Ensuring Wellness: Immunosensors Monitoring Food Quality and Pathogenicity Risks

Kamil Kuča • Harsh Kumar • Deepak Kala
Editors

Ensuring Wellness: Immunosensors Monitoring Food Quality and Pathogenicity Risks

Springer

Editors
Kamil Kuča
Centre for Basic and Applied Research
University of Hradec Králové
Hradec Králové, Czech Republic

Harsh Kumar
Centre of Advanced Technologies, Faculty of Science
University of Hradec Králové
Hradec Králové, Czech Republic

Deepak Kala
Polish Academy of Sciences
Institute of High Pressure Physics
Warsaw, Poland

ISBN 978-3-031-97476-2 ISBN 978-3-031-97477-9 (eBook)
https://doi.org/10.1007/978-3-031-97477-9

This Springer imprint is published by the registered company Springer Nature Switzerland AG
The registered company address is: Gewerbestrasse 11, 6330 Cham, Switzerland

Preface

Human society views food from different perspectives, yet it remains a valuable source of energy and nutrition. Because fraudulent activities can compromise the quality of food, consumers are always curious about quality. The significant development of analytical techniques, particularly protein-based sensing technology, has made it easy for food safety agencies to use them as point-of-care testing methods to detect various biological or chemical agents present in food. In addressing these issues, the development of "immunosensors" has emerged as a key area of research and innovation.

This book, *Ensuring Wellness: Immunosensors Monitoring Food Quality and Pathogenicity Risks*, intends to provide an in-depth review of strategies and technologies regarding the use of immunosensors in different food matrixes. The chapters cover the fundamental principles of immunosensors and role of nanomaterials and their applications in the detection of foodborne pathogens, plant pathogens, microbial toxins, chemicals/drugs as food adulterants.

This book aims to serve as a valuable resource for researchers, industry professionals, and students engaged in chemistry, microbiology, food science and nutrition, and agriculture. We extend our heartfelt gratitude to the contributors whose expertise and dedication have enriched this book, ensuring comprehensive coverage of the subject matter. We hope this book will inspire future research and innovation in the field of immunosensor technology, contributing to the creation of portable sensing platforms.

Hradec Králové, Czech Republic — Kamil Kuča
Hradec Králové, Czech Republic — Harsh Kumar
Warsaw, Poland — Deepak Kala

Contents

Introduction to Immunosensors: Principles and Applications in Food Safety

Anjuman Ayub, Daljeet Singh Dhanjal, Varun Sharma, Devinder Kumar Sharma, Reena Singh, and Chirag Chopra

Abstract

Immunosensors are analytical devices that combine the specificity of antibodies with various transduction methods to detect contaminants in food. They offer rapid, sensitive, on-site detection of pathogens, toxins, allergens, and chemical residues, enhancing food safety monitoring. This chapter provides an overview of immunosensor principles, including surface plasmon resonance, ellipsometry, optical waveguides, microgravimetric sensors, and their applications in food safety. Recent advancements, such as nanostructured materials and electrochemical detection techniques, have improved immunosensor performance, enabling real-time monitoring and early detection of foodborne hazards. The chapter also discusses current challenges, including the need for standardized protocols and the addressing of non-specific binding. It explores future directions for integrating immunosensors into food safety management systems.

Keywords

Immunosensors · Biosensors · Food safety · Electrochemical biosensors · Immunoassay · Food quality

A. Ayub · D. S. Dhanjal · R. Singh · C. Chopra (✉)
School of Bioengineering and Biosciences, Lovely Professional University, Jalandhar, Punjab, India
e-mail: chirag.18298@lpu.co.in

V. Sharma
NMC Genetics India Pvt Ltd, Gurugram, India

D. K. Sharma
Department of Physics, Govt. Gandhi Memorial Science College, Jammu, India

K. Kuča et al. (eds.), *Ensuring Wellness: Immunosensors Monitoring Food Quality and Pathogenicity Risks*, https://doi.org/10.1007/978-3-031-97477-9_1

1 Introduction

Ensuring food safety and quality is critical to public health and consumer protection. Detecting biological contaminants, such as pathogenic bacteria, and chemical contaminants, including pesticides, toxins, and veterinary drug residues, is crucial for maintaining food integrity (Pan et al. 2021). Traditional microbiological methods, which depend on culturing and counting viable bacterial cells on specific media, have long been the foundation of food safety testing (Harshavardhan et al. 2019). While these methods are highly sensitive and cost-effective, their dependence on microbial growth and visible colony formation renders them time-consuming, often requiring several days to yield results (Karachaliou et al. 2022). Similarly, chemical contaminants are typically detected using advanced separative techniques such as gas chromatography (GC) and high-performance liquid chromatography (HPLC), often coupled with sophisticated detectors. While effective, these techniques demand significant time, resources, and expertise, limiting their efficiency for real-time monitoring (Dinc et al. 2024). The pattern of food safety control has gradually shifted from end-product testing to process control methodologies over the past decade. End-product testing, though valuable, is often insufficient due to the heterogeneous distribution of contaminants and its inability to identify the root causes of contamination (Jiang et al. 2024). To address these limitations, a globally recognized Hazard Analysis and Critical Control Point (HACCP) system has emerged as a framework for identifying and managing specific hazards throughout the food production chain. The adoption of HACCP has highlighted the pressing need for rapid, sensitive, and precise detection methods that facilitate timely corrective measures and reduce risks to public health (Uzoigwe and Kongolo 2024). Biosensor technology has revolutionized food safety testing. Among the various types of biosensors, immunosensors have received notable recognition for their remarkable specificity and sensitivity (Dinc et al. 2024).

Immunosensors utilize biological recognition elements like nucleic acids, enzymes, or antibodies, alongside transducers to accurately detect a wide range of analytes in complex food matrices, often with minimal sample preparation (Pan et al. 2021). Their ability to deliver rapid results within minutes or hours makes them indispensable tools in modern food safety management. Immunosensors are categorized based on their signal transduction mechanisms into four primary types: electrochemical, optical, piezoelectric, and thermometric (Dinc et al. 2024). Each type provides a specific analytical requirement and often employs innovative labelling strategies to enhance detection capabilities. For example, electrochemical immunosensors use enzymatic or non-enzymatic labels for precise quantification, while optical immunosensors, particularly those based on Surface Plasmon Resonance (SPR), facilitate label-free detection of immunocomplexes (Dinc et al. 2024). Piezoelectric immunosensors, such as quartz crystal microbalances (QCM), offer another label-free approach with direct quantification capabilities. The adaptability of immunosensors is reflected in their assay formats, such as competitive and sandwich assays tailored to the molecular weight of the target analyte (Nsanzabera et al. 2024). Despite extensive research and development in immunosensor technology, its routine application in food analysis remains limited. However, recent advancements in electrochemical, optical, and piezoelectric immunosensors have demonstrated their transformative potential in food safety testing (Siavashy et al. 2024). The strategic immobilization of antibodies or antigens on transducers ensures high sensitivity, rapid detection, and reduced measurement time, further establishing immunosensors as invaluable and reliable tools for maintaining food quality and safety (Bhatlawande et al. 2024). This book chapter explores the principles underlying immunosensor technology, its design, signal transduction mechanisms, and assay formats. It illustrates the practical applications of immunosensors in food analysis, offering a comprehensive overview of developments in the field over the past 5 years. By connecting the gap between research and application, this chapter aims to serve as an essential guide for researchers, food safety professionals, and industry stakeholders, empowering them to harness the potential of immunosensors in safeguarding public health.

2 Biosensors and Immunosensors

Biosensors are innovative, cutting-edge analytical devices that combine biological recognition elements with signal transducers, allowing fast and accurate detection of analytes (Fig. 1). These devices rely on a biorecognition layer made up of single-stranded DNA, peptides, receptors, enzymes, or even living cells immobilized on a solid-state surface (Gangopadhyay et al. 2024). This biorecognition layer selectively binds to the target analyte and produces a signal that is subsequently converted into a measurable output by the transducer. Biosensors are distinguished by their ability to merge exceptional analytical specificity with the advanced processing capabilities of modern electronics for creating versatile and highly sensitive detection systems

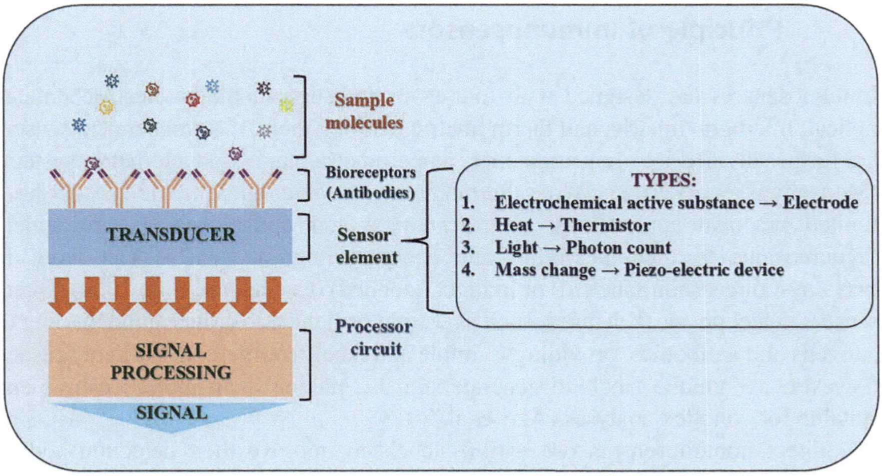

Fig. 1 Schematic illustration of an immunosensor detection and processing mechanism

suitable for a wide range of applications (Askari et al. 2025). There are broadly two major types of biosensors, i.e. biocatalytic and bioaffinity-based biosensors. Biocatalytic biosensors primarily use enzymes to catalyse biochemical reactions that produce a detectable or measurable signal (Bhatlawande et al. 2024).

In contrast, bioaffinity-based biosensors monitor binding events between specific biomolecules, such as antibodies, receptors, nucleic acids, or membranes, to recognize the target analyte. When antibodies or antibody fragments are used as the biological recognition element, the biosensor is classified as an immunosensor (Malvano et al. 2020). Immunosensors are explicitly designed to detect antigens through the highly specific and stable interaction between an antibody and its target antigen. Immunosensors represent a subset of biosensors and have gained significant attention due to their unparalleled specificity, sensitivity, and versatility. These devices integrate biological recognition molecules, such as antibodies or antigens, with signal transducers, enabling the detection of a wide range of analytes in complex matrices. Immunosensors have revolutionized diagnostic testing and food safety analysis, providing a platform for rapid and accurate monitoring (Ansari et al. 2025).

The process involves immobilizing antibodies or antigens onto a solid substrate, such as gold or silicon, ensuring accessibility for target analyte interaction. The target analyte binds to the immobilized element upon sample exposure, forming a stable antigen–antibody complex. This binding induces a physicochemical change at the sensor's surface, which the transducer converts into a measurable signal correlating to the analyte's concentration. The transducer type, optical, electrochemical, or piezoelectric, determines the signal manifestation, such as changes in light intensity, electrical current, or frequency. This integration enables rapid, sensitive, and specific detection, positioning immunosensors as indispensable tools in environmental monitoring, food safety, and medical diagnostics.

3 Principle of Immunosensors

Immunosensors are designed with four main detection methods: electrochemical, optical, microgravimetric, and thermometric. Among these, electrochemical sensors are further divided into potentiometric, amperometric, and conductometric systems (Martsenyuk et al. 2018). While thermometric and microgravimetric sensors have limited practical applications, electrochemical and optical sensors are widely favoured due to their flexibility and accuracy. Immunosensors can work in two ways: direct (non-labelled) or indirect (labelled) (Cancelliere et al. 2024). Direct sensors detect physical changes, such as weight or light shifts, during the binding of antigens and antibodies, providing a simple and label-free method. Indirect sensors, however, use unique labels to generate signals, making them more sensitive and suitable for complex analyses (Azri et al. 2018).

Indirect immunosensors use various labels to improve their detection ability, many of which come from traditional immunoassay methods. These labels include enzymes like peroxidase, glucose oxidase, and luciferase, which create measurable signals during chemical reactions (Harshavardhan et al. 2019; Nsanzabera et al. 2024). Other labels include electroactive compounds like ferrocene and fluorescent dyes such as rhodamine, fluorescein, and ruthenium complexes, which enhance signal strength. Advanced techniques like FRET (Fluorescence Resonance Energy Transfer), which uses light transfer between molecules, further improve accuracy, especially in systems like fibre-optic sensors. Thus, the development and innovation in immunosensors have increased their application across diverse fields like environmental monitoring, food safety testing, and diagnostics (Azri et al. 2018).

Even though indirect immunosensors offer exceptional sensitivity because of their advanced labels, but direct immunosensors are gaining attraction due to their simplicity and easy-to-use design. Direct sensors do not require labelling, which makes them cheaper and easier to use. This simplicity has immense potential, offering a practical alternative to more complex immunoassay systems (Nsanzabera et al. 2024). With ongoing research, direct immunosensors are expected to improve further, especially with the integration of nanotechnology and microfluidics. These advancements aim to create highly efficient, next-generation diagnostic tools that combine simplicity with high performance, making immunosensors a promising technology for the future of biosensing (Cancelliere et al. 2024; Askari et al. 2025).

4 Electrochemical Immunosensors

Electrochemical immunosensors are a specialized subgroup of electrochemical sensors that incorporate immunological components, such as antibodies or antigens, to achieve specific detection of target analytes (Baranwal et al. 2022). While both types of sensors operate on electrochemical principles, electrochemical immunosensors leverage the high specificity of antigen–antibody interactions to enhance selectivity. This integration allows for the precise detection of biomolecules, making electrochemical immunosensors particularly valuable in clinical diagnostics and bioanalysis (Mahato and Wang 2021). Recent research has focused on enhancing the sensitivity and selectivity of electrochemical immunosensors through various strategies: incorporating nanomaterials such as graphene oxide to improve electrochemical properties, leading to better detection limits and response times; functionalizing sensor surfaces with biomolecules like antibodies, aptamers, and peptides to enhance specificity towards target analytes, making them suitable for detecting biomolecules and animal cells, including cancer and stem cells; and utilizing advanced electrode materials such as glassy carbon electrodes and screen-printed carbon electrodes to improve sensitivity and analysis speed, particularly in pharmaceutical and biological sample analysis (Gao et al. 2023). These advancements have broadened the application scope of electrochemical immunosensors, making them invaluable tools in medical diagnostics, environmental monitoring, and food safety. The combination of quick response, high sensitivity, and miniaturization potential of these systems remains the key factor driving research and development in this field. Additionally, recent studies have demonstrated the use of electrochemical enzyme-linked immunoassays utilizing

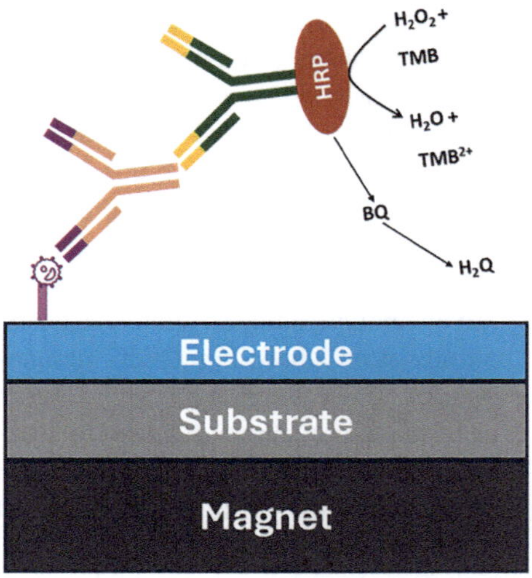

Fig. 2 Representation of an electrochemical enzyme-linked immunoassay using silica-coated iron oxide nanoparticles for antigen detection

antigens immobilized on silica-coated iron oxide nanoparticles, significantly improving both specificity and sensitivity in target analyte detection and expanding the potential of electrochemical immunosensors (Fig. 2) (Wang et al. 2023). The integration of nanomaterials, biomolecular modifications, and advanced electrode materials has significantly improved the performance of electrochemical immunosensors, expanding their applications and reinforcing their importance in various analytical fields.

4.1 Amperometric Immunosensors

Amperometric immunosensors detect current changes resulting from electrochemical reactions at a constant voltage. Direct detection is limited, as many protein analytes lack inherent electrochemical activity. Therefore, electroactive labels (often enzymes) are used to facilitate measurable reactions. Commonly used enzymes include horseradish peroxidase (HRP), glucose oxidase, and alkaline phosphatase, which catalyse reactions producing detectable electroactive species (Huang et al. 2024a). For example, HRP can catalyse reactions involving hydrogen peroxide, leading to measurable current changes. The Clark oxygen electrode, introduced in 1956, exemplifies this approach. It comprises an electrolyte chamber equipped with a silver/silver chloride reference electrode and a platinum cathode, separated from the sample by an oxygen-permeable membrane (Malvano et al. 2020). Oxygen reduction at the cathode generates a current proportional to the oxygen concentration, enabling analyte detection through enzymatic reactions that consume or produce oxygen (Kumar et al. 2025).

Despite the indirect nature of detection, amperometric immunosensors offer high sensitivity due to their linear response to analyte concentration, unlike the logarithmic response observed in potentiometric systems. However, challenges such as mass transport limitations of redox partners to the electrode surface and potential interferences from other electroactive species in the sample must be addressed to ensure accurate measurements (Fabiani et al. 2024). Recent advancements in nanomaterials have significantly enhanced the performance of amperometric immunosensors. Incorporating nanostructures, such as nanoparticles and nanocomposites, into sensor design has improved sensitivity and lowered detection limits. These nanomaterials provide an increased surface area for antibody immobilization and facilitate efficient electron transfer, leading to more robust and responsive sensors (Brusnitsyn et al. 2024).

Additionally, innovative immobilization techniques utilizing nanomaterials have been developed to meet the specific requirements of various electrochemical immunosensors. These methods enhance the stability and functionality of the immobilized antibodies, contributing to improved sensor performance. These advancements have significantly expanded the scope of amperometric immunosensors, allowing their application in various fields like environmental monitoring, food safety, and clinical diagnostics through the precise and sensitive detection of various analytes (Docherty et al. 2024).

4.2 Optical Immunosensors

Optical immunosensors are widely used in bioanalysis due to their advantages over other transducer techniques. They offer non-destructive operation, rapid signal generation, and precise detection. The integration of fibre-optic technologies and advanced optoelectronics has enhanced their versatility, making them valuable tools in clinical applications (Kourti et al. 2023). These sensors detect changes in light properties such as luminescence, fluorescence, refractive index, absorption, or scattering on interaction of light with sensing surfaces. Typically, photodiodes or photomultipliers are used as detectors. Applications range from label-free optical detection of immunological reactions to enzymatic reaction products and identification of labelled immunospecies. Fluorescent labels are commonly used, but bioluminescent and chemiluminescent labels are also utilized (Karachaliou et al. 2022).

Label-free optical sensors offer the advantage of real-time monitoring without additional labelling steps. However, they face challenges such as non-specific binding effects and reduced sensitivity to small analytes. To solve these challenges, researchers are investigating the incorporation of mass labels such as latex particles to strengthen and boost the signal strength. A notable example of innovation is the development of a nanoparticle-enhanced bidiffractive grating coupler immunosensor for detection of thyroid-stimulating hormones, which has demonstrated improved performance (Singh and Natarajan 2023). Optical immunosensors are potent bioanalysis tools, offering rapid and precise detection capabilities. Ongoing research aims to overcome current challenges, enhancing their sensitivity and broadening their applications in diagnostics and environmental monitoring (Petrou et al. 2022).

4.3 Microgravimetric Immunosensors

Microgravimetric immunosensors, particularly those utilizing quartz crystal microbalance (QCM) technology, are highly effective in monitoring antigen–antibody interactions. These sensors detect mass changes on the sensor surface resulting from the specific binding of antigens to immobilized antibodies, leading to measurable shifts in oscillation frequency (Alanazi et al. 2023). This method allows for real-time, label-free detection of biomolecular interactions. Recent research has showcased the use of QCM-based immunosensors in detecting aerosolized influenza viruses using specific antibodies anchored on the sensor surface. Upon exposure to aerosolized virus particles, a significant decrease in frequency was observed, indicating successful antigen–antibody binding. Control experiments with nebulized water and allantoic fluid confirmed that the frequency changes were due to specific interactions between the virus and the immobilized antibodies (Yan et al. 2024).

5 Advancements in Immunosensor Technologies for Enhanced Food Safety Monitoring

Recent advancements in biosensor technology have significantly enhanced food safety monitoring, ensuring that consumers receive products free from harmful contaminants. One notable development is the use of immunosensors, which detect specific substances through antigen–antibody interactions (Table 1) (Mandal and Banerjee 2022; Ansari et al. 2025). These devices have been instrumental in identifying bacterial pathogens responsible for foodborne illnesses, offering rapid and accurate detection methods that surpass traditional microbiological techniques (Fig. 3). In addition to immunosensors, electrochemical biosensors have gained prominence in food safety applications (Zhang et al. 2024). These sensors achieve high specificity in detecting contaminants by utilizing bioreceptors and biological molecules like antibodies or enzymes. Recent studies are now focused on improving bioreceptors to boost and enhance the reliability and sensitivity of biosensors, making them valuable tools for real-time food quality monitoring. The integration of hydrogel-based biosensors into food safety practices has also shown significant potential. Hydrogels are recognized for their broad surface area and exceptional biocompatibility; therefore, they are used for the detection of diverse food contaminants like biotoxins, pesticide residues, and pathogens. Their adaptability and mechanical stability make hydrogel-based biosensors suitable for various food matrices, enhancing the detection capabilities in complex food systems (Gangopadhyay et al. 2024).

Furthermore, the development of flexible biosensors has opened new avenues for food pathogen detection. These sensors, characterized by their adaptable materials and structures, can conform to different surfaces, enabling effective monitoring across diverse food products (Barrias et al. 2024). Their flexibility and sensitivity contribute to more efficient detection of contaminants, ensuring better food safety management. The incorporation of intelligent biosensors that leverage deep learning

Table 1 Recent advances in immunosensor applications for food safety

Immunosensor type	Detection method	Target analyte	Limit of detection (LOD) CFU/ml	Type of antigen–antibody reaction	Specific substrate detection	Application	Sensitivity	References
Electrochemical	Impedance spectroscopy	*Salmonella* spp.	10	Noncompetitive (Sandwich)	Lipopolysaccharides	Pathogen detection in poultry products	High	Silva et al. (2018)
Optical	Surface Plasmon resonance	*E. coli* O157:H7	5	Noncompetitive (Sandwich)	Outer membrane proteins	Monitoring of waterborne pathogens in beverages	High	Meira et al. (2024)
Piezoelectric	Quartz crystal microbalance	Listeria monocytogenes	15	Noncompetitive (Sandwich)	Internalin proteins	Detection in dairy products	High	Kourti et al. (2023) and Wang et al. (2024c)
Fluorescent	Fluorescence spectroscopy	Aflatoxin B1	0.1	Competitive	Aflatoxin B1 molecule	Mycotoxin detection in grains	High	Wang et al. (2024a)
Colourimetric	Gold nanoparticle aggregation	*Staphylococcus aureus*	20	Noncompetitive (Sandwich)	Protein A	Pathogen detection in meat products	High	Umapathi et al. (2021)
Electrochemical	Amperometry	Pesticide residues	0.05	Competitive	Specific pesticide molecules	Monitoring of pesticide levels in vegetables	High	Ng et al. (2016) and Umapathi

(continued)

Table 1 (continued)

Immunosensor type	Detection method	Target analyte	Limit of detection (LOD) CFU/ml	Type of antigen–antibody reaction	Specific substrate detection	Application	Sensitivity	References
								et al. (2022)
Optical	Photonic crystal resonance	*Campylobacter jejuni*	8	Noncompetitive (Sandwich)	Flagellin proteins	Detection in poultry products	High	Petrou et al. (2022)
Piezoelectric	Acoustic wave sensor	*Norovirus*	25	Noncompetitive (Sandwich)	Capsid proteins	Virus detection in seafood	High	Grasso et al. (2022)
Fluorescent	Quantum dot-based assay	Ochratoxin A Quantum	0.2	Competitive	Ochratoxin A molecule	Mycotoxin detection in wine	High	Huang et al. (2024a)
Colourimetric	Enzyme-linked assay	*Vibrio cholerae*	30	Noncompetitive (Sandwich)	Cholera toxin	Pathogen detection in water samples	High	Fabiani et al. (2024)

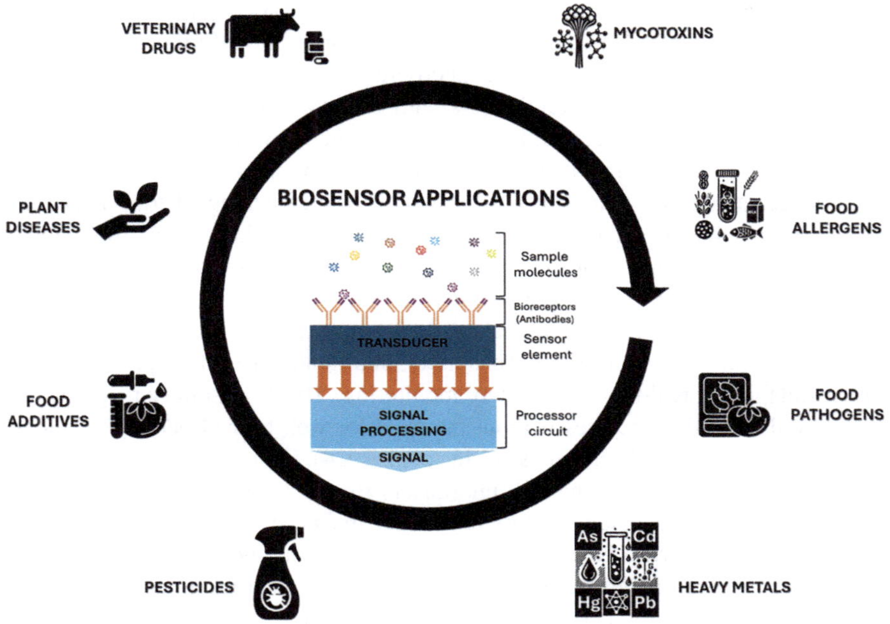

Fig. 3 Overview of biosensor technology application in food safety monitoring

algorithms represents a significant leap towards more intelligent food safety solutions (Chen et al. 2023). These advanced systems can analyse complex data patterns, facilitating real-time monitoring and classification of food products. Such intelligent biosensors enhance the ability to detect contaminants promptly, thereby improving overall food safety management practices. Lastly, the application of optical immunosensors has been explored for detecting bacteria in food matrices. These sensors utilize light-based detection methods to identify specific bacterial pathogens, offering a non-invasive and rapid approach to ensuring food safety (Huang et al. 2024b). Ongoing research and development in this area aim to provide more efficient and user-friendly devices for routine food safety assessments. The continuous evolution of biosensor technologies, including immunosensors, electrochemical, hydrogel-based, flexible, intelligent, and optical immunosensors, is playing a pivotal role in advancing food safety measures. These innovations provide faster, more accurate, and more reliable methods for detecting contaminants while safeguarding public health (Dinc et al. 2024).

6 Conclusion and Future Outlook

Immunosensors have emerged as vital tools in ensuring food safety, offering rapid, sensitive, and specific detection of contaminants such as pathogens, toxins, and allergens. Their integration into food safety protocols has significantly enhanced the monitoring and control of foodborne hazards, thereby protecting public health (Dinc et al. 2024). Recent advancements in nanotechnology have further improved the performance of immunosensors. The incorporation of nanomaterials has led to the development of diverse immunosensors with enhanced sensitivity and specificity, enabling the detection of contaminants at even lower concentrations (Bhatlawande et al. 2024). Electrochemical immunosensors have been particularly impactful in food safety analysis. Their ability to provide robust screening of food impurities, hazardous chemicals, and toxins has made them invaluable in ensuring food quality. The ease of miniaturization and the potential for integration into user-friendly formats further enhance their applicability in various settings.

Optical immunosensors have also shown promise in detecting foodborne pathogens. Their high selectivity, sensitivity, and rapid detection capabilities make them suitable for real-time food safety monitoring. These sensors can significantly improve diagnostic processes without the need for well-trained personnel, making them accessible for widespread use (Dinc et al. 2024). Despite these advancements, challenges remain in adapting immunosensors for food safety applications. Issues such as non-specific binding interference and the need for further validation in complex food matrices demand continuous research and development. Thus, overcoming these challenges is essential for the widespread implementation of immunosensors in food safety monitoring. Looking forward, the future of immunosensors in food safety is promising. Ongoing research aims to enhance their sensitivity, specificity, and ease of use (Chen et al. 2023). The integration of immunosensors into innovative food traceability systems is anticipated to revolutionize food safety and security by enabling timely responses to contamination events and facilitating continuous monitoring in real time. Immunosensors represent a transformative technology in the field of food safety. Their continued development and integration into food safety protocols hold the potential to significantly reduce the incidence of foodborne illnesses and enhance public health outcomes.

References

Alanazi N, Almutairi M, Alodhayb AN (2023) A review of quartz crystal microbalance for chemical and biological sensing applications. Sens Imaging 24:10

Ansari A, Siddique MU, Nayak AK, Askari VR, Zairov RR, Aminabhavi TM, Hasnain MS (2025) Biosensors: history and classifications. In: Hasnain MS, Nayak AK, Aminabhavi TM (eds) Fundamentals of biosensors in healthcare, 1st edn. Academic Press, London, pp 1–20

Azri FA, Sukor R, Selamat J et al (2018) Electrochemical immunosensor for detection of aflatoxin B1 based on indirect competitive ELISA. Toxins 10:196

Baranwal J, Barse B, Gatto G et al (2022) Electrochemical sensors and their applications: a review. Chemosensors 10:363

Barrias S, Ibáñez J, Fernandes JR, Martins-Lopes P (2024) The role of DNA-based biosensors in species identification for food authenticity assessment. Trends Food Sci Technol 145:104350

Bhatlawande AR, Ghatge PU, Shinde GU et al (2024) Unlocking the future of smart food packaging: biosensors, IoT, and nano materials. Food Sci Biotechnol 33:1075–1091

Brusnitsyn DV, Medyantseva EP, Ramazanova AN et al (2024) Capabilities of a supramolecular system based on hexamolybdenum cluster complexes in the determination of amitriptyline in human urine using amperometric immunosenors. J Anal Chem 79:726–732
Cancelliere R, Paialunga E, Grattagliano A, Micheli L (2024) Label-free electrochemical immunosensors: a practical guide. Trends Anal Chem 180:117949
Chen X, Yao C, Li Z (2023) Microarray-based chemical sensors and biosensors: fundamentals and food safety applications. Trends Anal Chem 158:116785
Dinc B, Demirci N, Üstünsoy R, Ertaş T (2024) Advances in rapid and sensitive biosensors for AF detection: toward enhanced food safety. Toxin Rev 43:48–62
Docherty N, Collins L, Pang S et al (2024) Cost-effective amperometric immunosensor for cardiac troponin I as a step towards affordable point-of-care diagnosis of acute myocardial infarction. Sens Bio-Sens Res 47:100725
Fabiani L, Fiore L, Fillo S et al (2024) Smartphone-assisted paper-based electrochemical immunosensor for SARS-CoV-2 detection in saliva. Bioelectrochemistry 156:108619
Gangopadhyay M, Chakravarti S, Sanyamat S, Dewanjee S (2024) Biosensors in agriculture: pioneering the future of precision farming and sustainability. In: Mathew R, Ajayan J (eds) Biosensors: developments, challenges and perspectives, 1st edn. Springer Nature, Singapore, pp 259–301
Gao F, Liu C, Zhang L et al (2023) Wearable and flexible electrochemical sensors for sweat analysis: a review. Microsyst Nanoeng 9:1–21
Grasso S, Di Loreto MV, Arienzo A et al (2022) Microbiological risk assessment of ready-to-eat leafy green salads via a novel electrochemical sensor. Chemosensors 10:134
Harshavardhan S, Rajadas SE, Vijayakumar KK et al (2019) Electrochemical immunosensors: working principle, types, scope, applications, and future prospects. In: Krishnaraj RN, Sani RK (eds) Bioelectrochemical interface engineering, 1st edn. Wiley, pp 343–369
Huang J, Xie Z, Luo S et al (2024a) A sandwich amperometric immunosensor for the detection of fowl adenovirus group I based on bimetallic Pt/Ag nanoparticle-functionalized multiwalled carbon nanotubes. Sci Rep 14:261
Huang L, Zhang C, Ye R et al (2024b) Capacitive biosensors for label-free and ultrasensitive detection of biomarkers. Talanta 266:124951
Jiang W, Tang Q, Zhu Y et al (2024) Research progress of microfluidics-based food safety detection. Food Chem 441:138319
Karachaliou C-E, Koukouvinos G, Goustouridis D et al (2022) Recent developments in the field of optical immunosensors focusing on a label-free, White Light Reflectance Spectroscopy-based immunosensing platform. Sensors 22:5114
Kourti D, Angelopoulou M, Petrou P, Kakabakos S (2023) Optical immunosensors for bacteria detection in food matrices. Chemosensors 11:430
Kumar A, Chandel R, Kaur R et al (2025) Biosensors: recent advancements in postharvest management of fruits. In: Irfan M, Kumar P (eds) Bioengineered fruit and vegetables, 1st edn. Apple Academic Press, Boca Raton, pp 253–280
Mahato K, Wang J (2021) Electrochemical sensors: from the bench to the skin. Sens Actuators B: Chem 344:130178
Malvano F, Pilloton R, Albanese D (2020) Label-free impedimetric biosensors for the control of food safety – a review. Int J Environ Anal Chem 100:468–491
Mandal D, Banerjee S (2022) Surface acoustic wave (SAW) sensors: physics, materials, and applications. Sensors 22:820
Martsenyuk VP, Klos-Witkowska A, Sverstiuk AS, Bihunyak TV (2018) On principles, methods and areas of medical and biological application of optical immunosensors. Med Inform Eng 2: 28–36
Meira DI, Barbosa AI, Borges J et al (2024) Recent advances in nanomaterial-based optical biosensors for food safety applications: Ochratoxin-A detection, as case study. Crit Rev Food Sci Nutr 64:6318–6360

Ng E, Nadeau KC, Wang SX (2016) Giant magnetoresistive sensor array for sensitive and specific multiplexed food allergen detection. Biosens Bioelectron 80:359–365

Nsanzabera F, Mwiseneza A, Irakoze E et al (2024) Emerging trends of immunosensors development for detection of food toxins. Turkish JAF Sci Tech 12:1046–1060

Pan R, Li G, Liu S et al (2021) Emerging nanolabels-based immunoassays: principle and applications in food safety. Trends Anal Chem 145:116462

Petrou P, Makarona E, Raptis I et al (2022) Monolithically integrated label-free optical immunosensors. Eng Proc 16:11

Siavashy S, Soltani M, Rahimi S et al (2024) Recent advancements in microfluidic-based biosensors for detection of genes and proteins: applications and techniques. Biosens Bioelectron X 19:100489

Silva NF, Magalhães JM, Freire C, Delerue-Matos C (2018) Electrochemical biosensors for Salmonella: state of the art and challenges in food safety assessment. Biosens Bioelectron 99: 667–682

Singh KRB, Natarajan A (2023) Molecularly imprinted polymer-based optical immunosensors. Luminescence 38:834–844

Umapathi R, Ghoreishian SM, Sonwal S et al (2022) Portable electrochemical sensing methodologies for on-site detection of pesticide residues in fruits and vegetables. Coord Chem Rev 453:214305

Umapathi R, Sonwal S, Lee MJ et al (2021) Colorimetric based on-site sensing strategies for the rapid detection of pesticides in agricultural foods: new horizons, perspectives, and challenges. Coord Chem Rev 446:214061

Uzoigwe DO, Kongolo D (2024) Integration of hazard analysis and critical control points HACCP with maintenance practices: enhancing food safety in the food and beverage industry; a review. Int J Latest Technol Eng Manag Appl Sci 13:88–101

Wang J, Cui X, Liang L et al (2024a) Advances in DNA-based electrochemical biosensors for the detection of foodborne pathogenic bacteria. Talanta 275:126072

Wang L, Pagett M, Zhang W (2023) Molecularly imprinted polymer (MIP) based electrochemical sensors and their recent advances in health applications. Sens Actuator Rep 5:100153

Wang S, Bao X, Liu M et al (2024c) Aptamer-based sensor for rapid and sensitive detection of ofloxacin in meat products. Sensors 24:1740

Yan S, Liu Q, Liang B et al (2024) Airborne microbes: sampling, detection, and inactivation. Crit Rev Biotechnol 11:1–35

Zhang Z, Hu J, Lin S et al (2024) Field effect transistor biosensors for healthcare monitoring. Interdiscip Med 2:e20240032

Nanoparticles as Catalysts for Immunosensor Innovation

Surbhi Sharma, Shagun Gupta, and Ankur Kaushal

Abstract

The advent of nanoparticles has transformed the field of biosensors, particularly immunosensors, which rely on specific interactions between antigens and antibodies. The physicochemical characteristics, such as tunable size, versatile functionalization, and high surface area, enable their integration into immunosensors. They have emerged as pivotal catalysts in the realm of immunosensor innovation, offering transformative potential in clinical diagnostics, environmental monitoring, and food safety. Nanoparticles, including carbon, gold, silver, and magnetic (MNP) materials, have been extensively studied for their catalytic properties, which can be harnessed to amplify detection signals. In addition, their superior electrical conductivity and antimicrobial properties have further contributed to the development of highly sensitive immunosensors. Different techniques, such as self-assembly, covalent binding, or electrostatic adsorption, have been employed for the integration of nanoparticles into immunosensors, involving surface functionalization with biological recognition elements, such as antibodies and aptamers. Additionally, the use of nanoparticles allows for the design of multianalyte sensors, which is crucial for comprehensive diagnostic applications. Nanoparticle-catalyzed immunosensors hold a prominent future, with ongoing research focusing on enhancing their stability, reproducibility, and real-time monitoring capabilities. Continued innovation and research in this field will undoubtedly lead to more robust, reliable, and versatile immune-sensing platforms, ultimately benefiting a wide range of scientific and industrial applications.

S. Sharma · S. Gupta · A. Kaushal (✉)
Department of Bio-Sciences and Technology, Maharishi Markandeshwar (Deemed to be) University, Mullana, Ambala, India

K. Kuča et al. (eds.), *Ensuring Wellness: Immunosensors Monitoring Food Quality and Pathogenicity Risks*, https://doi.org/10.1007/978-3-031-97477-9_2

Keywords

Nanoparticles · Immunosensor · Magnetic nanoparticles

1 Introduction

Analytical devices that are used to detect antigen (Ag) and antibody (Ab) interactions and form a stable complex include immunosensors. The immobilization of either antigen or antibody can be performed on the surface for detection (Felix and Angnes 2018). Immunosensors amalgamate the precision of immunological assays with transducers that aid in converting the interaction into a measurable signal in the form of piezoelectric, optical, and electrochemical signals, allowing sensitive, rapid, and selective target analyte detection (Aydin et al. 2021). The applications of these immunosensors are vast in various fields, particularly in food safety, environmental monitoring, medical diagnostics, and pharmaceutical research (Soria et al. 2011; Gallardo-Carreño et al. 2020; Soleymani and Golsanamluo 2021). Immunosensors play a crucial role in the healthcare industry in the detection of biomarkers for testing drug levels in blood, saliva, or any other biological fluid for diseases, hormones, infectious agents, etc., and many health-related conditions, such as cardiovascular diseases, cancer, and infectious diseases, including coronavirus disease-19, HIV, etc., can be diagnosed with the help of these sensors (Li et al. 2022). Moreover, food quality and safety can be monitored by detecting contaminants and allergens in food. Additionally, various pollutants, pathogens, and toxins in soil or water enable onsite and real-time determination of environmental factors (Gavrilaş et al. 2022). Owing to their fast detection with more precision even in complex samples, immunosensors have received tremendous attention in diagnostics, allowing early disease detection that further facilitates timely treatment and improves patient outcomes. Miniaturization makes decentralized and point-of-care testing easier, thus enabling remote settings with limited resource requirements to be reached. The potential of multianalyte detection in comprehensive disease profiling and monitoring has increased. Despite having many advantages over traditional methods, immunosensors still impose certain limitations related to their effectiveness, sensitivity, and specificity in different applications. The quality of the Ag–Ab interaction and transduction mechanisms can limit the sensitivity, as very low concentrations of pathogens in biological samples can make detection difficult by producing very low signals, resulting in undetectable or false results. The measurement of low-abundance analytes is often repressed by the signal-to-noise ratio. The performance of the immunosensor is strongly affected by background interference within the sample and sensor material, which produces low signals (Matysiak-Brynda et al. 2018). Another issue that may arise is the cross-reactivity of antibodies with other structurally similar compounds that can affect the specificity of immunosensors, resulting in false-positive results. Complex matrices such as blood or serum address this issue, where similarly structured molecules interfere with the detection of the target molecule. Despite these challenges, advanced materials such as quantum dots, nanoparticles, and nanocomposites aid in reducing cross-reactivity issues, improve signal amplification, and pave the way for the detection of low concentrations of analytes, leading to the enhanced performance of the sensor (Bilal et al. 2020).

As powerful enhancers in the field of immunosensing, nanoparticles have advanced the performance and sensitivity of sensors because of their high catalytic activity, ease of functionalization, high surface area, and good electrical conductivity. The sensitivity and accuracy of the immunosensors are directly improved with the application of nanoparticles. The ability of these materials to amplify signals because of their catalytic behavior plays a major role in enhancing performance. Signal amplification of the Ag–Ab interaction can be harnessed by the use of metallic nanoparticles such as platinum, silver, and gold (Agrawal et al. 2022; Zhang et al. 2024). For example, nanoparticles work as catalysts in electrochemical sensors by catalyzing reactions that produce electrochemical signals and detect even trace amounts of analytes. The increased surface area of nanoparticles such as carbon nanotubes, graphene, and metal oxides allows more antibody/antigen attachment to the surface, thus increasing antibody loading and enabling better entrapment and interaction of the analyte and high signal transduction. Moreover, easy functionalization of biomolecules on the sensor surface can also be performed to improve selectivity within complex matrices. Some challenges, such as stability, regulatory considerations, and reusability, still remain (Asal et al. 2019). Nanoparticles also have a role in environmental monitoring, biofuel production, the food industry, as well as in diagnostics (Mishra et al. 2021; Sabarees et al. 2023; Pandit et al. 2023; Yadav et al. 2023; Bose et al. 2023; Manjunathan et al. 2024). This chapter discusses those limitations and highlights future directions, including their role in AI, data analysis, and the development of hybrid nanoparticles with multifunctional capabilities. Overall, this chapter briefly discusses the catalytic properties of nanoparticles, representing a breakthrough in the innovation of immunosensors and providing new, fast, and reliable opportunities for highly sensitive diagnostic platforms in various applications.

2 Types of Nanoparticles Used in Immunosensors

An increase in the performance of the immunosensors is a major contribution of nanoparticles possessing unique properties for refinement of selectivity, stability, and sensitivity. Various metals, metal oxides, quantum dots, and carbon-based nanoparticles are discussed below.

2.1 Metallic Nanoparticles

The performance of immunosensors is significantly enhanced by the signal amplification of various nanoparticles, such as silver (Ag), gold (Au), and platinum (Pt), primarily because of their catalytic properties in redox reactions (Begum et al. 2022).

Each of the nanoparticles contributes distinctly to increasing the performance of the sensors. Properties such as biocompatibility, easy functionalization, efficient bioelement immobilization in AuNPs, and optical properties make them ideal for colorimetric and electrochemical detection (Yadav et al. 2023; Nangan et al. 2023; Manjunathan et al. 2024). Pakchin et al. developed an immunosensor for the detection of carcinoma antigen 125 by combining chitosan-gold nanoparticles with CNTs and GO using lactate oxidase as a probe. This sandwich of an immunosensor was able to detect 0.002 U/mL of the analyte, surpassing traditional ELISA tests (Samadi Pakchin et al. 2018). On the other hand, AgNPs possess excellent conductivity and catalytic ability for signal amplification in electrochemical immunosensors. Dorozhko et al. conjugated AgNPs with immunoglobulin G to develop an immunosensor through a simple adsorption method. They essentially demonstrated the use of silver NPs in electrochemical immunoassays and as an alternative to enzyme-labeled conjugates (Dorozhko et al. 2023). Owing to their thermal and chemical stability, PtNPs, along with their superior catalytic activity, mainly in hydrogen peroxide reactions, are most suitable for use in electrochemical immunosensors. Khan et al. explored luminol-functionalized platinum nanoparticles loaded on graphene sheets for the detection of prostate-specific antigens. This electrochemiluminescent immunosensor was able to detect 0.3 pg/mL analyte, with a linear range between 1 pg/mL and 10 ng/mL (Khan et al. 2019).

2.2 Metal Oxide Nanoparticles

Like metallic NPs, metal oxide nanoparticles such as iron oxide NPs (Fe_2O_3 and Fe_3O_4), cerium oxide, etc., have attracted attention in the development of immunosensors. The magnetic and peroxidase-like activity of iron oxides assist in signal amplification and analyte isolation, whereas the flexibility and antioxidant properties of cerium oxide increase sensitivity and stability. The ability of cerium oxide to switch between its two oxidation states (Ce^{3+} and Ce^{4+}) plays a major role in its catalytic action, allowing it to mimic enzymes in redox reactions (Manoj et al. 2021).

2.3 Quantum Dots

Quantum dots possess both optical and electrical properties that make them effective for both types of sensors. The size-dependent electronic properties of quantum dots enable them to mimic natural enzymes in redox reactions, leading to enhanced catalytic amplification of the immunosensor signal. High catalytic activity also aids in fast electron transfer, thereby improving the detection limit (Kaushal et al. 2022). Multiplexed detection through QDs is supported by their optical properties, such as high quantum yield and size-tunable fluorescence, simultaneously allowing multiple target analyses with various fluorescent signals. Kuang et al. developed a sandwich immunosensor for the detection of *Salmonella* using magnetic nanoparticles and QDs (Shen and Gao 2015; Liu et al. 2019). Other than immunosensors, these luminescent nanoparticles also play a role in the detection of food pathogens and in diagnostics (Srivastava et al. 2021; Nangan et al. 2023). All the properties of QDs play crucial roles in effective and sensitive diagnosis and in early detection with rapid and precise analyte determination.

2.4 Carbon-Based Nanoparticles

Nanomaterials such as graphene oxide, fullerenes, and carbon nanotubes have emerged as excellent and highly effective matrices for the development of immunosensors and their performance. Graphene oxide, a single carbon-layered structure containing oxygen functional groups, aids in its dispersion in aqueous media and allows easy functionalization (Kala et al. 2021). These 2D sheets provide an excellent platform for biomolecule immobilization because their high surface area-to-volume ratio and oxygen groups help with conjugation with antibodies, proteins, DNA, etc. Sharma et al. developed an electrochemical immunosensor for the detection of 25-hydroxy vitamin D_3 using graphene oxide as a matrix. The immunosensor displayed a detection limit as low as 5.96 fg/μL (Sharma et al. 2024). Another class of carbon-based nanomaterials involves single-walled and multiwalled carbon nanotubes (SWCNTs and MWCNTs) that offer high mechanical strength, high surface area, and excellent electrical conductivity (Natrayan et al. 2022). A significant increase in the rate of electron transfer can be observed between the sensor surface and the analyte, resulting in faster and accurate measurements. Yang et al. developed an immunosensor based on optical immunodetection of a major foodborne disease pathogen, staphylococcal enterotoxins, via CNTs. The antibody–nano mixture was adsorbed onto the polycarbonate film, which increased the sensitivity up to sixfold compared with that of their optical-based immunological assays (Yang et al. 2011). On the other hand, for fullerenes, spherical structures act as efficient electron acceptors and serve as electron transport materials for enhancing immunosensor performance. Pan et al. developed a piezoelectric quartz crystal immunosensor modified with fullerene C_{60}–antibody conjugates for the detection of hemoglobin and human IgG. The response was partially reversible, implying a chemical reaction between C60 and the two antibodies. The sensor demonstrated a detection limit below 10–4 mg/mL for both antibodies (Pan and Shih 2004). Overall, these carbon nanoparticles are revolutionizing the field of immunosensors, as they help enhance the electrochemical properties of the sensor. Improved catalytic activity, electron transfer efficiency, and immobilization of biomolecules can be observed for the development of rapid, stable, and sensitive sensors.

3 Catalytic Properties of Nanoparticles in Immunosensors

Enhancing electron transfer along with facilitating signal amplification are the main roles played by nanoparticles in immunosensors and directly lead to increased sensitivity and improved performance. Focusing on catalyzing electron transfer is crucial in immunosensors, especially when biomolecule detection relies on redox reactions for measuring signals or interactions. The increased mobility of electrons also plays an integral role in catalysis. CNTs and GO exhibit excellent electron mobility; as a result, more efficient transfer of charge can be observed. The increased mobility aids in improving the response time and ensures minimal analyte concentration detection (Voon and Sam 2018; Zhang et al. 2020). In addition to the catalytic properties of Au and Pt nanoparticles, they reduce H_2O_2 in electrochemical sensors. Likewise, Fe nanoparticles act as peroxidase-like catalysts, enhancing redox reactions. NPs also mimic some natural enzymes and catalyze redox reactions. A signal boost is provided as nanoparticles promote charge transfer between the analyte and the electrode surface.

Signal amplification is another factor that can be improved by increasing electron transfer in nanoparticles, thereby increasing the number of electrons generated during each measurement. A molecule and its reaction with an antibody generate a signal that is further enhanced. Overall, the conductive nature, surface area, and catalytic ability of nanoparticles ensure that efficient electron flow enables low analyte detection because of the amplified signal (Wang and Pang 2015; Li et al. 2018).

Some nanoparticles exhibit enzyme-like activities and mimic the catalytic behavior of natural enzymes; such nanoparticles are known as nanozymes (Fig. 1). Nanozymes include metals, metal oxides, and carbon-based nanoparticles. They provide a stable and cost-effective approach for enhancing the performance of immunosensors. The peroxidase-like activity of nanozymes enables the breakdown of H_2O_2 into H_2O and oxygen, resulting in a measurable signal. Electrochemical immunosensors address this phenomenon, where a current or voltage signal is produced from decomposition. Common nanoparticles that exhibit this property are AuNPs, CeO_2, FeO4, etc. One such example of enzyme mimicking was developed by Fareeha et al. in the design of optical sensors. They reported colloidal AuNPs/nanoplatelets of graphene/chitosan nanozyme composites possessing peroxidase-mimicking properties to carry out porcine gelatin detection. Nanozyme I, in the presence of 3,3′5,5′-tetramethylbenzidine and H_2O_2, produced a colorimetric response with a detection limit of 86.42 pg/mL (Arshad et al. 2024). By catalyzing redox reactions and amplifying signals generated during the interaction of bioelements and analytes, nanozymes enhance the sensitivity of immunosensors. Other nanoparticles exhibit oxidase-like activity, in which substrate oxidation in the presence of oxygen is catalyzed by NPs, and they often produce H_2O_2 as a byproduct. Nanomaterials such as CuO, AuNPs, and CeO possess this behavior. As a result, they amplify the signal by facilitating the generation of H_2O_2 from organic molecules or analyte oxidation (Ndolomingo et al. 2020).

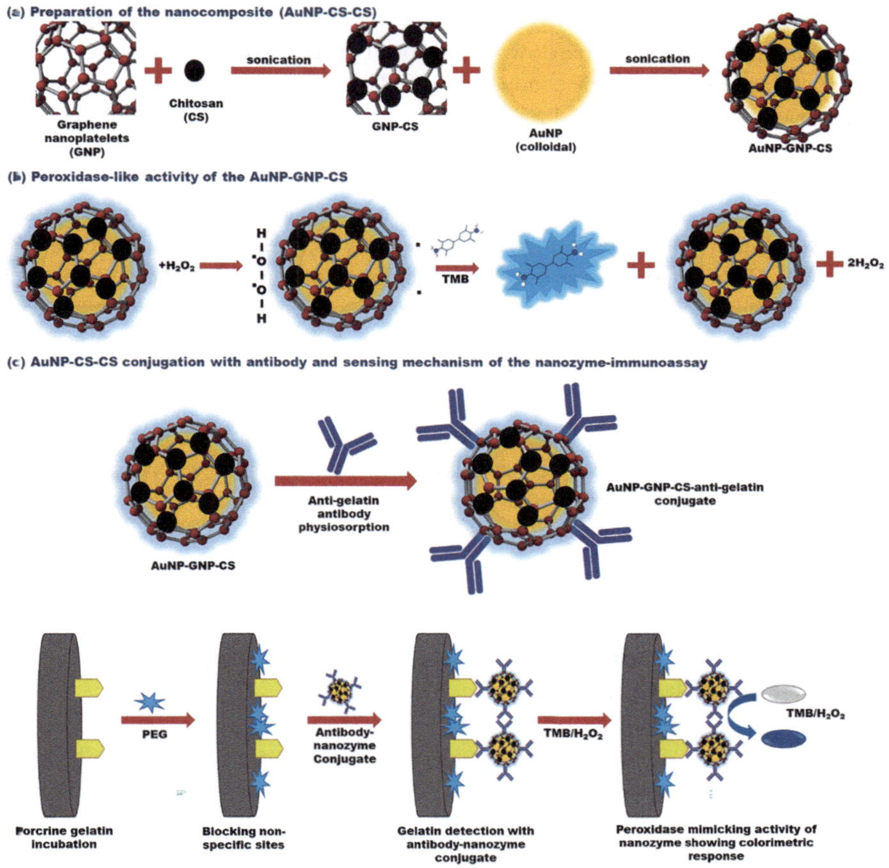

Fig. 1 Nanozyme-based colorimetric detection of porcine gelatin immunosensors. The figure shows the preparation of colloidal AuNPs, graphene nanoplatelets, and chitosan nanocomposites, and the use of this nanozyme instead of HRP for catalysis. Furthermore, the 2D nanonetwork increases the number of active sites, increasing the sensitivity of the sensor

Redox cycling and signal amplification are two main mechanisms by which nanoparticles enhance the performance of immunosensors. The repeated oxidation and reduction of an analyte or redox mediator in an electrochemical reaction involves redox cycling, which is particularly common in metallic nanoparticles such as PtNPs or AuNPs. The interaction of the antigen and the antibody on the sensor surface involves a mediator that can cause a redox reaction to occur. A mediator can be any electroactive species, such as H_2O_2 or any other. Multiple active sites for electron transfer and increased contact areas between the analyte and the surface are provided by the nanoparticles. This cycling of oxidation and reduction mechanisms often amplifies the original signal multiple-fold, thus enabling the detection of trace amounts of analytes (Cao et al. 2020). Conductive nanomaterials such as graphene or CNTs and metallic NPs efficiently improve electron transfer at the surface of the sensor. In electrochemical immunosensors, these conductive materials allow easy electron flow throughout the redox reaction. Owing to their larger surface area and high conductance, CNTs and graphene oxide, along with their nanocomposites, are used for improving signal amplification (Naresh and Lee 2021).

Fluorescence and surface plasmon resonance also enhance the signal in optical immunosensors, as they provide an additional layer of signal enhancement (Srivastava et al. 2021). NPs interact with light, which results in shifts in the case of SPR and better fluorescence due to the increased intensity of scattered or emitted light generated in response to Ag–Ab interactions upon binding. NPs concentrate light at a particular surface, which leads to signal enhancement. By encoding signal transduction, nanomaterials improve the performance of immunosensors through various mechanisms, such as catalytic amplification, nanozyme catalysis, electron transfer enhancement, and redox cycling. The incorporation of nanoparticles in sensors makes them more stable, sensitive, and reliable for the detection of a wide range of analytes in medical diagnosis, food safety, and environmental monitoring applications (Darwish et al. 2018).

4 Mechanism of Action of Nanoparticles in Signal Transduction and Amplification for Immunosensor Innovation

NPs can directly catalyze chemical reactions on the sensor surface, substantially increasing the response and sensitivity. Catalysis can be direct, label-based, nanozyme-based, or combined with other polymers for synergetic effects. Owing to their high surface energy and area, nanoparticles efficiently provide a suitable environment for catalytic reactions when antigen–antibody binding takes place. Basically, the target analyte triggers the redox reaction at the surface of the sensor, and the nanoparticles provide active sites for the reaction to occur more efficiently by interacting with the substrates, thus accelerating the rate of the reaction for faster signal generation. Various metal oxides catalyze the decomposition of H_2O_2 or other substrates in a manner similar to how peroxidase enzymes work. The production of electrochemical signals is catalyzed by these nanozymes, for example, in the generation of current by an electrochemical immunosensor. Figure 2 shows the enzyme-mimicking mechanism of nanozymes in a sandwich system for the detection of analytes. It first describes the catalysis of H_2O_2 via the HRP enzyme, and subsequently, the nanocomposite PdNPs/NH_2-ZIF-67 acts as a nanozyme for hydrogen peroxide catalysis. PdNPs are efficient nanozymes for the oxide–peroxide reaction, resulting in enhanced electrochemical signals and sensitivity (Polli et al. 2024). Nanoparticles also act as intermediates to shuttle electrons from the surface of the electrode to the redox species, leading to a faster reaction rate and amplified intensity of the signal. Certain nanoparticles can also promote dehydration or hydrolysis reactions that can be linked to antigen–antibody interactions. Hartati et al. constructed an immunosensor using PEGylated CeO nanoparticles for the detection

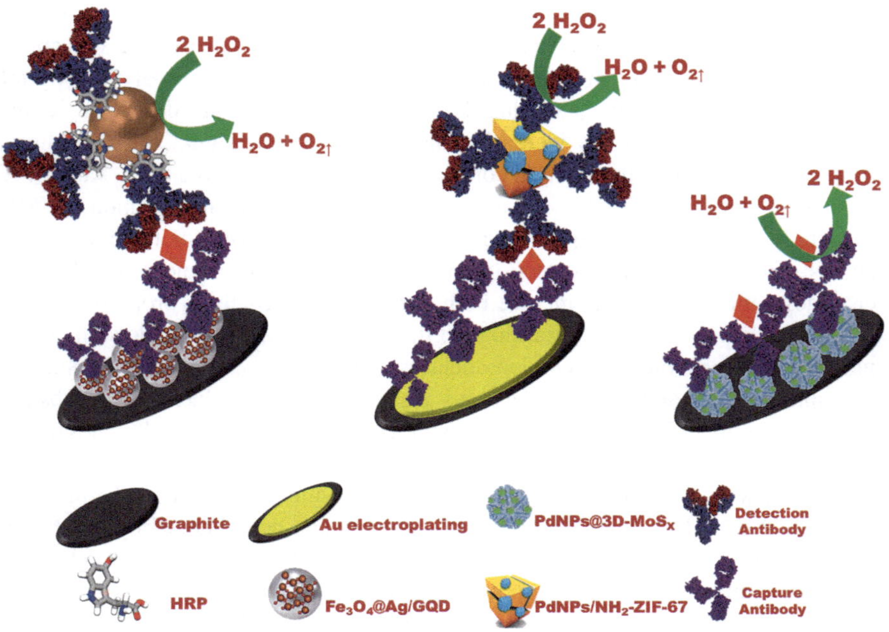

Fig. 2 Role of nanoparticles in electrochemical immunosensors with HRP-labeled detection antibodies that are capable of catalyzing the H_2O_2 reaction of its mediator in the electrochemical cell and a nanozyme-mimicking enzyme with the catalytic property of H_2O_2 instead of the HRP enzyme. PdNP efficiency in catalysis via a peroxide–oxide reaction

of Herceptin. The electrochemical immunosensor platform was fabricated with a CeO–anti-HER2 biconjugate on a screen-printed paper electrode. The sensor was able to detect 34.9 pg/mL HER2 in a human serum sample (Hartati et al. 2020). Direct catalysis can function effectively over a wide variety of environmental conditions, such as temperature, pH, and humidity, and is stable, which makes it ideal for monitoring the stability of enzymes. The cost-effectiveness and versatility of these methods aid in their practical application in various fields. Tan et al. developed a label-free immunosensor for the detection of hepatitis B surface antigens using PtPd nanocubes on MoS_2 via in situ oxidation polymerization. This nanozyme exhibited peroxidase-like activity, synergetic amplification, and good conductivity, and was able to detect 10.2 fg/mL of hepatitis B (Tan et al. 2019).

Label-based catalysis of nanoparticles utilizes NPs as tags or labels conjugated to recognition elements or specific antibodies for analyte detection. A cascade of catalytic events is triggered by analytes, with NPs facilitating redox reactions or nanozyme activity. The labeled NPs also participate in redox cycling, where the analyte induces a chemical reaction and labeled NPs catalyze the process by providing additional active sites for electron transfer. On the other hand, nanozymes catalyze the breakdown of substrates to electrochemically detectable products for effective binding of the analyte to its corresponding bioelements. Du et al. developed an ultrasensitive immunosensor for the electrochemical detection of cortisol via the use of Au nanozymes fabricated via an HRP-labeled anti-cortisol antibody. This single-atom nanozyme showed an excellent response toward cortisol and was able to detect its concentration in human saliva, with a detection limit of 0.48 pg/mL (Du et al. 2024). For optical sensing, NP labels facilitate localized SPR or fluorescence in immunosensors by increasing the intensity of the signal and providing a proportional signal to the analyte concentration. Wang et al. prepared Au@Cu_2O core-shell nanocatalysts with peroxidase-like activity for efficient oxidation of TMB. The study proved the concept of alternative click-chemistry-promoted colorimetric analysis for small molecule detection (Wang et al. 2023). The synergistic effects of enzymes with nanoparticles include both enzymatic and catalytic properties, which facilitate a large surface area for enzyme function, generate detectable signals, and increase the reaction rate. Oxidoreductases containing Cu or Fe ions, such as tyrosinase, HRP, and laccase, serve as electrocatalysts for facilitating substrate redox reactions. When supplemented with external oxidants, small redox-active molecules such as ferrocene and TMB can function as mediators to facilitate electron transfer between the sensing electrode and the enzyme's active site (Thamilselvan and Kim 2024). The surface properties of sensors can be enhanced by combining nanoparticles with some polymers to improve bioelement stability and provide better signal amplification. Conductive polymers such as polypyrrole and polyaniline can create stable and responsive surfaces for analyte detection. Linting et al. developed an immunosensor for aflatoxin B1 detection based on graphene oxide, AuNPs, and 2,5-di-(2-thienyl)-1-pyrrole-1-(p-benzoic acid) conjugates electrodeposited on the electrode surface (Linting et al. 2012). The sensor was stable for up to 26 weeks, with a detection limit of less than 1 fM. This method was successfully applied in practical applications for aflatoxin B1 determination in spiked food samples. Quantum-based signal amplification uses semiconductor nanoparticles with unique optical properties such as tunable size, size-dependent fluorescence, etc. In contrast, QDs produce fluorescence signals proportional to the analyte concentration that is produced when the QDs are excited by light, and at particular wavelengths, fluorescence is emitted that can be used to track or quantify the analyte. Electrochemical and optical sensors have various applications in food safety and quality control, medical diagnostics, environmental monitoring, and point-of-care testing.

5 Recent Advances, Challenges, and Future Directions in Nanoparticle-Catalyzed Immunosensors

Over the years, nanoparticle integration into immunosensor platforms has led to fundamental improvements in detection sensitivity, even at very low analyte concentrations. Properties such as high catalytic activity, high surface area, and the ability to enhance signals have drastically contributed to the development of ultra-sensitive immunosensors with a wide range of applications. For example, Zhou et al. built 3D-PdPtCuRu mesoporous nanospheres with improved catalytic activity for signal amplification for the detection of circulating tumor cells. The biosensor was able to detect 10 cells/mL for model A549 cells in spiked serum samples (Zhou et al.

2023). Carbon-based nanoparticles are known to improve the performance of electrochemical sensors because of their electron conductivity, surface area, and ease of functionalization. Nanoparticle-catalyzed immunosensors have revolutionized the field of ultrasensitive detection. Along with exploiting the unique properties of quantum dots, graphene oxide, and CNTs, researchers have developed highly specific and sensitive sensors with broad applications ranging from medical to agricultural fields. Advancements in POCTs have led to the emergence of lateral flow assays because of their simplicity, low cost, and easy handling. One such example is the pregnancy test, which uses colloidal AuNPs as labels for the detection of human chorionic gonadotropin hormone from urine samples (Zhong et al. 2021). Some other examples of onsite colorimetric detection based on gold and silver nanoparticles are for the detection of heavy metals in various samples (Ali et al. 2023). They have also played a critical role in the development of immunosensors for the detection of respiratory diseases such as COVID-19, tuberculosis, and influenza. Rapid and early diagnosis with immediate action has proven to be important (Bardhan et al. 2021; Maghoul et al. 2022; Nikolaev et al. 2024). Recently, portable nanoPCR devices have used GO and AuNPs for DNA amplification and detection to facilitate amplification and better sensitivity (Anuradha and Bhatia 2021). In addition to these advancements, lab-on-a-chip microfluidic devices process minute volumes of sample liquids, and upon NP integration, these systems enhance diagnostic performance and facilitate automated sample preparation with minimal operator intervention (Ali et al. 2022). Wearable glucose sensors utilize AuNPs functionalized with glucose oxidase to detect glucose levels in sweat and blood or interstitial fluids. The electrochemical response allows real-time measurements (Mohapatra et al. 2018).

Stability and reusability issues remain a challenge for immunosensors when they are exposed to complex biological environments such as serum, blood, or saliva. Nanoparticle aggregation, cross-reactivity with nonspecific targets, and leaching of toxic substances remain challenges that potentially affect the biocompatibility and performance of the sensor.

Designing robust nanomaterials that possess better biocompatibility and chemical stability with mechanical resilience is essential for improving the performance of immunosensors in challenging environments. The integration of smart devices for real-time monitoring and data analysis can improve the long-term usability of sensors and their applications in various fields other than immunosensors, such as drug delivery, therapeutics, etc. (Yeganeh et al. 2022; Kumar et al. 2022; Panwar et al. 2022; Raina et al. 2023; Mohanty et al. 2023; Pramanik et al. 2023; Yadav et al. 2023; Worku et al. 2024). The mass production of cost-effective nanoparticle-based sensors makes them applicable for widespread use in POCT, and advancements in surface functionalization and chemistry with the use of smart coatings can address issues related to aggregation, fouling, and degradation (Vashist et al. 2015; Polli et al. 2023).

6 Conclusion

Advancements in immunosensors have led to the emergence of nanoparticles playing a pivotal role, especially as catalysts enabling ultrasensitive, rapid, and specific analyte detection. NPs aid in signal amplification in optical, electrochemical, and magnetic immunosensors, thereby assisting direct or label-based catalysis. With these advancements, extremely low analyte detection is possible, enabling various applications in portable devices and point-of-care diagnosis. Immunosensors play critical roles in medical diagnosis, food safety, and environmental monitoring, and the incorporation of nanoparticles, whether metallic, metal oxide, quantum dot, or carbon-based, facilitates electron transfer and enhances signals, allowing improved performance. Silver, gold, or iron oxide nanoparticles possess unique redox and catalytic properties that increase the overall performance of the immunosensors.

References

Agrawal N, Saxena R, Singh L, Saha C, Kumar S (2022) Recent advancements in plasmonic optical biosensors: a review. ISSS J Micro Smart Syst 11:31–42. https://doi.org/10.1007/s41683-021-00079-0

Ali MA, Hu C, Yttri EA et al (2022) Recent advances in 3D printing of biomedical sensing devices. Adv Funct Mater 32:2107671. https://doi.org/10.1002/adfm.202107671

Ali S, Chen X, Shi W et al (2023) Recent advances in silver and gold nanoparticles-based colorimetric sensors for heavy metal ions detection: a review. Crit Rev Anal Chem 53:718–750. https://doi.org/10.1080/10408347.2021.1973886

Anuradha, Bhatia T (2021) Novel nanomaterials in forensic investigations: a review. Mater Today Proc 50:1071–1079. https://doi.org/10.1016/j.matpr.2021.07.466

Arshad F, Nurul Azian Zakaria S, Uddin Ahmed M (2024) Nanohybrid nanozyme based colourimetric immunosensor for porcine gelatin. Food Chem 438:137947. https://doi.org/10.1016/j.foodchem.2023.137947

Asal M, Özen Ö, Şahinler M et al (2019) An overview of biomolecules, immobilization methods and support materials of biosensors. Sens Rev 39:377–386. https://doi.org/10.1108/SR-04-2018-0084

Aydin M, Aydin EB, Sezgintürk MK (2021) Advances in immunosensor technology. Adv Clin Chem 102:1–62. https://doi.org/10.1016/bs.acc.2020.08.001

Bardhan NM, Jansen P, Belcher AM (2021) Graphene, carbon nanotube and plasmonic nanosensors for detection of viral pathogens: opportunities for rapid testing in pandemics like COVID-19. Front Nanotechnol 3:733126. https://doi.org/10.3389/fnano.2021.733126

Begum SJP, Pratibha S, Rawat JM et al (2022) Recent advances in green synthesis, characterization, and applications of bioactive metallic nanoparticles. Pharmaceuticals 15:455. https://doi.org/10.3390/ph15040455

Bilal M, Barani M, Sabir F et al (2020) Nanomaterials for the treatment and diagnosis of Alzheimer's disease: an overview. NanoImpact 20:100251. https://doi.org/10.1016/j.impact.2020.100251

Bose R, Jayawant M, Raut R et al (2023) Cyclodextrin nanoparticles in targeted cancer theranostics. Front Pharmacol 14:1218867. https://doi.org/10.3389/fphar.2023.1218867

Cao L, Han GC, Xiao H et al (2020) A novel 3D paper-based microfluidic electrochemical glucose biosensor based on rGO-TEPA/PB sensitive film. Anal Chim Acta 1096:34–43. https://doi.org/10.1016/j.aca.2019.10.049

Darwish NT, Sekaran SD, Alias Y et al (2018) Immunofluorescence–based biosensor for the determination of dengue virus NS1 in clinical samples. J Pharm Biomed Anal 149:591–602. https://doi.org/10.1016/j.jpba.2017.11.064
Dorozhko E, Kazachinskaia E, Kononova Y et al (2023) Electrochemical immunoassay of antibodies using freshly prepared and aged conjugates of silver nanoparticles. Talanta 253: 124028. https://doi.org/10.1016/j.talanta.2022.124028
Du Y, Guo M, Chen Y et al (2024) Ultrasensitive cortisol electrochemical immunosensor amplifying by Au single-atom nanozymes and HRP enzymes. Anal Chim Acta 1303:342462. https://doi.org/10.1016/j.aca.2024.342462
Felix FS, Angnes L (2018) Electrochemical immunosensors – a powerful tool for analytical applications. Biosens Bioelectron 102:470–478. https://doi.org/10.1016/j.bios.2017.11.029
Gallardo-Carreño I, Moreno-Paz M, Aguirre J et al (2020) A multiplex immunosensor for detecting perchlorate-reducing bacteria for environmental monitoring and planetary exploration. Front Microbiol 11:590736. https://doi.org/10.3389/fmicb.2020.590736
Gavrilaş S, Ştefan UC, Perţa-Crişan S et al (2022) Recent trends in biosensors for environmental quality monitoring. Sensors 22:1513. https://doi.org/10.3390/s22041513
Hartati YW, Letelay LK, Gaffar S et al (2020) Cerium oxide-monoclonal antibody bioconjugate for electrochemical immunosensing of HER2 as a breast cancer biomarker. Sens Bio-Sensing Res 27:100316. https://doi.org/10.1016/j.sbsr.2019.100316
Kala D, Sharma TK, Gupta S et al (2021) Graphene oxide nanoparticles modified paper electrode as a biosensing platform for detection of the htra gene of o. Tsutsugamushi. Sensors 21:4366. https://doi.org/10.3390/s21134366
Kaushal A, Kumar A, Kala D et al (2022) AuNPs-/c-MWCNT-/GQDs-modified electrode-based genosensor for early diagnosis of rheumatic heart disease. Chem Pap 76:4621–4626. https://doi.org/10.1007/s11696-022-02163-7
Khan MS, Zhu W, Ali A et al (2019) Electrochemiluminescent immunosensor for prostate specific antigen based upon luminol functionalized platinum nanoparticles loaded on graphene. Anal Biochem 566:50–57. https://doi.org/10.1016/j.ab.2018.11.010
Kumar A, Devi M, Kumar M et al (2022) Silicon nanostructures and nanocomposites for antibacterial and theranostic applications. Sensors Actuators A Phys 347:113912. https://doi.org/10.1016/j.sna.2022.113912
Li F, Guo Y, Wang X et al (2018) Multiplexed aptasensor based on metal ions labels for simultaneous detection of multiple antibiotic residues in milk. Biosens Bioelectron 115:7–13. https://doi.org/10.1016/j.bios.2018.04.024
Li Z, Zhang J, Huang Y et al (2022) Development of electroactive materials-based immunosensor towards early-stage cancer detection. Coord Chem Rev 471:214723. https://doi.org/10.1016/j.ccr.2022.214723
Linting Z, Ruiyi L, Zaijun L et al (2012) An immunosensor for ultrasensitive detection of aflatoxin B 1 with an enhanced electrochemical performance based on graphene/conducting polymer/gold nanoparticles/the ionic liquid composite film on modified gold electrode with electrodeposition. Sensors Actuators B Chem 174:359–365. https://doi.org/10.1016/j.snb.2012.06.051
Liu M, Sun X, Liao Z et al (2019) Zinc oxide end-capped Fe3O4@mSiO2 core-shell nanocarriers as targeted and responsive drug delivery system for chemo-/ions synergistic therapeutics. Drug Deliv 26:732–743. https://doi.org/10.1080/10717544.2019.1642419
Maghoul A, Simonsen I, Rostami A et al (2022) An optical modeling framework for coronavirus detection using graphene-based nanosensor. Nanomaterials 12:2868. https://doi.org/10.3390/nano12162868
Manjunathan J, Revathi M, Sowmya H et al (2024) Recent advancements in nanotechnological approaches for pollution monitoring and environmental sustainability. Clean Techn Environ Policy 26:3667–3683. https://doi.org/10.1007/s10098-023-02676-z
Manoj D, Manigandan R, Rajendran S et al (2021) Self-assembled dendrite-like 3D-CeO2 nanostructures for non-enzymatic vitamin B2 sensor. Mater Lett 295:129834. https://doi.org/10.1016/j.matlet.2021.129834

Matysiak-Brynda E, Wagner B, Bystrzejewski M et al (2018) The importance of antibody orientation in the electrochemical detection of ferritin. Biosens Bioelectron 109:83–89. https://doi.org/10.1016/j.bios.2018.02.063

Mishra A, Bhatt R, Bajpai J et al (2021) Nanomaterials based biofuel cells: a review. Int J Hydrog Energy 46:19085–19105. https://doi.org/10.1016/j.ijhydene.2021.03.024

Mohanty S, Panwar V, Anoop G et al (2023) Eco-friendly phyllanthus emblica-based ionic polymer composite for enhanced mechanical, electrical, and wearable sensing performance. Mater Chem Phys 303:127791. https://doi.org/10.1016/j.matchemphys.2023.127791

Mohapatra J, Ananthoju B, Nair V et al (2018) Enzymatic and non-enzymatic electrochemical glucose sensor based on carbon nano-onions. Appl Surf Sci 442:332–341. https://doi.org/10.1016/j.apsusc.2018.02.124

Nangan S, Kanagaraj K, Kaarthikeyan G et al (2023) Sustainable preparation of luminescent carbon dots from syringe waste and hyaluronic acid for cellular imaging and antimicrobial applications. Environ Res 237:116990. https://doi.org/10.1016/j.envres.2023.116990

Naresh V, Lee N (2021) A review on biosensors and recent development of nanostructured materials-enabled biosensors. Sensors (Switzerland) 21:1–35. https://doi.org/10.3390/s21041109

Natrayan L, Arul Kumar PV, Dhanraj JA et al (2022) Synthesis and analysis of impregnation on activated carbon in multiwalled carbon nanotube for Cu adsorption from wastewater. Bioinorg Chem Appl 2022:7470263. https://doi.org/10.1155/2022/7470263

Ndolomingo MJ, Bingwa N, Meijboom R (2020) Review of supported metal nanoparticles: synthesis methodologies, advantages and application as catalysts. J Mater Sci 55:6195–6241. https://doi.org/10.1007/s10853-020-04415-x

Nikolaev VV, Lepekhina TB, Alliluev AS et al (2024) Quantum dot-based nanosensors for in vitro detection of mycobacterium tuberculosis. Nanomaterials 14:1553. https://doi.org/10.3390/nano14191553

Pan NY, Shih JS (2004) Piezoelectric crystal immunosensors based on immobilized fullerene C60-antibodies. Sensors Actuators B Chem 98:180–187. https://doi.org/10.1016/j.snb.2003.09.034

Pandit C, Banerjee S, Pandit S et al (2023) Recent advances and challenges in the utilization of nanomaterials in transesterification for biodiesel production. Heliyon 9:e15475. https://doi.org/10.1016/j.heliyon.2023.e15475

Panwar V, Anoop G, Sharma M et al (2022) Sugarcane liquid-generated silver nanoparticles connected ionic polymer nanocomposite for enhanced electrical and wearable sensing signals. Mater Today Chem 26:101195. https://doi.org/10.1016/j.mtchem.2022.101195

Polli F, D'Agostino C, Zumpano R et al (2023) ASu@MNPs-based electrochemical immunosensor for vitamin D3 serum samples analysis. Talanta 251:123755. https://doi.org/10.1016/j.talanta.2022.123755

Polli F, Simonetti F, Surace L et al (2024) Nanoparticles in electrochemical immunosensors – a concept and perspective. Chem Electro Chem 11:e202300408. https://doi.org/10.1002/celc.202300408

Pramanik B, Sar P, Bharti R et al (2023) Multifactorial role of nanoparticles in alleviating environmental stresses for sustainable crop production and protection. Plant Physiol Biochem 201:107831. https://doi.org/10.1016/j.plaphy.2023.107831

Raina N, Shiekh PA, Kumar A et al (2023) Oxygen-generating nanomaterials for chronic wound management. Nanomedicine 18:845–847. https://doi.org/10.2217/nnm-2023-0090

Sabarees G, Tamilarasi GP, Velmurugan V et al (2023) Emerging trends in silk fibroin based nanofibers for impaired wound healing. J Drug Deliv Sci Technol 79:103994. https://doi.org/10.1016/j.jddst.2022.103994

Samadi Pakchin P, Ghanbari H, Saber R et al (2018) Electrochemical immunosensor based on chitosan-gold nanoparticle/carbon nanotube as a platform and lactate oxidase as a label for detection of CA125 oncomarker. Biosens Bioelectron 122:68–74. https://doi.org/10.1016/j.bios.2018.09.016

Sharma S, Gupta S, Saini AK et al (2024) Portable graphene oxide modified immunosensor for the detection of vitamin D3. Microchem J 197:109745. https://doi.org/10.1016/j.microc.2023.109745

Shen W, Gao Z (2015) Quantum dots and duplex-specific nuclease enabled ultrasensitive detection and serotyping of Dengue viruses in one step in a single tube. Biosens Bioelectron 65:327–332. https://doi.org/10.1016/j.bios.2014.10.060

Soleymani J, Golsanamluo Z (2021) Advanced materials for immunosensing of pharmaceutical and drug compounds. ImmunoAnalysis 1:5. https://doi.org/10.34172/ia.2021.05

Soria S, Berneschi S, Brenci M et al (2011) Optical microspherical resonators for biomedical sensing. Sensors 11:785–805. https://doi.org/10.3390/s110100785

Srivastava S, Singh L, Kaushik V et al (2021) Electrically controlled nanophotonic slot structure based on photocatalytic nanocomposite for optical detection of foodborne pathogens. J Light Technol 39:6670–6677. https://doi.org/10.1109/JLT.2021.3104409

Tan Z, Dong H, Liu Q et al (2019) A label-free immunosensor based on PtPd NCs@MoS2 nanoenzymes for hepatitis B surface antigen detection. Biosens Bioelectron 142:111556. https://doi.org/10.1016/j.bios.2019.111556

Thamilselvan A, Kim MI (2024) Recent advances on nanozyme-based electrochemical biosensors for cancer biomarker detection. TrAC-Trends Anal Chem 177:117815. https://doi.org/10.1016/j.trac.2024.117815

Vashist SK, Luppa PB, Yeo LY et al (2015) Emerging technologies for next-generation point-of-care testing. Trends Biotechnol 33:692–705. https://doi.org/10.1016/j.tibtech.2015.09.001

Voon CH, Sam ST (2018) Physical surface modification on the biosensing surface. In: Nanobiosensors for biomolecular targeting, pp 23–50. https://doi.org/10.1016/B978-0-12-813900-4.00002-6

Wang X, Pang G (2015) Amplification systems of weak interaction biosensors: applications and prospects. Sens Rev 35:30–42. https://doi.org/10.1108/SR-03-2014-629

Wang PY, Wang WJ, Li WP (2023) Serine-modified au@Cu2O Core-Shell nanoparticles for catalysis-mediated colorimetric detection of small molecules. ACS Appl Nano Mater 6:15651–15662. https://doi.org/10.1021/acsanm.3c02468

Worku LA, Tadesse MG, Bachheti A et al (2024) Experimental investigations on PVA/chitosan and PVA/chitin films for active food packaging using Oxytenanthera abyssinica lignin nanoparticles and its UV-shielding, antimicrobial, and antiradical effects. Int J Biol Macromol 254:127644. https://doi.org/10.1016/j.ijbiomac.2023.127644

Yadav S, Chamoli S, Kumar P et al (2023) Structural and functional insights in polysaccharides coated cerium oxide nanoparticles and their potential biomedical applications: a review. Int J Biol Macromol 246:125673. https://doi.org/10.1016/j.ijbiomac.2023.125673

Yang M, Javadi A, Gong S (2011) Sensitive electrochemical immunosensor for the detection of cancer biomarker using quantum dot functionalized graphene sheets as labels. Sensors Actuators B Chem 155:357–360. https://doi.org/10.1016/j.snb.2010.11.055

Yeganeh FE, Yeganeh AE, Far BF et al (2022) Synthesis and characterization of tetracycline loaded methionine-coated NiFe2O4 nanoparticles for anticancer and antibacterial applications. Nano 12:2286. https://doi.org/10.3390/nano12132286

Zhang X, Yu Y, Shen J et al (2020) Design of organic/inorganic nanocomposites for ultrasensitive electrochemical detection of a cancer biomarker protein. Talanta 212:120794. https://doi.org/10.1016/j.talanta.2020.120794

Zhang M, Xu F, Cao J et al (2024) Research advances of nanomaterials for the acceleration of fracture healing. Bioact Mater 31:368–394. https://doi.org/10.1016/j.bioactmat.2023.08.016

Zhong ZT, Song LB, Li CQ et al (2021) Lateral flow biosensor for universal detection of various targets based on hybridization chain reaction amplification strategy with pregnancy test strip. Sens Actuators B: Chem 337:129778. https://doi.org/10.1016/j.snb.2021.129778

Zhou X, Bai D, Yu H et al (2023) Detection of rare CTCs by electrochemical biosensor built on quaternary PdPtCuRu nanospheres with mesoporous architectures. Talanta 253:123955. https://doi.org/10.1016/j.talanta.2022.123955

Detection of Foodborne Pathogens in Milk and Milk-Based Products Using Immunosensors

Rachna Poria, Renu Poria, Ankur Kaushal, and Shagun Gupta

Abstract

Foodborne illness associated with pathogens, toxins, and other contaminants can cause serious health problems that are dangerous for humans. Owing to the presence of sufficient nutrients in dairy products, they are recommended on a regular basis. However, these nutrients promote the development of pathogens that lead to foodborne illness. Over the past few years, researchers have focused more on determining the presence of live foodborne pathogens in dairy products. Biosensors, which can convert biological, chemical, or biochemical signals into measurable electrical signals, are innovative methods for quickly identifying foodborne pathogens in the food industry. One of the most widely used types of sensors is the immunosensor, which is mainly used in food safety, environmental monitoring, and diagnostic areas. Researchers have developed various rapid, accurate, and affordable biosensor-based analytical tools to identify milk-borne pathogens that could help the dairy industry reduce costs and improve production management. The current chapter aims to provide an overview of various nanomaterial-based immunosensors for the detection of several foodborne bacterial toxins and contaminants, emphasizing their principles, benefits, and drawbacks in terms of sensitivity, multiplexing capability, and simplicity in milk samples. It also focuses on the application of these sensors for the determination of pathogenic bacteria in food samples.

Keywords

Foodborne pathogens · Immunosensor · Milk and milk-based products

R. Poria · R. Poria · A. Kaushal · S. Gupta (✉)
Department of Bio-Sciences and Technology, Maharishi Markandeshwar Engineering College, Maharishi Markandeshwar (Deemed to be) University, Mullana, Ambala, India

K. Kuča et al. (eds.), *Ensuring Wellness: Immunosensors Monitoring Food Quality and Pathogenicity Risks*, https://doi.org/10.1007/978-3-031-97477-9_3

1 Introduction

Milk and milk-based products, including proteins, calcium, vitamins, and fats, which are crucial for human health and development, are fundamental to human nutrient requirements. All age groups of humans use these products, ranging from milk and cheese to yogurt and butter, which are widely consumed (Górska-Warsewicz et al. 2019). However, the nutrient richness of dairy products makes them valuable and similarly produces an environment highly conducive to the growth of harmful microorganisms, including foodborne pathogens such as *Salmonella, Listeria monocytogenes, and E. coli.* These pathogens can easily infect dairy products during milking, processing, packaging, or even during storage if proper hygiene is not performed (Yeganeh et al. 2022). Bacterial replication is the most critical issue since most types of dairy products are characterized by high levels of moisture and a neutral pH (Górska-Warsewicz et al. 2019). In the past decade, societies around the world have observed a trend toward increased outbreaks related to dairy foods, which may be attributed to products made from raw milk or contaminated dairy products. Such outbreaks have numerous consequences for public health, resulting in severe forms of gastrointestinal diseases and sometimes even death among the entire population; however, these outbreaks specifically target the most vulnerable groups of people, including children, elderly individuals, and immunocompromised individuals (Grace et al. 2020). The increase in these incidences requires better techniques for tracking the contamination of these foods with disease-causing organisms, ensuring that they remain safe for consumption. Current methods used for the detection of pathogens in dairy products are crucial for preventing foodborne disease, but these methods have limitations that hamper the monitoring process (Yunus 2019). Conventional methods such as culture-based and PCR methods and enzyme-linked immunosorbent assays are also known as ELISAs. However, these methods have limitations, such as being slow and time-consuming, ranging from several hours to days to produce results (Priyanka et al. 2016). Culture methods are believed to be highly accurate; however, growing bacteria in a laboratory requires a substantial amount of time and may require safety measures. PCR and ELISA are faster than antigen tests, but these tests are more challenging and time-consuming and still require highly qualified personnel, instruments, and procedures, increasing the cost of the tests (Law et al. 2015). In addition, these approaches often lack the ability to detect a low level of contamination, which is crucial for analyzing the potential of an outbreak since few pathogenic cells may proliferate and infect. This is challenging with dairy foods, where foodborne pathogens can be present in very small amounts but can still be dangerous. The fact that detection methods require rapid, sensitive, and highly accurate tools is therefore critical for ensuring food safety in the dairy industry (Chen et al. 2017). The delivery of these demands suggests that biosensors that are based on another technology can provide a probable solution. Electrochemical biosensors, including immunosensors, enhance pathogen detection, thus enabling a rapid response to potential contamination (Kumar et al. 2022). They also have high sensitivity, which is important for detecting pathogen levels and is essential for early intervention and preventing outbreaks (Cesewski and

Johnson 2020). These modern tools are essential for improving food safety, decreasing the probability of contamination or threats in dairy products, and enhancing public health in the face of evolving challenges in the dairy sector. Biosensors are analytical devices that incorporate biological materials, including enzymes, antibodies, or nucleic acids, and a physicochemical transducer to convert biological interactions into measurable signals (Manjunathan et al. 2024). In the context of food safety, biosensors should be used to detect the pathogens present in food, utilizing certain biological markers, such as antigens or toxins (Curulli 2021). These sensors are highly important for the detection of pathogens and offer rapid, accurate, and ultrasensitive analyses of food samples. Although accurate and efficient, conventional techniques are often time-consuming and labor-intensive, although biosensors can provide real-time results, enabling quicker responses to contamination events. Especially in industries such as dairy, determining the presence of certain pathogens is crucial for preventing contaminated products from reaching consumers, thus reducing the risk of foodborne illnesses (Nath 2024). Biosensors used in food analysis help increase food safety and quality assurance because they enable constant or periodic testing of food products at various stages of production, processing, and distribution. They can detect all levels of pathogen concentrations; this enables high-level accuracy, which is very important since initial action must be initiated. In addition, they are convenient to use, compact, and capable of field use, such as on dairy farms, in processing plants, and in supermarkets (Kakimova et al. 2023). In addition, this capability ensures the enhancement of general quality control because producers can meet high safety indicators to minimize product recalls and losses. By integrating biosensors into routine testing procedures, the dairy industry can ensure a safer food supply, lower costs associated with contamination, and provide greater consumer confidence in the safety of dairy products (Xu et al. 2021).

2 Biosensor Technology in Food Safety

2.1 Overview of Biosensor Types

Biosensors are analytical devices that convert a biological response into a quantifiable signal by detecting specific biological or chemical substances. They typically consist of three core components: a biological recognition element, for example, enzymes, antibodies or DNA, a transducer, and a signal processor (Mehrotra 2016). In certain cases, a direct interaction of the target analyte, such as a pathogen or a toxin, with the biological recognition element leads to a specific biochemical event (Fig. 1). The transducer converts the biochemical signal into an electrical, thermal, optical, or mechanical signal (Wang et al. 2023). The signal is then amplified, usually by using NPs, and a quantitative form of analysis that could lead to the identification or quantification of the substance of interest is generated (Vikal et al. 2023). It helps in the rapid detection of harmful contamination in foods, providing an efficient way to ensure food safety in industries, including dairy and other food production sectors (Bharti et al. 2024).

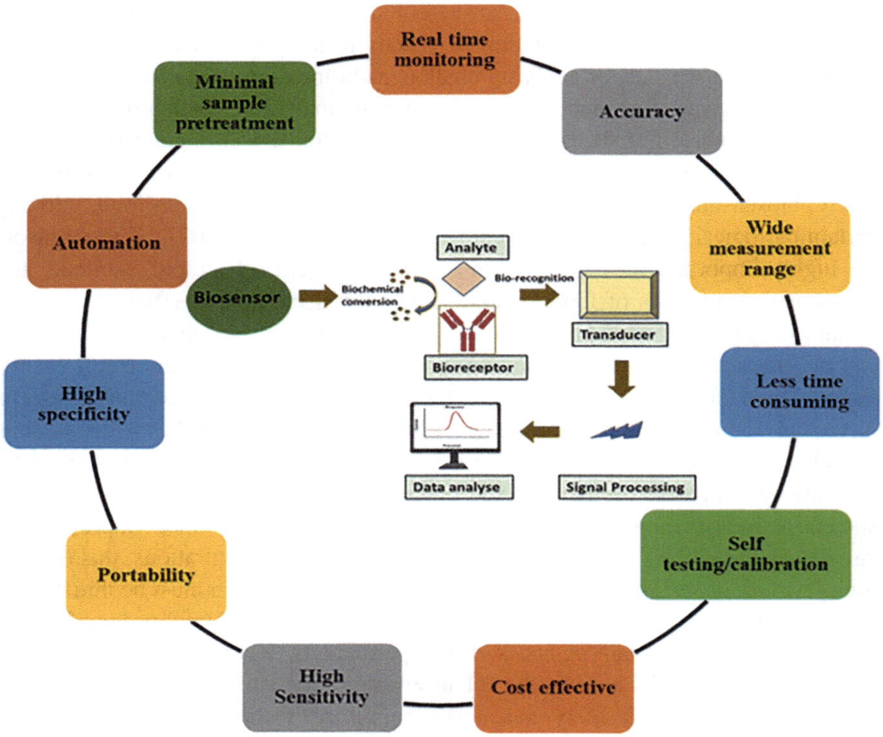

Fig. 1 Applications and unique characteristics of biosensors

2.2 Focus on Immunosensors: Definition, Working Mechanism, and Applications

Immunosensors are biosensors that utilize the immunological reactions of antibodies and antigens to detect target materials, such as pathogenic microorganisms, toxins, or allergens. They function by incorporating an antibody (or antigen) as the biological recognition element, which attaches to the specific counterpart (antigen or antibody) that is present in the sample (Janik-Karpinska et al. 2022). This signal can be measured by an electrochemical, optical, or piezoelectric transducer on the basis of the type of immunosensor. For example, when a change in an electrochemical immunosensor occurs, the reaction being measured is a change in electrical signals such as current or voltage. Owing to the exact nature of antibody–antigen binding, immunosensors are very selective, which makes them very helpful for applications involving food safety (Hassanpour et al. 2018). They are widely applied in the detection of pathogens *such as E. coli, Salmonella*, and *Listeria*, as well as harmful toxins and allergens in food products. These sensors are highly fast and specific for detecting the quality and safety of foods, and they help to avoid contamination outbreaks and the regulation of foods in industry. Because of its adaptability, versatility also allows for applications in other fields, such as clinical diagnostics and environmental monitoring (Begum et al. 2022; Ali et al. 2020).

2.3 Advantages of Biosensors for the Detection of Foodborne Pathogens in Dairy Foods

Biosensors are helpful for the detection of foodborne pathogens in dairy foods and are useful tools for ensuring food safety. Some of the benefits that have been described include their ability to detect lower concentrations of foodborne pathogens such as *E. coli, Salmonella*, and *Listeria* in the complex matrices of dairy products, providing accurate results with minimal false positives (Nnachi et al. 2022). Additionally, biosensors facilitate rapid detection and real-time monitoring, reducing the time required for pathogen identification compared with traditional methods, which often take several hours or even days. This speed is important for preventing contaminated products from reaching consumers. The portability and onsite use of biosensors have increased their usefulness because dairy companies can perform testing on farms, food processing plants, or distribution centers without the use of complex laboratory equipment (Biswas et al. 2024). This flexibility simplifies the detection process and hence makes it faster. In conclusion, biosensors are most often affordable because they require less manpower, fewer resources, and less money than traditional detection methods do, making them suitable for constant monitoring in the dairy industry. The idea is that all these benefits accrue to biosensors and can be deemed a valuable asset in the task of protecting dairy products from contamination (Castillo-Henríquez et al. 2020).

2.4 Immunosensors for the Detection of Pathogens

Immunosensors, coupled with highly specific interactions between antigens and antibodies, which are pathogens in food products, are widely used in sensory technology because of their accuracy. The fundamental operation of the mechanism involves contact of the antibody-based biological recognition element with its original antigen, which is situated on the surface of target microorganisms such as *Escherichia coli* or *Listeria monocytogenes* (Janik-Karpinska et al. 2022). This antigen–antibody interaction leads to a biological response that is detected by the immunosensor in this study. The sensor transducer transforms this particular biological event to yield a signal, which is usually electrical in nature and is directly proportional to the amount of pathogen detected. For example, in an electrochemical immunosensor, the interaction between the analyte and the corresponding antibody changes one or many of the electrical characteristics, such as current or voltage, and therefore enables a pathogen to be detected and quantified (Bhatia et al. 2024). The conversion of biological signals into electrical signals allows rapid detection of pathogens with a high degree of accuracy, making immunosensors valuable for monitoring food safety and quality (Daliri et al. 2019).

2.5 Nanomaterials in Immunosensors

Most nanomaterials, especially immunosensors, have focused on biosensing, and their chemical and physical properties are unique (Pramanik et al. 2023). Nanomaterials, which include nanoparticles, nanowires, nanotubes, and quantum dots, are characterized by dimensions of up to 100 nanometers or smaller in at least one dimension and exhibit promising optical, electrical, and catalytic behaviors (Joshi et al. 2023; Neethirajan et al. 2017). These materials are very useful for how biosensors operate since high surface-area-to-volume ratios improve the sensitivity and signal-to-noise ratio of biosensors (Masanam et al. 2022; Rauf et al. 2022). It is particularly useful in immunosensors, in which changes in the antibody–antigen reaction are used to test for biomolecules because such signals can be boosted by nanomaterials, including at very low concentrations (Bahadır and Sezgintürk 2017). Furthermore, nanomaterials tend to possess high stability, which increases the stability and longevity of immunosensors for various applications in biological matrices. The use of almost all types of nanomaterials in immunosensors has led to the development of higher sensitivity, detection limits, and signal amplification (Ansari 2023). For example, metals such as gold and silver are highly preferred because of their increased optical features, including LSPR, which amplifies signals from biomolecular reactions, real-time detection, and label-free detection (Djearamane et al. 2022). Graphene and carbon nanotubes have large surface-to-area ratios and better electrical conductivity (Mishra et al. 2021; Chakroborty et al. 2023; Upadhyay et al. 2023), which is beneficial for increasing the signal-to-noise ratio as well as increasing the electron transfer rates in electrochemical immunosensors. These materials are also quite flexible, making them adaptable to various biosensor configurations (Wang and Hu 2016). Another type of semiconductor nanomaterial consists of quantum dots because of their optical properties, such as high photostability and ability to facilitate fluorescence emission. These properties make QDs ideal for use in optical immunosensors that involve the use of fluorochromes for tagging, offering solutions of high sensitivity and multiplexing, enabling not only single but also multiple targets to be detected. When utilized in immunosensors, these nanomaterials enable optimization of sensor sensitivity and enable the development of sensors with low detection limits, especially for clinical biomarkers, environmental samples, or toxins and foodborne pathogens. Owing to their large surface area and ability to functionalize them in a variety of ways, the nanomaterials are optimal for biomolecule attachment because they provide dependable and selective identification of the target analytes. Therefore, the incorporation of nanomaterials such as metal nanoparticles, carbon-based nanomaterials, and QDs in immunosensors has led to tremendous changes in the sensing platform, which has increased the sensitivity and stability of selective biomolecular detection for further enhanced biosensing applications (Mohamadi et al. 2017).

3 Nanomaterial-Based Sensors for Foodborne Pathogens in Milk and Milk-Based Products via Immunosensors

3.1 *Escherichia coli*

E. coli O157:H7 is one of the most dangerous serotypes among *E. coli* strains and is famous for its toxins, which affect the mucous membrane of the intestines (Some et al. 2021). It can cause severe illnesses, including stomach cramps, anemia, kidney failure, and colon bleeding, through an infective dose of one hundred bacterial cells. It can be transmitted to humans through the consumption of raw or undercooked ground meat products and raw milk. Specific foods linked to outbreaks of *E. coli* O157:H7 are made up of ingredients that are lightly cooked, including hamburgers, milk, and dairy products such as certain types of cheese and yogurt, dried cured salami, and unpasteurized apple juice. In 2013, C.K. Joung developed a novel nanoporous membrane-based impedimetric immunosensor for the detection of bacterial pathogens in whole milk. This method involves the functionalization of a preexisting alumina nanoporous membrane with hyaluronic acid (HA), which helps eliminate most of the nonspecific binding of biomolecules to other cells but allows the antibodies to bind. *Escherichia coli* O157:H7 was chosen because it is one of the most pathogenic foodborne bacteria. The sensor measured shifts in ionic impedance through the nanopores due to antibody–pathogen interactions via NIC derived from impedance plots. A regression equation was developed to correlate NIC at 1 kHz with *E. coli* O157:H7 concentrations ranging from 10 to 10^{-5} cfu/mL, achieving a detection limit as low as 10 cfu/mL. Furthermore, the immunosensor successfully detected *E. coli* O157:H7 in samples enriched with whole milk, with a limit of detection of 83.7 cfu/mL and 95% probability (Joung et al. 2013). Another study introduced a disposable amperometric immune-sensing strip for the rapid detection of *Escherichia coli* O157:H7, which employs an indirect sandwich enzyme-linked immunosorbent assay with two antibodies. For electrochemical analysis, the carbon electrodes were functionalized with a primary antibody specific to *E. coli* O157:H7. The electrode was functionalized with a primary antibody specific to *E. coli* O157:H7, intact *E. coli* O157:H7 cells, and a secondary *E. coli* O157:H7-specific antibody conjugated with horseradish peroxidase (HRP). Hydrogen peroxide served as the substrate for HRP, whereas ferrocene dicarboxylic acid (FeDC) acted as the mediator. The tested current of the strips increased, and the highest response current (RC) was achieved when 13 nm AuNPs were coated on the working electrode. With this synthesis of AuNPs and FeDC, the RC increased by 13.1-fold. The SPCE-based strips demonstrated specificity for *E. coli* O157:H7 via PCR, with improved detection sensitivity and specificity within a range of 10^2 to 10^7 cfu/mL. The detection limits were also found to be approximately 6 cfu/strip in PBS buffer and 50 cfu/strip in milk (Lin et al. 2008). In 2018, E.C. Cho et al. fabricated a composite electrode with improved performance through the integration of peptide nanotubes. These nanotubes were prepared with cyclic peptides built with D- and L-amino acid monomers. Afterwards, the peptide nanotubes were cast on a screen-printed carbon paste electrode made through a sol–gel procedure. Antibodies specific to *E. coli*

O157:H7 were immobilized on the surface of peptide nanotubes. The anti-*E. coli* O157:H7 antibodies bound to the PN–SPCE effectively captured *E. coli* O157:H7 from the samples via antigen–antibody interactions. The response current was subsequently measured via cyclic voltammetry (Cho et al. 2008). This work presented a highly sensitive electrochemical immuno-ASV based on core-shell Cu@Au NPs as labels for the rapid detection of *Escherichia coli*. The characteristics of the Cu@Au NPs before and after antibody attachment were further investigated via transmission electron microscopy (TEM). In this assay, the Cu@Au-labeled antibodies were bound to *E. coli* attached to a surface coated with a polystyrene (PS)/ITO chip. Upon oxidation, the Cu@Au NPs disintegrated and made Cu^{2+} ions available in the solution, which were estimated at a GC/Nafion/Hg electrode via ASV. This modified electrode enhanced the Cu^{2+} detection sensitivity, with a detection limit as low as 9.0×10^{-12} mol/L. As the Cu@Au NP labels are present only when the antibody binds to *E. coli*, the content of Cu^{2+} ions is coupled with the amount of *E. coli*. The assay presented a limit of detection of 30 cfu/mL for *E. coli*, and the entire process took less than 2 h. Furthermore, with the incorporation of preenrichment in the electrochemical immunoassay, the detection limit of *E. coli* in the surface water samples obtained in this study was 3 cfu/10 mL (Zhang et al. 2009).

3.2 *Salmonella*

This bacterial infection is caused primarily by *Salmonella* species, including *Salmonella enteritidis* and *Salmonella typhimurium*. According to World Health Organization (WHO) data, this disease affects millions of people globally each year, resulting in approximately 100,000 fatalities. The common symptoms of salmonellosis include fever, abdominal cramps, diarrhea, nausea, and vomiting (Valdés et al. 2009; Naresh and Lee 2021). A highly sensitive electrochemical immunoassay for the rapid detection of *Salmonella typhi* (*S. typhi*) in biological samples has been designed via the use of gold nanoparticle labels combined with anodic stripping voltammetry. Specific monoclonals for the *S. typhi* isotype were first bound to polystyrene microwells for adherence, which then bound the *S. typhi* bacteria. A polyclonal antibody–colloidal gold conjugate was subsequently used to attach itself to the bacterial surface. A copper-enhancer solution, which included ascorbic acid and copper (II) sulfate, was then added. Ascorbic acid was used to bleach the copper (II) ion to copper (0), which was subsequently deposited on the gold nanoparticle labels. The remaining copper dissolved in nitric acid was assayed via anodic stripping voltammetry of the copper ions (Fig. 2). The amount of copper formed was proportional to the density of the gold nanoparticle tags, which itself was directly proportional to the *S. typhi* bound to the antibody–gold conjugate. The anodic stripping peak current increased as the concentration of *S. typhi* increased and was correlated with the range of 1.30×10^2 cfu/mL to 2.6×10^3 cfu/mL on a base 10 logarithmic scale, with an LOD of $\approx$ 98.9 cfu/mL. The interactions of key experimental conditions, such as the concentration of copper and time of interaction

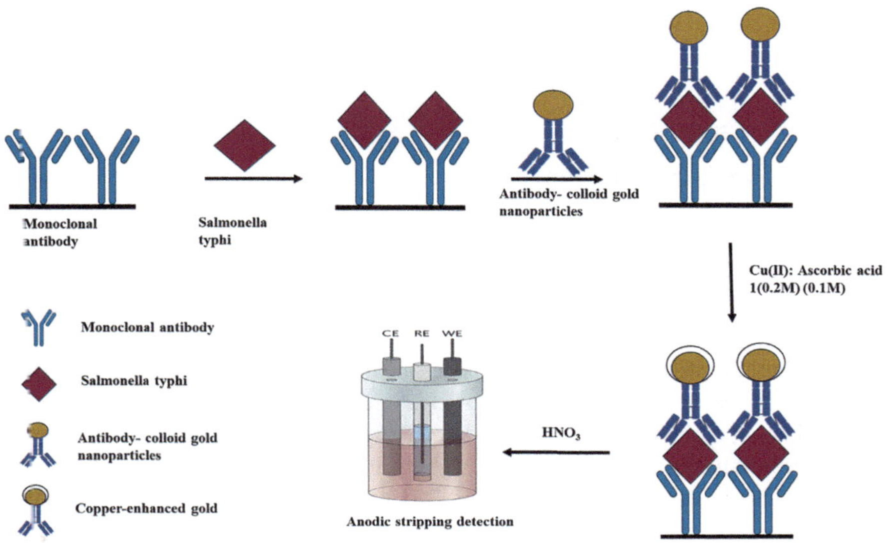

Fig. 2 Schematic diagram of the *S. typhi* immunoreaction

between *S. typhi* and antibodies, were studied. The present study employed this method for the successful detection of *S. typhi* in human serum, thus improving the sensitivity and reliability of nanoparticle-based immunoassays (Dungchai et al. 2008). Another study demonstrated that a new capacitive immunosensor for the detection of *Salmonella* spp. was fabricated through ethylene diamine grafting and self-assembled gold nanoparticle monolayers on a GC electrode. In this method of antibody conjugation to gold nanoparticles, McAbs specific for *Salmonella* spp. were adsorbed to the gold nanoparticles. The presence of *Salmonella* in the interaction with the McAbs was monitored by EIS, with the anti-GalNAc antibody concentration identified by a change in capacitance. Thus, the immunosensor was characterized by a linear dependence of the response on the *Salmonella* concentration within the range of 1.0×10^2 to 1.0×10^5 cfu/mL, with a detection limit of 1.0×10^2 cfu/mL. Sensor stability was retained for up to 8 weeks and began to decline to 85.3% of its initial stability at 4 months. Furthermore, the acyl component could also be regenerated by using a glycine-HCl buffer solution. An immunosensor was successfully used to detect *Salmonella* spiked in commercially processed pork samples (Zhang et al. 2009).

Semiconductor quantum dots are relatively new types of fluorophores that have been found to be very useful in pathogen detection. In this study, quantum dots were employed in immunoassays to identify *Salmonella typhimurium* in chicken wash water. Magnetic beads coated with anti-*Salmonella* antibodies were used to stain *Salmonella* cells, followed by the binding of biotin-tagged secondary antibodies. Finally, streptavidin-coated quantum dots that specifically recognize biotin in solution were incubated to create a detectable fluorescence signal. The FI increased with increasing *Salmonella* cell count and was linearly related to the FI and cell count in the range of 10^3–10^7 cfu/mL, with a detection limit of 10^3 cfu/mL. The model was fitted by FI = 198.6 Log N − 639.03, with a coefficient of determination of 0.96 (Yang et al. 2009a, b).

3.3 *Listeria monocytogenes*

Listeria monocytogenes is an aerobic, nonsporulating, gram-positive bacterium responsible for the disease listeriosis. This pathogen is recognized as the most toxigenic of all foodborne pathogens and is one of the three main causes of fatalities resulting from foodborne bacteria (Kakkar et al. 2023). Grossman et al. introduced a new method for identifying *Listeria monocytogenes* through the utilization of 50 nm antibody-coated superparamagnetic particles in an aqueous medium. An oscillation magnetization orients the particles, and a very sensitive SQUID identifies magnetic rejuvenation after the removal of the magnetization. The detection limit is approximately 5.6 million target cells in 20 μL, with further optimization to ~230 *L. monocytogenes* cells. Time-resolved measurements also reveal the binding rate between particles and bacteria (Grossman et al. 2004).

4 Nanomaterial-Based Sensors for the Detection of Bacterial and Food Toxins

4.1 *Staphylococcal enterotoxin*

Staphylococcal enterotoxins (SEs), which constitute 21 heat-stable proteins derived from *Staphylococcus aureus*, cause foodborne illnesses when they are consumed through contaminated foods. SE exposure results in symptoms that include nausea, vomiting, diarrhea, and decreased appetite even at low concentrations of 20–100 ng/person. Furthermore, they are linked to atopic eczema, rheumatoid arthritis, and toxic shock syndrome (Yang et al. 2008, 2009a). Compared with traditional detection methods such as ELISA, these methods are high throughput and tend to be sensitive to production rate optimization. To resolve this, Yang et al. (2008) introduced CNTs to directly immobilize them on a glass slide to build a novel optical immunosensor for the detection of SEs on the basis of an inverse method of sample antibodies, HRP-conjugated secondary antibodies, and tracking of HRP fluorescence. The proposed sandwich immunosensor is 6–8-fold more sensitive than existing methods and has a detection range of 0.1–100 ng/mL and a detection limit of 0.1 ng/mL. Later, in another study, the same group reported a GNP-based ECL immunosensor for SE detection in foods (Yang et al. 2009a). The first anti-SE antibody was anchored to the GNPs, and subsequently, the GNPs were adhered to the polycarbonate. A sandwich-type ELISA was used for ECL detection of the antirabbit IgG-HRP conjugates. This method has a detection limit of 0.01 ng/mL, which, in fact, is tenfold more sensitive than both ELISA and other immunosensors based on CNTs (Yang et al. 2008, 2009a). Similarly, GNPs are less toxic and easier to prepare than CNTs are and do not need to be functionalized with acid.

4.2 Ricin

Ricin, derived from castor bean *(Ricinus communis),* is a potent threat to bioterrorism because of its illicit application. It becomes toxic upon interaction with β-d-galactosamine or β-d-galactose receptors on host cells (Kaittanis et al. 2010; Uzawa et al. 2008). For the identification of ricin (Uzawa et al. 2008), a colorimetric assay was designed for the synthesis of gold nanoparticles conjugated with a β-glucosylceramide thiol. If ricin (RCA60) or the equivalent molecule agglutinin RCA120 is added, the color of the nanoparticles changes from red to purple. Other plant lectins or noncarbohydrate-binding proteins, such as Con A, PVA, BSA, or γ-globulin, resulted in no color change. This assay can identify concentrations as low as 3.3 μg/mL within 10 min or 1.7 μg/mL within 30 min. Another study used surface plasmon resonance enhanced by microfluidic devices fabricated from polydimethylsiloxane, etc., for the detection of ricin. By binding ganglioside-monosialic acid 1 receptor-containing lipid vesicles to a gold-nanoglassified surface via streptavidin, they obtained a detection range of 0.83–5.83 nM. It is less sensitive than other methods, but making changes to design and signal amplification may normally increase the sensitivity and multiplexing nature of the system (Uzawa et al. 2008).

4.3 Cholera Toxin (CT)

CT produced by *Vibrio cholerae* causes watery diarrhea in patients with cholera infections. AB5 is an oligomeric complex holoenzyme of bacterial toxins that consists of one A subunit and five B subunits (Kaittanis et al. 2010). In accordance with Viswanathan et al. (2006), another method for the detection of CT was reported: an electrochemical immunosensor with the use of liposome amplification. The method included the conjugation of an anti-CT-B monoclonal antibody to PEDOT on a CNT film and a sandwich assay of the released potassium ferrocyanide via square-wave stripping voltammetry. The detection limit of the developed method was 10^{-16} g/mL, and the working range of the method was approximately 10^{-14} to 10^{-7} g/mL. Another study by GNPs on ganglioside lipid bilayers reported that the sensitivity of these probes was 100-fold greater than that of fluorescent immunoassays in the range of 10 molecules to 100 micromolar and 10 micromolar to 100 nanomolar, respectively (Minke et al. 1999). Schofield et al. (2007) introduced a colorimetric detection method by conjugating thiolate-lactose derivatives on GNPs, where binding to the CT-B subunit aggregation causes the color to shift from red to purple. The detection limit was 3 μg/mL.

4.4 Brevetoxins

K. brevis contains toxins called *brevetoxins* (BTXs); they are sodium channel blockers that cause neurotoxic shellfish poisoning with symptoms of muscle pain, nausea, and respiratory irritation (Arnich et al. 2021). To quickly detect BTX-B in food, an electrochemical immunosensor was designed by Tang et al. (2011). They captured and stabilized a BTX-B–BSA conjugate on GNP-decorated polyamidoamine dendrimers (GNP–PAADs). The sensor response, using a competitive immunoassay with HRP-labeled anti-BTX antibodies, met a detection limit of 0.01 ng/mL and a working range of 0.03–8 ng/mL. GNPs and PAADs yield conductivity and biomolecule surface coverage, hence improving the performance of the sensor (Tang et al. 2011).

4.5 Shiga Toxin

Shiga-like toxins produced by *E. coli* O157:H7 resemble cholera toxins and adhere to the globotriose (Pk) blood group antigen, which has a trisaccharide (Kaittanis et al. 2010; Chien et al. 2008). Chien et al. (2008) presented an SPR competition assay with the help of glyconanoparticles (GNPs) immobilized by Pk derivatives. A greater chain length of GNPs and longer Pk chains improved the binding specificity and resulted in a more flexible ligand interacting with the toxin. This led to the formation of a glyconanoparticle chip-based assay of glycoantibodies. With respect to detection, the color changed from purple to brown within 5 min when toxin-producing bacteria were present, which showed that the method was highly selective and sensitive, with a detection range of 1200–7200 U/μL (Nagy et al. 2008).

5 Conclusion

This chapter focuses on the promising potential of nanomaterials in food analysis applications. Nanomaterial sensors achieve this through the interaction of a biological component with the target species, leading to the formation of a detectable signal, which enables rapid detection of food contaminants to allow preventive measures after timely detection. Sensors are rapid, highly sensitive, simple to operate, and portable for applications in the field. However, challenges such as interference from real-sample analysis, reproducibility, and treatment of the toxicity of nanomaterials remain issues to be solved (Kirthi et al. 2022). Additionally, with respect to gold nanoparticles (GNPs), the great potential of other nanomaterials, such as quantum dots (QDs), carbon nanotubes (CNTs), metal oxide/metal nanoparticles, nanowires, and nanorods, for food analysis still remains to be explored.

References

Ali AA, Altemimi AB, Alhelfi N et al (2020) Application of biosensors for detection of pathogenic food bacteria: a review. Biosensors 10:58. https://doi.org/10.3390/bios10060058

Ansari MA (2023) Nanotechnology in food and plant science: challenges and future prospects. Plan Theory 12:2565. https://doi.org/10.3390/plants12132565

Arnich N, Abadie E, Amzil Z et al (2021) Guidance level for Brevetoxins in French shellfish. Mar Drugs 19:520. https://doi.org/10.3390/md19090520

Bahadır EB, Sezgintürk MK (2017) Biosensor technologies for analyses of food contaminants. In: Nanobiosensors. Elsevier, pp 289–337

Begum SJP, Pratibha S, Rawat JM et al (2022) Recent advances in green synthesis, characterization, and applications of bioactive metallic nanoparticles. Pharmaceuticals 15:455. https://doi.org/10.3390/ph15040455

Bharti A, Jain U, Chauhan N (2024) Progressive analytical techniques utilized for the detection of contaminants attributed to food safety and security. Talanta Open 10:100368. https://doi.org/10.1016/j.talo.2024.100368

Bhatia D, Paul S, Acharjee T et al (2024) Biosensors and their widespread impact on human health. Sens Int 5:100257. https://doi.org/10.1016/j.sintl.2023.100257

Biswas SK, Islam MS, Jia F et al (2024) Flexible biosensors for food pathogen detection. Adv Electron Mater 10. https://doi.org/10.1002/aelm.202300898

Castillo-Henríquez L, Brenes-Acuña M, Castro-Rojas A et al (2020) Biosensors for the detection of bacterial and viral clinical pathogens. Sensors 20:6926. https://doi.org/10.3390/s20236926

Cesewski E, Johnson BN (2020) Electrochemical biosensors for pathogen detection. Biosens Bioelectron 159:112214. https://doi.org/10.1016/j.bios.2020.112214

Chakroborty S, Pal K, Nath N et al (2023) Sustainable synthesis of multifunctional nanomaterials from rice wastes: a comprehensive review. Environ Sci Pollut Res 30:95039–95053. https://doi.org/10.1007/s11356-023-29235-9

Chen J-Q, Regan P, Laksanalamai P et al (2017) Prevalence and methodologies for detection, characterization and subtyping of Listeria monocytogenes and L. ivanovii in foods and environmental sources. Food Sci Human Wellness 6:97–120. https://doi.org/10.1016/j.fshw.2017.06.002

Chien Y, Jan M, Adak AK et al (2008) Globotriose-functionalized gold nanoparticles as multivalent probes for Shiga-like toxin. ChemBioChem 9:1100–1109. https://doi.org/10.1002/cbic.200700590

Cho EC, Choi J-W, Lee M et al (2008) Fabrication of an electrochemical immunosensor with self-assembled peptide nanotubes. Colloids Surf A: Physicochem Eng Asp 313–314:95–99. https://doi.org/10.1016/j.colsurfa.2007.04.154

Curulli A (2021) Electrochemical biosensors in food safety: challenges and perspectives. Molecules 26:2940. https://doi.org/10.3390/molecules26102940

Daliri F, Aboagye AA, Kyei-Baffour V et al (2019) Immunosensors for food safety: current trends and future perspectives. J Food Hyg Saf 34:509–518. https://doi.org/10.13103/JFHS.2019.34.6.509

Djearamane S, Loh ZC, Lee JJ et al (2022) Remedial aspect of zinc oxide nanoparticles against Serratia Marcescens and Enterococcus Faecalis. Front Pharmacol 13. https://doi.org/10.3389/fphar.2022.891304

Dungchai W, Siangproh W, Chaicumpa W et al (2008) Salmonella typhi determination using voltammetric amplification of nanoparticles: a highly sensitive strategy for metalloimmunoassay based on a copper-enhanced gold label. Talanta 77:727–732. https://doi.org/10.1016/j.talanta.2008.07.014

Górska-Warsewicz H, Rejman K, Laskowski W et al (2019) Milk and dairy products and their nutritional contribution to the average polish diet. Nutrients 11. https://doi.org/10.3390/nu11081771

Grace D, Wu F, Havelaar AH (2020) MILK symposium review: foodborne diseases from milk and milk products in developing countries—review of causes and health and economic implications. J Dairy Sci 103:9715–9729. https://doi.org/10.3168/jds.2020-18323
Grossman HL, Myers WR, Vreeland VJ et al (2004) Detection of bacteria in suspension by using a superconducting quantum interference device. Proc Natl Acad Sci 101:129–134. https://doi.org/10.1073/pnas.0307128101
Hassanpour S, Baradaran B, Hejazi M et al (2018) Recent trends in rapid detection of influenza infections by bio and nanobiosensor. TrAC Trends Anal Chem 98:201–215. https://doi.org/10.1016/j.trac.2017.11.012
Janik-Karpinska E, Ceremuga M, Niemcewicz M et al (2022) Immunosensors—the future of pathogen real-time detection. Sensors 22:9757. https://doi.org/10.3390/s22249757
Joshi NC, Gururani P, Bhatnagar P et al (2023) Advances in metal oxide-based nanocatalysts for biodiesel production: a review. ChemBioEng Rev 10:258–271. https://doi.org/10.1002/cben.202200019
Joung C-K, Kim H-N, Lim M-C et al (2013) A nanoporous membrane-based impedimetric immunosensor for label-free detection of pathogenic bacteria in whole milk. Biosens Bioelectron 44:210–215. https://doi.org/10.1016/j.bios.2013.01.024
Kaittanis C, Santra S, Perez JM (2010) Emerging nanotechnology-based strategies for the identification of microbial pathogenesis. Adv Drug Deliv Rev 62:408–423. https://doi.org/10.1016/j.addr.2009.11.013
Kakimova Z, Orynbekov D, Zharykbasova K et al (2023) Advancements in nano bio sensors for food quality and safety assurance – a review. Potravin Slovak J Food Sci 17:728–747. https://doi.org/10.5219/1903
Kakkar S, Gupta P, Kumar N et al (2023) Progress in fluorescence biosensing and food safety towards point-of-detection (PoD) system. Biosensors 13:249. https://doi.org/10.3390/bios13020249
Kirthi AV, Kumar G, Pant G et al (2022) Toxicity of nanoscaled zero-valent iron particles on tilapia, Oreochromis mossambicus. ACS Omega 7:47869–47879. https://doi.org/10.1021/acsomega.2c05696
Kumar A, Devi M, Kumar M et al (2022) Silicon nanostructures and nanocomposites for antibacterial and theranostic applications. Sensors Actuators A Phys 347:113912. https://doi.org/10.1016/j.sna.2022.113912
Law JW-F, Ab Mutalib N-S, Chan K-G et al (2015) Rapid methods for the detection of foodborne bacterial pathogens: principles, applications, advantages and limitations. Front Microbiol 5. https://doi.org/10.3389/fmicb.2014.00770
Lin Y-H, Chen S-H, Chuang Y-C et al (2008) Disposable amperometric immunosensing strips fabricated by Au nanoparticles-modified screen-printed carbon electrodes for the detection of foodborne pathogen Escherichia coli O157:H7. Biosens Bioelectron 23:1832–1837. https://doi.org/10.1016/j.bios.2008.02.030
Manjunathan J, Revathi M, Sowmya H et al (2024) Recent advancements in nanotechnological approaches for pollution monitoring and environmental sustainability. Clean Techn Environ Policy 26:3667–3683. https://doi.org/10.1007/s10098-023-02676-z
Masanam HB, Perumal G, Krishnan S et al (2022) Advances and opportunities in nanoimaging agents for the diagnosis of inflammatory lung diseases. Nanomedicine 17:1981–2005. https://doi.org/10.2217/nnm-2021-0427
Mehrotra P (2016) Biosensors and their applications – a review. J Oral Biol Craniofac Res 6:153–159. https://doi.org/10.1016/j.jobcr.2015.12.002
Minke WE, Roach C, Hol WGJ et al (1999) Structure-based exploration of the ganglioside GM1 binding sites of Escherichia coli heat-labile enterotoxin and cholera toxin for the discovery of receptor antagonists. Biochemistry 38:5684–5692. https://doi.org/10.1021/bi982649a
Mishra A, Bhatt R, Bajpai J et al (2021) Nanomaterials based biofuel cells: a review. Int J Hydrog Energy 46:19085–19105. https://doi.org/10.1016/j.ijhydene.2021.03.024

Mohamadi E, Moghaddasi M, Farahbakhsh A et al (2017) A quantum-dot-based fluoroassay for detection of food-borne pathogens. J Photochem Photobiol B Biol 174:291–297. https://doi.org/10.1016/j.jphotobiol.2017.08.005
Nagy JO, Zhang Y, Yi W et al (2008) Glycopolydiacetylene nanoparticles as a chromatic biosensor to detect Shiga-like toxin producing Escherichia coli O157:H7. Bioorg Med Chem Lett 18:700–703. https://doi.org/10.1016/j.bmcl.2007.11.055
Naresh V, Lee N (2021) A review on biosensors and recent development of nanostructured materials-enabled biosensors. Sensors 21:1109. https://doi.org/10.3390/s21041109
Nath S (2024) Advancements in food quality monitoring: integrating biosensors for precision detection. Sustain Food Technol 2:976–992. https://doi.org/10.1039/D4FB00094C
Neethirajan S, Ahmed SR, Chand R et al (2017) Recent advances in biosensor development for foodborne virus detection. Nanotheranostics 1:272–295. https://doi.org/10.7150/ntno.20301
Nnachi RC, Sui N, Ke B et al (2022) Biosensors for rapid detection of bacterial pathogens in water, food and environment. Environ Int 166:107357. https://doi.org/10.1016/j.envint.2022.107357
Pramanik B, Sar P, Bharti R et al (2023) Multifactorial role of nanoparticles in alleviating environmental stresses for sustainable crop production and protection. Plant Physiol Biochem 201:107831. https://doi.org/10.1016/j.plaphy.2023.107831
Priyanka B, Patil R, Dwarakanath S (2016) A review on detection methods used for foodborne pathogens. Indian J Med Res 144:327. https://doi.org/10.4103/0971-5916.198677
Rauf A, Abu-Izneid T, Khalil AA et al (2022) Nanoparticles in clinical trials of COVID-19: an update. Int J Surg 104:106818. https://doi.org/10.1016/j.ijsu.2022.106818
Schofield CL, Field RA, Russell DA (2007) Glyconanoparticles for the colorimetric detection of cholera toxin. Anal Chem 79:1356–1361. https://doi.org/10.1021/ac061462j
Some S, Mondal R, Mitra D et al (2021) Microbial pollution of water with special reference to coliform bacteria and their nexus with environment. Energy Nexus 1:100008. https://doi.org/10.1016/j.nexus.2021.100008
Tang D, Tang J, Su B et al (2011) Gold nanoparticles-decorated amine-terminated poly (amidoamine) dendrimer for sensitive electrochemical immunoassay of brevetoxins in food samples. Biosens Bioelectron 26:2090–2096. https://doi.org/10.1016/j.bios.2010.09.012
Upadhyay S, Mir RA, Kumar N et al (2023) Reusing waste plastic bottle to reduce MoO3 into carbon-supported MoO2 nanoparticles for efficient water electrolysis. Surf Interfaces 42:103297. https://doi.org/10.1016/j.surfin.2023.103297
Uzawa H, Ohga K, Shinozaki Y et al (2008) A novel sugar-probe biosensor for the deadly plant proteinous toxin, ricin. Biosens Bioelectron 24:923–927. https://doi.org/10.1016/j.bios.2008.07.049
Valdés MG, Valdés González AC, García Calzón JA et al (2009) Analytical nanotechnology for food analysis. Microchim Acta 166:1–19. https://doi.org/10.1007/s00604-009-0165-z
Vikal S, Gautam YK, Kumar A et al (2023) Bioinspired palladium-doped manganese oxide nanocorns: a remarkable antimicrobial agent targeting phyto/animal pathogens. Sci Rep 13:14039. https://doi.org/10.1038/s41598-023-40822-1
Viswanathan S, Wu L, Huang M-R et al (2006) Electrochemical Immunosensor for cholera toxin using liposomes and poly(3,4-ethylenedioxythiophene)-coated carbon nanotubes. Anal Chem 78:1115–1121. https://doi.org/10.1021/ac051435d
Wang Y, Hu S (2016) Applications of carbon nanotubes and graphene for electrochemical sensing of environmental pollutants. J Nanosci Nanotechnol 16:7852–7872. https://doi.org/10.1166/jnn.2016.12762
Wang B, Huang D, Weng Z (2023) Recent advances in polymer-based biosensors for food safety detection. Polymers (Basel) 15:3253. https://doi.org/10.3390/polym15153253
Xu L, Bai X, Bhunia AK (2021) Current state of development of biosensors and their application in foodborne pathogen detection. J Food Prot 84:1213–1227. https://doi.org/10.4315/JFP-20-464
Yang M, Kostov Y, Rasooly A (2008) Carbon nanotubes based optical immunodetection of Staphylococcal Enterotoxin B (SEB) in food. Int J Food Microbiol 127:78–83. https://doi.org/10.1016/j.ijfoodmicro.2008.06.012

Yang M, Kostov Y, Bruck HA, Rasooly A (2009a) Gold nanoparticle-based enhanced chemiluminescence immunosensor for detection of Staphylococcal Enterotoxin B (SEB) in food. Int J Food Microbiol 133:265–271. https://doi.org/10.1016/j.ijfoodmicro.2009.05.029

Yang G-J, Huang J-L, Meng W-J et al (2009b) A reusable capacitive immunosensor for detection of Salmonella spp. based on grafted ethylene diamine and self-assembled gold nanoparticle monolayers. Anal Chim Acta 647:159–166. https://doi.org/10.1016/j.aca.2009.06.008

Yeganeh FE, Yeganeh AE, Far BF et al (2022) Synthesis and characterization of tetracycline loaded methionine-coated NiFe2O4 nanoparticles for anticancer and antibacterial applications. Nanomaterials 12:2286. https://doi.org/10.3390/nano12132286

Yunus G (2019) Biosensors: an enzyme-based biophysical technique for the detection of foodborne pathogens. In: Enzymes in food biotechnology. Elsevier, pp 723–738

Zhang X, Geng P, Liu H et al (2009) Development of an electrochemical immunoassay for rapid detection of E. coli using anodic stripping voltammetry based on Cu@Au nanoparticles as antibody labels. Biosens Bioelectron 24:2155–2159. https://doi.org/10.1016/j.bios.2008.11.019

Detection of Antibiotics in Meat and Meat-Based Products Using Immunosensors

Harsh Kumar, Shivani Guleria, Kanika Dulta, Richard Cimler, Sivakumar Manickam, and Kamil Kuča

Abstract

Consumers are always curious about the quality of meat and their products they are consuming. There are lots of adulterants being added to the meat that not only compromise the quality but also create serious health issues. Antibiotics are generally used to treat various types of microbial infections in the livestock sector. In addition, some antibiotics also act as growth promoters for meat animals' growth. Although many food safety agencies operating in different countries set the maximum residue limits for drugs found in the different body parts of meat animals, besides this, these agencies usually ban many veterinary drugs/antibiotics for use as growth promoters. Immunosensor technology has gained popularity as point-of-care testing, especially for detecting fraudulent

H. Kumar · R. Cimler
Centre of Advanced Technologies, Faculty of Science, University of Hradec Kralove, Hradec Kralove, Czech Republic
e-mail: richard.cimler@uhk.cz

S. Guleria
Department of Biotechnology, TIFAC-Centre of Relevance and Excellence in Agro and Industrial Biotechnology (CORE), Thapar Institute of Engineering and Technology, Patiala, India

K. Dulta
Department of Food Technology, School of Applied and Life Sciences, Uttaranchal University, Dehradun, India

S. Manickam
Petroleum and Chemical Engineering, Faculty of Engineering, Universiti Teknologi Brunei, Bandar Seri Begawan, Brunei
e-mail: manickam.sivakumar@utb.edu.bn

K. Kuča (✉)
Centre for Basic and Applied Research, University of Hradec Kralove, Hradec Kralove, Czech Republic
e-mail: kamil.kuca@uhk.cz

K. Kuča et al. (eds.), *Ensuring Wellness: Immunosensors Monitoring Food Quality and Pathogenicity Risks*, https://doi.org/10.1007/978-3-031-97477-9_4

activities in meat and their products. Our chapter well describes the immobilization technology for the designing of immunosensors as well as details about the developed immunosensors' application in the detection of antibiotics in the meat and their products.

Keywords

Health · Meat quality · Antibody-based detection · Biosensor technology · Drugs

1 Introduction

The reason humans are omnivores is their meat consumption dates back around 2.6 million years. Meat is a popular source of protein in the average person's diet. Additionally, it has been shown that it is possible to reduce the burden on meat intake by strengthening certain lentils (Aimon et al. 2021). Since 1961, the amount of global meat production has increased by more than four times. Pork is the most consumed meat, even if the availability of poultry is growing quickly. In different countries, the amount of meat consumed depends on a number of factors, such as the culture and economics of the nation. More meat is consumed in nations with greater per capita incomes than in those with lower incomes, as has been shown. The world generates around 340 million metric tons of meat annually, three times as much as 50 years ago. Meat consumption is predicted to increase to 460–570 million metric tons by 2050 on average (Ali and Alsayeqh 2022).

Nevertheless, recent reports have shown that consistent use of antimicrobial agents in large amounts could result in the accumulation of antimicrobial drug residues in the organs and muscles of the animal (Sanz et al. 2015). Consumption of animal products with these residues (intake through meat and their products) may cause serious health risks such as hypersensitivity reactions and antibiotic resistance development amongst the consumers. FAO/WHO has reported that the presence of antimicrobial residues in edible animal-based food products has surpassed the permissible levels (very high) in developing countries (Falowo and Akimoladun 2019).

Solid-state devices known as immunosensors are a crucial type of biosensor. They are distinguished by the creation of a signal that can be detected using a transducer as a result of an immunochemical reaction (Cristea et al. 2015; Khan et al. 2020; Chen et al. 2023c). In a process similar to immunoassays, they often serve as affinity biosensors, in which antibodies' molecular recognition of a particular antigen results in the creation of a stable immunocomplex (Cristea et al. 2015; Nie et al. 2019). However, they are different from immunoassays in that the immunological complex may now be detected and quantified label-free, thanks to modern transducer technology advancements (Luppa et al. 2001). Immunosensors may be roughly divided into four categories based on the transduction mode: (a) electrochemical, such as amperometric, voltammetric, potentiometric, and impedimetric; (b) optical; (c) microgravimetric; and (d) thermometric immunosensors (Muratova et al. 2015;

Cristea et al. 2015; Kim and Park 2021; Naresh and Lee 2021; Chen et al. 2023a, b, c). These immunosensors may be used as indirect or labeled (using a label that is observable with the antibody or antigen in conjunction) or direct (determining the immunochemical response directly by measuring the physical alterations brought on by the development of immunocomplexes) (Cristea et al. 2015). Although these terms are related to tracer-based detection, the words "direct" and "indirect" take on different connotations when applied to immunoassays. Whether antibody binding is detected directly determines the difference. An example of this would be the attachment of an enzyme-labeled antibody to an immobilized coated conjugate or the attachment of an enzyme tracer to an antibody that has been immobilized (Lim and Ahmed 2016). In contrast, identification in an indirect technique occurs only after a further binding step. For example, the first antibody coupled to a fixed coating compound can be marked with an enzyme-labeled secondary antibody (Cristea et al. 2015).

Here, we described significant developments in antibody-based sensors that go beyond their conventional use in immunity-related scenarios. Different types of immunosensors and how they work are not particularly covered in this chapter. Rather, our main goal is to shed light on the immobilization strategies for the development of immunosensors, and finally, we focused on immunosensor applications in the detection of antibiotics in meat and their products. Lastly, we briefed about the challenges with immunosensor development.

2 Immobilization of Antibody (Ab) for the Development of Immunosensor

A bio/immunosensor's performance is contingent upon three key elements: (i) its capacity to immobilize biological molecules, or recognition elements, while preserving their inherent activity; (ii) the availability of the relevant analyte in solution to the recognition element; and (iii) low levels of nonspecific adsorption onto the solid substrate. The interface's or sensor surface's physicochemical characteristics are crucial for achieving the best possible antibody (Ab) immobilization and preventing nonspecific adsorption (Kim and Herr 2013). As mentioned before, one of the most crucial processes during an immunosensor construction is the immobilization of Abs onto the sensor surface without losing their immunological activity and specificity. The immobilization step affects the detection limit, sensitivity, and overall performance of the immunosensor (Trilling et al. 2013). Ab, an asymmetric macromolecule, may adhere to a solid surface in a variety of ways. Using the constant fragment to immobilize the sensor can optimize its performance in the (Fc) region, and the Ab's binding site is completely open for interaction with the appropriate antigen. When the Ab is immobilized on the sensor surface via the antigen-binding sites, which include the variable heavy (VH) and constant regions (CH) regions, the Ab's binding activity is either reduced or completely removed (Conroy et al. 2009). Furthermore, there are several intermediate phases of immobilization that can either result in less access of the Ab-binding site to the cognate antigen or associate the Ab close to this location with a reduction in binding and, consequently, sensor-related functionality. Additionally, the degree of bonding between the Ab and the surface may require cautious handling. If the antibody is designed to have extremely distinct points of attachment, such as when using Ab fragments like single-chain fragment variable (scFv), single-chain antibody (scAb), or Fabs with tags like biotin, polyhistidine, or modified thiol spots for linking, this can be accomplished. When chemical crosslinking is used, such as to -carboxyl groups (COOH) or amino groups (NH2), this is more difficult to regulate since many crosslinks might lower the total Ab-binding ability, for example, by blocking the Ab-binding site's access (Sharma et al. 2016).

When Abs are immobilized on a solid surface, favorable orientation and little structural alteration can immediately lead to excellent performance of immunosensors, with enhanced performance metrics up to 200 times greater than when immobilization is confused (Trilling et al. 2013). Certain reactive molecules on the surface and on the Ab interact to affect Abs' arrangement on sensor surfaces. To improve the sensor surface's suitability for immobilizing Ab using a particular arrangement, several methods have been documented in the literature. These include functionalizing the surface containing certain groupings, like periodate, maleimide, succinimide ester, glutaraldehyde, or carbodiimide, or embellishing it with nanostructures (Kierny et al. 2012). There are two main ways that immunosensors can be produced, namely, covalent and non-covalent immobilization. A schematic illustration of the many immobilization techniques utilized to immobilize Abs on a sensor surface may be seen in Fig. 1.

2.1 Immobilization Without Covalent Bonding

Simple non-covalent forces, such as van der Waals forces, hydrophobic contacts, and electrostatic or ionic connections, can adsorb Abs onto the sensor surface. Adsorption is a straightforward technique that falls into one of two categories: specifically, physisorption or chemisorption. Physisorption is chiefly based on hydrophobic contacts and weak van der Waals forces, chemisorption chiefly on an electrostatic relationship between the surface and the Abs. Another non-covalent method is Ab entrapment into conducting polymer sheets, as shown by John et al. (1991), whereby pyrrole that had been galvanostatically polymerized was used to entrap Abs for albumin in human serum on a platinum wire substrate. Several popular conducting polymers, including polyacetylene, polythiophene, polyaniline, polyindole, and polypyrrole, have been widely employed in the production of immunosensors because of their remarkable biocompatibility in aqueous solutions that are neutral. For immobilizing biomolecules, polyquinone is another form of smart material. It offers coupling and transduction capabilities in addition to outstanding biocompatibility, obviating the necessity for an extra redox label on the Ab or in solution. Sun et al. (2011) created a "label-free" immunosensor based on a silica sol-gel-entrapped Ab for the detection of carbofuran, which demonstrated a detection limit of 0.33 ng/ml. Around the Ab, silica sol-gel creates a biocompatible

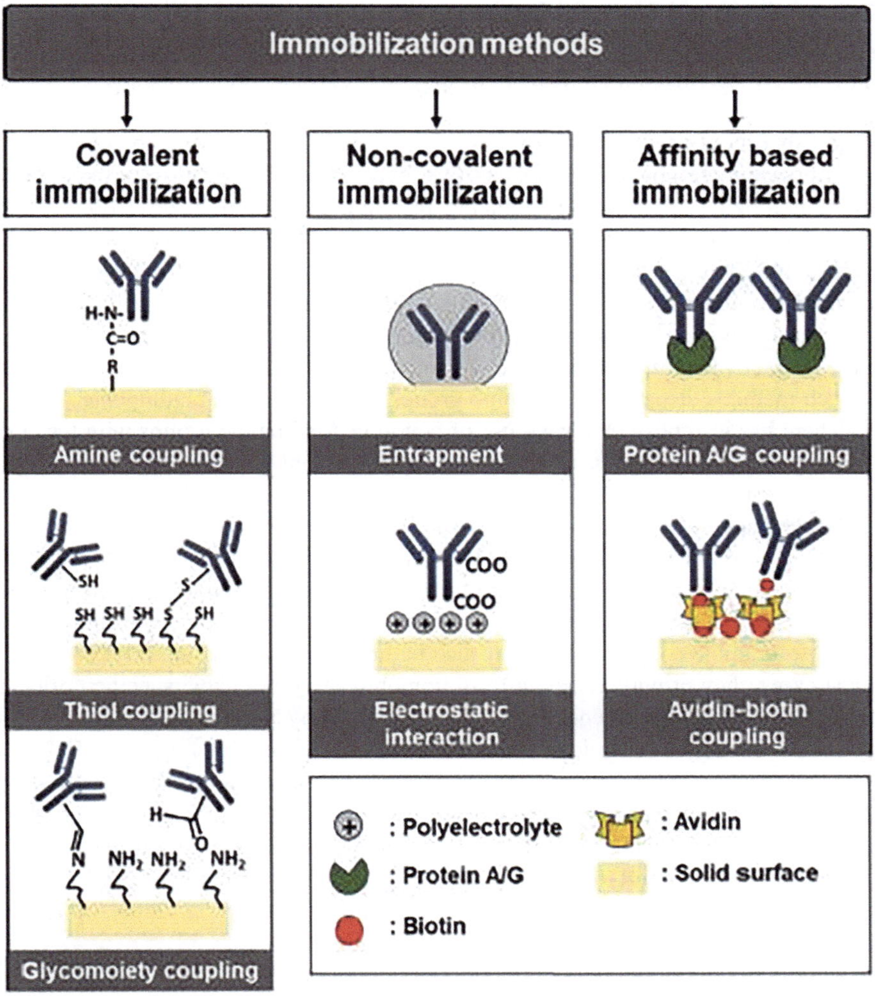

Fig. 1 Schematic presentation of antibody immobilization techniques (Kumar et al. 2024a Licence Number 5896610014401)

milieu thanks to its silicate network. Despite the improved attachment, the burying of the active sites may reduce the sensitivity of an entrapment-based immunosensor, leading to low immunological activity or identification efficiency. Although the simplicity of these untreated coupling techniques makes them popular, restrictions on mass transfer across membranes or gels limit their practical use, and they provide only a limited amount of control over the Abs' orientation. This might limit the Fc region's ability to immobilize in a proper and desired way, making the antigen-binding sites less accessible.

2.2 Covalent Immobilization

The most researched immobilization technique is covalent immobilization of Abs on different solid surfaces, which makes immunosensors easier to store and reuse over time (Sharma et al. 2016). Covalent binding refers to the surface modification to afford reactive groups like amino, carboxy, hydroxy, or thiol groups for the purpose of immobilizing Abs later on. Ionizing radiation graft co-polymerization, photochemical grafting, plasma gas discharge, and chemical modification are some of the surface modification methods that have been developed. Due to the amine groups' comparatively easy access found in the side chains of lysine amino acids, targeting these groupings in order to couple Abs to the sensor surface is a commonly utilized method of covalent immobilization. The activity of Abs might be decreased, though, if such methods also react with amino groups at or around the binding site.

There has also been extensive use of coupling Abs' amine groups with terminal carboxylated gold surfaces. In order to create a covalent bond, reactive succinimide esters are attached to the primary amine groups on Abs, which are created when carboxylic acid reacts with 1-Ethyl-3-(3-dimethylaminopropyl) carbodiimide (EDC) and N-Hydroxysuccinimide (NHS) (Sharma et al. 2016).

The excessive number of lysine groups in the Ab causes amine-coupling chemistry to typically result in unpredictable orientation and less homogeneous binding. In contrast to amine coupling, Abs may also be immobilized by using thiol groups; in most cases, they provide more uniform immobilization or enable specified orientation of Abs, especially when Abs have been modified by adding thiol groups through molecular biology or chemical means. The process of immobilizing Abs on a thiol-grouped sensor surface usually involves the exchange of thiols on the Ab with active disulfides that have been added to the surface. It has been demonstrated that, when immobilized in a specific orientation, Ab fragments using amine groups have a 20-fold higher antigen-binding capacity than randomly immobilized Abs (Trilling et al. 2013).

2.3 Affinity-Based Immobilization

The method's shortcomings, including incomplete denaturation and unclear orientation, make covalent immobilization less than ideal, even if it is robust. Therefore, other nonphysical immobilization techniques were devised, such as the use of biotin–avidin or immobilizing intermediary binding proteins, such as Protein A or G, first on the sensor surface and then capturing Abs. These proteins encourage Ab immobilization through the sensor surface's Fc region because of their excellent binding specificity and affinity for the Fc region of many Abs. A novel, straightforward, and quick method of immobilizing Ab was presented by de Juan-Franco et al. (2013). It made use of a fusion protein called the Protein A–gold-binding domain (PAG). When the biotin-tagged Abs are coupled with metallic and functionalized polymeric surfaces, they may readily be trapped on avidin, streptavidin, or genetically produced combinations of the two proteins. In contrast to an entrapment-based immunosensor, Barton et al. (2009) obtained much greater sensitivities using an avidin–biotin interaction-based immunosensor.

3 Detection of Antibiotics Using Immunosensors

Since past few years, researchers have gained interest in immunosensors, as these tend to detect antibiotics in meat and meat products (Chomthong et al. 2024; Kumar et al. 2023, 2024b; Shah et al. 2022; Stevenson et al. 2019; Pan et al. 2017a). Table 1 displays the applications of diverse immunosensors to detect antibiotics present in meat samples.

It was possible to quickly identify tilmicosin (TMC) residues in pork by creating a label-free electrochemical immunosensor (Chen et al. 2024). To create the immunosensor, a high-affinity monoclonal antibody against TMC was immobilized on an rGO-PEI-Ag (reduced graphene oxide-polyethyleneimine-silver) nanocomposite-modified electrode (Fig. 2). The rGO-PEI-Ag nanocomposites that were produced demonstrated high conductivity and redox activity. The immunosensor's linear range was 0.01–100 ng/ml, its low detection limit (LOD) was 0.0013 ng/ml, and its recoveries ranged from 92.77% to 100.02% in pork. Scientists have devised an immunosensor based upon rGO, using a fabricated electrode in conjugation with anti-Penicillin antibody (rGO/Pen-Ab) and used it for sensitive determination of Penicillin G antibiotic in case of spiked meat samples (Shah et al. 2022). Their studies revealed that the developed immunosensor displayed an LOD of 0.724 pM and showed negligible cross-reactivity toward other β-lactam antibiotics (ampicillin and cephalexin). Recently, Aymard et al. (2022) have fabricated a dual electrochemical sensor for the sensitive and rapid detection of enrofloxacin (EF) in meat. Immobilization of an anti-quinolone antibody was carried out on the carbon-based screen-printed dual electrodes by coupling carbodiimide. The next step included the synthesis of a novel electrochemical probe by conjugation of amino-ferrocene (FC) and difloxacin (DF). The principle for detection included competitive binding of the conjugate and free antibiotic on the immobilized antibodies. This immunosensor allowed the detection of EF with an LOD value of 0.003 μg/ml in case of spiked pork muscle samples. The proposed immunosensor was stable at 4 °C for a minimum of 1 month and showed good specificity toward other fluoroquinolones. A nonenzymatic and photothermal immunosensing assay for the detection of enrofloxacin (ENR) in animal meats was developed by Li et al. (2021) using black phosphorus-platinum (BP-Pt) two-dimensional nanomaterials. Results showed that BP-Pt with peroxidase activity worked as a nonenzymatic and photothermal immunosensing assay (PT-NISA) without a secondary antibody. The developed method showed an LOD of 0.063–1.56 μg/kg with sensitivities of 12.9–35.9 μg/kg in fish, beef, and chicken samples. Yu et al. (2020) developed a carboxylated single-walled carbon nanotubes/chitosan (SWNTs-COOH/CS) composite to create a sensitive and selective electrochemical immunosensor for cephalexin (CEX) measurement on a glassy carbon electrode (GCE) surface. The competitive immunoreaction of CEX-antibody and

Table 1 Detection of antibiotics/drugs from the meat and their products using antibody-based sensors

Type of sensor	Type of antibody	Target antibiotic/drug	Type of meat sample	Detection limit	References
Electrochemical	Anti-norfloxacin and anti-enrofloxacin (monoclonal)	Norfloxacin, enrofloxacin	Fish (spiked)	2.02 ng/ml (norfloxacin), 1.70 ng/ml (enrofloxacin)	Chomthong et al. (2024)
Electrochemical	Anti-colistin (monoclonal)	Colistin	Chicken liver (spiked)	0.07 μg/kg	Kumar et al. (2024b)
Electrochemical	Anti-cephalexin (monoclonal)	Cephalexin	Meat (spiked)	0.53 fm	Kolhe et al. (2023)
Electrochemical	Anti-amoxicillin (monoclonal)	Amoxicillin	Fish (spiked)	0.5 pM	Kumar et al. (2023)
Electrochemical	Anti-colistin (monoclonal)	Colistin	Chicken liver (spiked)	0.85 μg/kg	Kumar et al. (2022)
Electrochemical	Anti-ceftiofur (monoclonal)	Ceftiofur	Turkey meat (spiked)	0.10 ng/ml	Stevenson et al. (2019)
Electrochemical	Anti-norfloxacin (monoclonal)	Norfloxacin	Pork (spiked)	0.38 μg/L	Liu et al. (2018)
Optical	Anti-sulfaquinoxaline, sulfachloropyridazine, sulfamethoxypyridazine, and sulfamethoxazole (polyclonal)	Sulfaquinoxaline, sulfachloropyridazine, sulfamethoxypyridazine, and sulfamethoxazole	Chicken muscles (spiked), beef (spiked), and fish (spiked)	66.9 ng/ml (Chicken muscles), 63.3 ng/ml (beef), 62.7 ng/ml (fish)	Pan et al. (2017a)
Optical	Anti-norfloxacin (monoclonal)	Norfloxacin	Chicken, beef, pork, pork kidney, pork liver, sea bass, shrimp, squid (all spiked)	10 pg/ml	Hu et al. (2015)
Electrochemical	Anti-kanamycin (monoclonal)	Kanamycin	Chicken liver (spiked)	5.74 pm/ml	Wei et al. (2012)
Gravimetric	Anti-chloramphenicol (monoclonal)	Chloramphenicol	Chicken muscles (spiked)	0.2 ng/ml	Karaseva and Ermolaeva (2012)

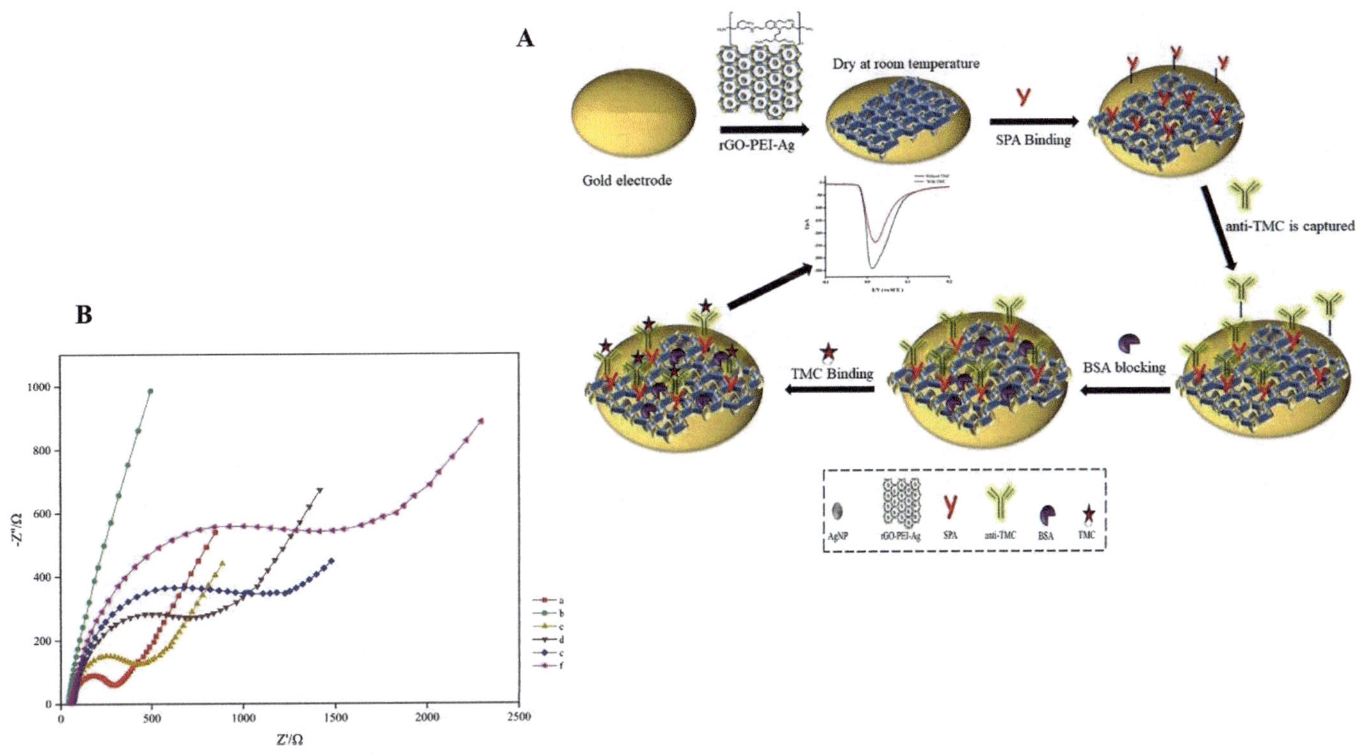

Fig. 2 (**a**) Schematic presentation of tilmicosin (TMC) electrochemical immunosensor (**b**) Electrochemical impedance spectroscopy (EIS): Bare gold electrode (a), rGO-PEI-Ag-Nf (b), SPA/GO-PEI-Ag-Nf (c), anti-TMC/SPA/rGO-PEI-Ag-Nf (d), BSA/anti-TMC/SPA/rGO-PEI-Ag-Nf (e), and TMC/BSA/anti-TMC/SPA/rGO-PEI-Ag-Nf, (f) in 0.1 M PBS with 0.1 M KCl (Chen et al. 2024, Licence Number 5896620719515)

CEX allowed for efficient measurement of the free CEX in solution. The electrochemical immunosensor provided an outstanding response for CEX under ideal circumstances. With an S/N of 3, the detection limit was 45.7 ng/ml and the linear range was 1–800 ng/ml. This method was applied to determine CEX in beef, prawn, and fish samples and obtained a recovery range from 75.39% to 87.27%.

A label-free and reproducible surface plasmon resonance (SPR) type immunosensor was developed by Pan et al. (2017b) to detect harmful enrofloxacin (ENRO) incidence in animal-based foods. Under optimized conditions, the prepared SPR immunosensor displayed a good linear response against ENRO with an LOD (IC_{15}: 1.2 ng/ml) and high sensitivity (IC_{50}: 3.8 ng/ml). The proposed immunosensor was then validated for favorable performance in the detection of ENRO residue in typical animal-based foods after pretreatment with a simple matrix, as well as precision (relative standard deviation ($n = 3$) of 1.8%–4.6%), sensitivity (IC15 $\leq$ 8.4 ng/ml), and acceptable accuracy (recovery of 84.3%–96.6%). Each SPR chip could be reused for a minimum of 100 times and with good stability during the analysis. Piezoelectric immunosensors on the basis of homologous and group-specific antibodies have been developed for detecting penicillin G, ampicillin, and the total content of penicillin antibiotics (Karaseva and Ermolaeva 2014). In order to create the sensor's receptor coating, penicillin G or ampicillin hapten-protein conjugates were immobilized on a polypyrrole film that was electropolymerized and glutaraldehyde-activated. Calculations have been made for the cross-reactivity coefficients and affinity constants. This allowed for the estimation of the polyclonal and monoclonal antibodies' specificity and affinity. In the concentration ranges of 2.5–250.0 ng/ml for penicillin G, 2.5–500.0 ng/ml for ampicillin, and 1–500 ng/ml for the group of penicillins, the calibration curves are linear. The detection thresholds of 0.8 and 3.9 ng/ml are below the maximum residue limit (MRL) set for antibiotics like penicillins. The sensors' ability to identify penicillins in liver, beef, and pig was examined.

4 Challenges with Immunosensors

Storage condition has an important impact on the stability of the immunosensor, and antibody preservation is a serious challenge. Thaw/freeze stresses may be quite effective at reducing the effectiveness of antibodies linked to enzymes. These stresses encompass diverse factors such as cold denaturation, crystallization of buffer components or non-buffer components, ice formation, phase separation, thawing time, redistribution of solutes, and pH fluctuation (Ma et al. 2020). Antibodies have been shown to aggregate as a result of thawing and freezing, a phenomenon related to intricate alterations in the solute and solvent's chemical and physical characteristics brought on by these stressors (Desai et al. 2017).

5 Conclusion

Majority of the studies used the covalent immobilization technique for anchoring the antibodies, especially amine and glycomoiety couplings. Moreover, those immunosensors using gold nanoparticles rather than thiol coupling worked efficiently. In terms of antibiotic detection from the meat samples with immunosensors, the majority of studies have been done with spiked samples with good recovery rates. Some of the studies have not well defined the nature of meat samples they have taken. In a nutshell, we can say that biosensing technology is very time-saving and effortless, particularly in food sectors.

References

Aimon SS, Liaquat A, Aamir SH et al (2021) Blend of flour with *Vigna radiata* and *Vigna mungo* used in muffins in order to increase nutritional properties. Agrobiol Rec 3:29–35

Ali S, Alsayeqh AF (2022) Review of major meat-borne zoonotic bacterial pathogens. Front Pub Health 10:1045599

Aymard C, Kanso H, Serrano MJ et al (2022) Development of a new dual electrochemical immunosensor for a rapid and sensitive detection of enrofloxacin in meat samples. Food Chem 370:131016

Barton AC, Collyer SD, Davis F et al (2009) Labeless AC impedimetric antibody-based sensors with pgml^{-1} sensitivities for point-of-care biomedical applications. Biosens Bioelectron 24: 1090–1095

Chen H, Li Y, Song Y et al (2023a) A sandwich-type electrochemical immunosensor based on spherical nucleic acids-templated Ag nanoclusters for ultrasensitive detection of tumor biomarker. Biosens Bioelectron 223:115029

Chen H, Zhang J, Huang R et al (2023b) The applications of electrochemical immunosensors in the detection of disease biomarkers: a review. Molecules 28:3605

Chen L, Leng YK, Qiu S et al (2023c) Ultrahigh-sensitivity label-free singlemode-tapered no core-singlemode fiber immunosensor for Listeria monocytogenes detection. Sens Actuators B Chem 376:132930

Chen Y, Zhang J, Liu R et al (2024) A highly sensitive electrochemical immunosensor based on rGO-PEI-ag nanocomposites for the detection of tilmicosin. Food Chem 461:140009

Chomthong K, Kunpatee K, Pimpitak U et al (2024) Label-free simultaneous detection of quinolone antibiotic residues using an origami paper–based electrochemical immunosensor. Sens Actuators B Chem 410:135667

Conroy PJ, Hearty S, Leonard P et al (2009) Antibody production, design and use for biosensor-based applications. Semin Cell Dev Biol 20:10–26

Cristea C, Florea A, Tertiş M et al (2015) Immunosensors. In: Rinken R (ed) Biosensors-micro and nanoscale applications. Intech Open, London, pp 165–202

de Juan-Franco E, Caruz A, Pedrajas JR et al (2013) Site-directed antibody immobilization using a protein A–gold binding domain fusion protein for enhanced SPR immunosensing. Analyst 138: 2023–2031

Desai KG, Pruett WA, Martin PJ et al (2017) Impact of manufacturing-scale freeze-thaw conditions on a mAb solution. BioPharm Int 30:30–36

Falowo AB, Akimoladun OF (2019) Veterinary drug residues in meat and meat products: occurrence, detection and implications. Vet Med Pharm 3:1–18

Hu G, Sheng W, Zhang Y et al (2015) A novel and sensitive fluorescence immunoassay for the detection of fluoroquinolones in animal-derived foods using upconversion nanoparticles as labels. Anal Bioanal Chem 407:8487–8496

John R, Spencer M, Wallace GG et al (1991) Development of a polypyrrole-based human serum albumin sensor. Anal Chim Acta 249:381–385
Karaseva NA, Ermolaeva TN (2012) A piezoelectric immunosensor for chloramphenicol detection in food. Talanta 93:44–48
Karaseva NA, Ermolaeva TN (2014) Piezoelectric immunosensors for the detection of individual antibiotics and the total content of penicillin antibiotics in foodstuffs. Talanta 120:312–317
Khan MZH, Hasan MR, Hossain SI et al (2020) Ultrasensitive detection of pathogenic viruses with electrochemical biosensor: state of the art. Biosens Bioelectron 166:112431
Kierny MR, Cunningham TD, Kay BK (2012) Detection of biomarkers using recombinant antibodies coupled to nanostructured platforms. Nano Rev 3:17240
Kim D, Herr AE (2013) Protein immobilization techniques for microfluidic assays. Biomicrofluidics 7:41501
Kim JY, Park M (2021) Recent progress in electrochemical immunosensors. Biosensors 11:360
Kolhe P, Roberts A, Gandhi S (2023) Fabrication of an ultrasensitive electrochemical immunosensor coupled with biofunctionalized zero-dimensional graphene quantum dots for rapid detection of cephalexin. Food Chem 398:133846
Kumar H, Valko M, Alomar SY et al (2022) Electrochemical immunosensor for the detection of colistin in chicken liver. 3 Biotech 12:190
Kumar R, Lakshmi GBVS, Dhiman TK et al (2023) Label-free and highly sensitive immunosensor based on L-cysteine capped vanadium disulfide quantum dots for amoxicillin detection. Microchem J 195:109433
Kumar H, Dhalaria R, Guleria S et al (2024a) Immunosensors in food, health, environment, and agriculture: a review. Environ Chem Lett 22:2573–2605
Kumar H, Kumar D, Valko M et al (2024b) Development of an electrode immunosensor using carbon nanofibres-gold nanoparticles-mercaptopropionic acid-polyethylenimine for chicken liver containing colistin. J Agric Food Res 15:100990
Li S, Wen W, Guo J et al (2021) Development of non-enzymatic and photothermal immuno-sensing assay for detecting the enrofloxacin in animal derived food by utilizing black phosphorus-platinum two-dimensional nanomaterials. Food Chem 357:129766
Lim SA, Ahmed MU (2016) Electrochemical immunosensors and their recent nanomaterial-based signal amplification strategies: a review. RSC Adv 6:24995–25014
Liu B, Li M, Zhao Y et al (2018) A sensitive electrochemical immunosensor based on PAMAM dendrimer-encapsulated Au for detection of norfloxacin in animal-derived foods. Sensors 18:1946
Luppa PB, Sokoll LJ, Chan DW (2001) Immunosensors-principles and applications to clinical chemistry. Clin Chim Acta 314:1–26
Ma H, Ó'Fágáin C, O'Kennedy R (2020) Antibody stability: a key to performance-analysis, influences and improvement. Biochimie 177:213–225
Muratova IS, Kartsova LA, Mikhelson KN (2015) Voltammetric vs. potentiometric sensing of dopamine: advantage sand disadvantages, novel cell designs, fundamental limitations and promising options. Sens Actuators B Chem 207:900–906
Naresh V, Lee N (2021) A review on biosensors and recent development of nanostructured materials-enabled biosensors. Sensors 21:1109
Nie R, Xu X, Cui X et al (2019) A highly sensitive capillary-based immunosensor by combining with peroxidase nanocomplex-mediated signal amplification for detection of procalcitonin in human serum. ACS Omega 4:6210–6217
Pan M, Li S, Wang J et al (2017a) Development and validation of a reproducible and label-free surface plasmon resonance immunosensor for enrofloxacin detection in animal-derived foods. Sensors 17:1984
Pan M, Wang X, Wang J et al (2017b) Stable and sensitive detection of sulfonamide residues in animal-derived foods using a reproducible surface plasmon resonance immunosensor. Food Anal Methods 10:2027–2035

Sanz D, Razquin P, Condón S et al (2015) Incidence of antimicrobial residues in meat using a broad spectrum screening strategy. Eur J Nutr Food Saf 5:156–165

Shah M, Kolhe P, Roberts A et al (2022) Ultrasensitive immunosensing of Penicillin G in food samples using reduced graphene oxide (rGO) decorated electrode surface. Colloids Surf B Biointerfaces 219:112812

Sharma S, Byrne H, O'Kennedy RJ (2016) Antibodies and antibody-derived analytical biosensors. Essays Biochem 60:9–18

Stevenson HS, Shetty SS, Thomas NJ et al (2019) Ultrasensitive and rapid-response sensor for the electrochemical detection of antibiotic residues within meat samples. ACS Omega 4:6324–6330

Sun X, Du S, Wang X et al (2011) A label-free electrochemical immunosensor for carbofuran detection based on a sol-gel entrapped antibody. Sensors 11:9520–9531

Trilling AK, Beekwilder J, Zuilhof H (2013) Antibody orientation on biosensor surfaces: a minireview. Analyst 138:1619–1627

Wei Q, Zhao Y, Du B et al (2012) Ultrasensitive detection of kanamycin in animal derived foods by label-free electrochemical immunosensor. Food Chem 134:1601–1606

Yu W, Sang Y, Wang T et al (2020) Electrochemical immunosensor based on carboxylated single-walled carbon nanotube-chitosan functional layer for the detection of cephalexin. Food Sci Nutr 8:1001–1011

Detection of *Escherichia coli* O157:H7 in Different Food Matrices

Usha, Shagun Gupta, Rupak Nagraik, and Ankur Kaushal

Abstract

Food poisoning causes significant discomfort to many people each year, with one of the main causes being the bacterium *Escherichia coli* O157:H7. Detecting pathogenic *Escherichia coli* O157:H7 in various types of food is crucial for ensuring food safety. Immunosensors have shown great potential as rapid, sensitive, and specific diagnostic tools for this particular illness because of their ability to use specific antigen–antibody interactions. This chapter provides a thorough examination of the progress and use of immunosensors for detecting *E. coli* O157:H7 in different types of food, such as dairy products, meat, fruits, and vegetables. This article provides a comprehensive elucidation of the basic concepts that form the basis of immunosensor function, with a particular focus on various transduction mechanisms, including optical, electrochemical, and piezoelectric approaches. This research investigates the potential of nanomaterials to reduce detection limits and increase sensor sensitivity. Additionally, it examines the difficulties associated with preserving the stability and functionality of antibodies under different nutritional circumstances. Case examples exemplify the pragmatic use of immunosensors and demonstrate their effectiveness in real-life situations. This chapter compares immunosensors to standard microbiological procedures, highlighting the benefits of immunosensors in terms of speed, mobility, and onsite testing capabilities. This discussion focuses on potential advancements in immunosensor technology, including the integration of wireless systems and the utilisation of multiplexed platforms for

Usha · S. Gupta · A. Kaushal (✉)
Department of Bio-Sciences and Technology, Maharishi Markandeshwar (Deemed to be) University, Mullana, Ambala, India

R. Nagraik
Department of Biotechnology, Graphic Era (Deemed to be University), Dehradun, Uttarakhand, India

K. Kuča et al. (eds.), *Ensuring Wellness: Immunosensors Monitoring Food Quality and Pathogenicity Risks*, https://doi.org/10.1007/978-3-031-97477-9_5

concurrent illness detection. These advancements have the potential to greatly improve food safety and monitor public health.

Keywords

Escherichia coli O157:H7 · Immunosensor · Food matrices

1 Introduction

In particular, in young, aged, and immunosuppressed people, *E. coli* O157:H7 is one of the most hazardous foodborne pathogens, as it can cause severe diseases, including inflammation, stomach discomfort, diarrhoea, haemolytic uremic syndrome, haemorrhagic colitis, and even death (Addis and Sisay 2015). The consumption of contaminated food and water is linked to epidemics of foodborne infections caused by *E. coli* O157:H7. Reducing illness outbreaks depends on the early identification of *E. coli* O157:H7 in food and water, thereby guaranteeing public health and preventing financial losses (Scallan et al. 2011). Many tests have been developed recently to monitor and manage *E. coli* O157:H7. The gold standard for bacterial detection, conventional culture and colony counting techniques, requires several days to obtain results, which is intensive and time-consuming. Previous conventional detection techniques are not universally relevant to present detection demands, given the 'zero tolerance' detection target for *E. coli* O157:H7. Many analytical and sensing techniques have been developed to meet the demands of quick, sensitive, and specific detection to overcome this obstacle. Optical detection is one such method. For the detection of *E. coli* in a watery environment, Silva et al. reviewed the advantages and drawbacks of polymerase chain reaction (PCR)-related technologies (Mendes Silva and Domingues 2015). Moreover, the development of isothermal amplification methods has increased the feasibility and usability of nucleic acid amplification techniques in real-world detection by eliminating the need for costly temperature-regulated equipment (Lv et al. 2020).

Owing to their advantages, including faster detection times and reduced detection limits, many nanomaterials are now utilised as fluorescent, colourimetric, and surface-enhanced Raman scattering (SERS) tags for low-concentration detection of *E. coli* O157:H7 (Duan et al. 2021). Nevertheless, optical detection frequently requires sample pretreatment, which complicates the experimental process and hence influences the general analytical efficiency. The main detection method for the electrochemical detection of *E. coli* O157:H7 involves the use of nanomodified electrodes to track the electrochemical signals directly produced by the target bacterium (Güner et al. 2017). Nevertheless, the lack of capacity is shown in the identification of complicated sample matrices; thus, the introduction of nanoparticles as electrochemical labels for aided detection will help to increase detectability in actual samples (Zhong et al. 2019). Razmi et al. examined, in terms of materials and concepts, the latest developments in electrochemical biosensors for *E. coli* O157:H7 detection (Razmi et al. 2020). Furthermore, the methods employed for the detection of *E. coli* O157:H7 in food and water include photothermal, piezoelectric (Prasad et al. 2022), and magnetoresistive effect-based sensing methods (Wu et al. 2017). These methods have shown good detection results. These detection techniques, along with lateral flow immunoassays (LFIAs) and microfluidic chips, are especially suitable for point-of-care and onsite testing.

While earlier studies focused on certain particular technologies for the detection of *E. coli* O157:H7, the above studies demonstrated that the detection techniques for *E. coli* O157:H7 are diverse and complex (Zhao et al. 2018). Furthermore, Rani et al. highlighted the relevance of the partial approaches applied for *E. coli* O157:H7 detection in point-of-care routine diagnostics (Rani et al. 2021). From the standpoint of detection techniques alone, there is no methodical and thorough analysis of several methods for *E. coli* O157:H7 detection. This paper presents a comprehensive assessment of the most recent *E. coli* O157:H7 detection techniques from three angles—optical, electrochemical, and other detection methods—based on the transduction mechanism of the detection signal. In addition to their ability to sense, their antimicrobial activity can be monitored by using different nanoparticles (Trache et al. 2022; Mitra et al. 2022; Begum et al. 2022; Kumar et al. 2022; Nangan et al. 2023). The evaluation mostly evaluates the performance differences between several detection techniques and critically examines the benefits and drawbacks of every technology.

2 Food Matrices and *E. coli* O157:H7 Contamination

Escherichia coli O157:H7 is alarming in food products, as every particular item causes particular problems in terms of identification. Bad processing, transportation, and retail techniques often expose fresh food to polluted water, soil, or airborne pollutants. Pathogens have been demonstrated to be conveyed via irrigation via nonportable water, contact with faeces, and insufficient fault-free washing and handling throughout the supply chain. Owing to their juicy characteristics and natural coatings or skins that shield bacteria such as *E. coli*, which prevent identification and isolation, vegetables, including spinach, tomato, and other fruits, provide attractive media for pathogen detection. Nevertheless, the diversity of produce aggravates this issue since it has unique microbial flora that make pathogen detection more difficult, and the need for techniques that are able to quickly and sufficiently separate *E. coli* O157:H7 from the rest of the contaminated flora is highly important (Besser 1993) (Fig. 1).

In particular, in beef and poultry, meat is sometimes considered a high-risk matrix for the potential occurrence of *E. coli* O157:H7. Ground beef and other raw meats are widely known to carry illnesses from dangerous handling or contact with contaminated surfaces or equipment during and after the slaughtering and processing stages. Furthermore, meat contamination should not be handled correctly or if the temperature used for cooking is not appropriate. Given that the meat structure of proteins, lipids, and fibres can complicate detection, identifying *E. coli* O157:H7 in meat products is challenging. In particular, fat content might interfere with the

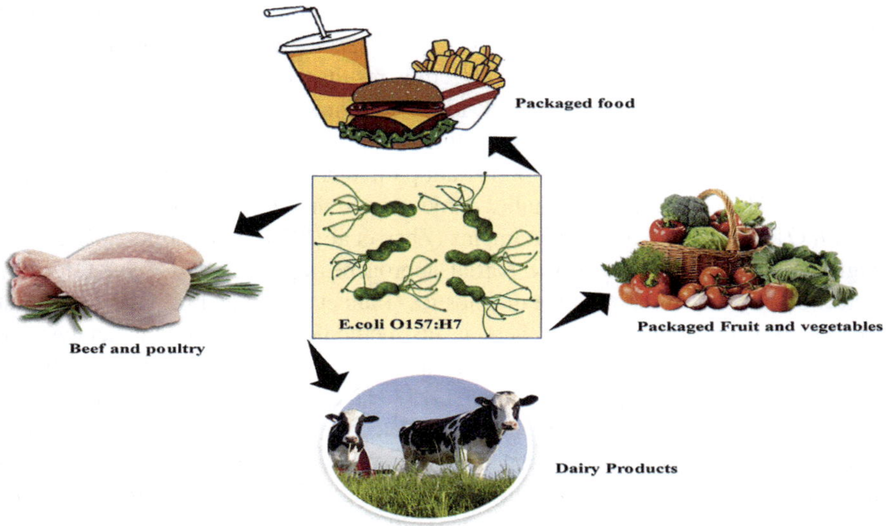

Fig. 1 Distribution and survival of *Escherichia coli* in different food matrices

binding capacity of the antibody or interact undesirably with the surface of the sensing element, resulting in the identification of no bacteria despite their existence (false negatives) or reducing the detection sensitivity. In addition, preservatives and other processing aids used in processed types of meat, such as sausage and deli meat, which seem to hide the organisms from identification, are needed (Eyi and Arslan 2012).

In particular, *E. coli* O157:H7 exposures are usually encountered from unpasteurised milk or insufficiently pasteurised foods, and dairy products also present the possibility of contamination. Possible methods of contamination during milking or processing include contact with contaminated milk, inadequate pasteurisation techniques, and filthy surroundings. Moreover, the high fat and protein contents of dairy products make even the identification of infections difficult, as these elements tend to bind to antibodies, therefore influencing the specificity and sensitivity of conventional detection techniques. Moreover, various other natural microflowers, such as lactic acid bacteria and several other forms, make it challenging to separate and pinpoint harmful bacteria in dairy products. Therefore, properly adjusting or creating islands for the identification of *E. coli* O157:H7 bacteria in dairy products must be used to reduce matrix effects and improve the degree of sensitivity and specificity in detection (Sidari and Caridi 2011).

The rise of processed and packaged food products presents special difficulties in terms of pathogen detection. These products run the danger of cross-contamination in manufacturing, packaging, and distribution systems, particularly in cases with several constituent origins. Packaged salads, ready-to-eat meals, frozen foods, and other sources of cross-contamination of raw ingredients or even inside procedures (Worku et al. 2024). Moreover, the strong mixture of preservatives, flavouring agents, colouring agents, etc., found in processed meals can also interfere with immunosensor detection, resulting in nonspecific binding and/or non-detectability. Furthermore, the processing techniques used include high-temperature heating, freezing, and particularly canning, which may alter the shape of the pathogens, making their separation from other quiet microorganisms challenging. Because of the variations in the chemical and physical properties influencing pathogens and sensors, all these idiosyncrasies require the development of detection methods for every processed food (Delaquis et al. 2007).

3 Principles of Immunosensors in Pathogen Detection

In pathogen detection, immunological-based sensors depend on a series of immunoassay techniques involving the interaction of a certain sort of biochemistry that protects other pathogens from becoming *E. coli* O157:H7 (Nesakumar et al. 2021). The most recommended method for detecting colour or fluorescence upon pathogen binding is the enzyme-linked immunosorbent assay (ELISA), which uses antibodies connected to an enzyme that catalyses a change, resulting in a detectable reaction. Within the same category, lateral flow assays (LFAs) are a form of immunoassay that also enable quick onsite testing, typically in the form of test strips. This is accomplished by the sample flowing across a reaction membrane where antibodies against a particular pathogen are immobilised, and a line or signal develops (Shah and Maghsoudlou 2016) (Fig. 2). These assays are therefore perfect for points of care, and other tests that need quick findings. Using magnetically retrievable solid supports covered with particular antibodies, immunomagnetic separation (IMS) captures the target organism from a complex matrix and thereafter proceeds with detection via other techniques, such as PCR or immunoassays. It has the advantage of being used for food samples with low pathogen levels, therefore improving detection sensitivity (Wang et al. 2020).

The main characteristic of any immunosensor is antibody and antigen interaction: antibodies bind to particular antigens set off by a pathogen, therefore allowing pathogen detection. This binding is frequently coupled with additional processes that convert a biological event into a measurable signal. Most electrical immunosensor signal transduction methods also involve optical and fluorescence-based equivalents. In electrochemical immunosensors, pathogen and antibody inclusion change the electrical properties of the sensor, e.g. current or impedance, which may be measured when an antigen binds to the opposite part of the sensor. For example, optical immunosensor surface plasmon resonance (SPR) depends on light pattern changes resulting from molecule attachment to a surface and provides label-free detection in real time. High sensitivity in pathogen detection, even at extremely low levels of contamination, is possible in this class of sensors by means of fluorescent dyes or nanoparticles that light up when bound to the target pathogen (Aydin et al. 2021).

Among others, infrared sensors abound in variety and quantity in relation to the detection of a particular pathogen. Every orthodox view differs in the criteria of

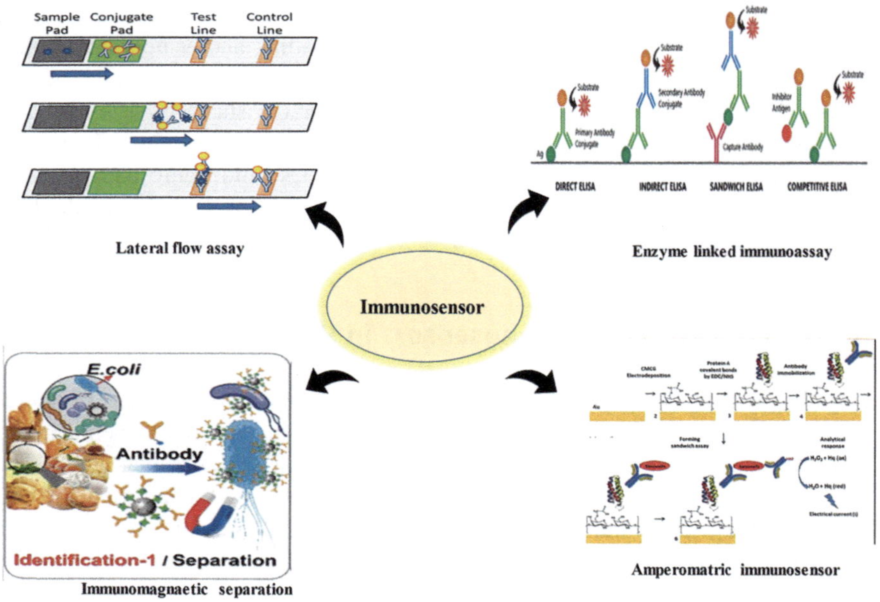

Fig. 2 Classification and mechanisms of immunosensors: advances in detection technologies for biomolecular recognition

sensitivity, quickness, or use. The right choice for pyrexia of unknown cause research is SPR-based immunosensors, as they have a suitable degree of sensitivity and real-time action detection. Since these electrochemical immunosensors are small and user-friendly and take little time to produce the required results except in the laboratory, they are used for food safety applications. This approach is especially relevant for diagnostic techniques that require onsite equipment. These kinds of sensors have excellent sensitivity and a low detection limit for pathogens; therefore, such devices are used in fields where high-sensitivity detection is necessary, such as in the clinical and food safety domains. From this point of view, all these kinds of immunosensors, together with the progressive development of nanotechnology and signal amplification, will improve pathogen detection performance during highly complicated food processes and improve frontal food safety with increased speed and accuracy (Gopinath et al. 2014).

4 Development of Immunosensors for *E. coli* O157:H7 Detection

Selecting the antibodies, immobilising them onto the sensor substrate, and investigating the amplification processes constitute many crucial stages in the process of developing immunosensors to identify *Escherichia coli* O157:H7. The type of antibodies used, i.e. monoclonal or polyclonal, significantly influences the sensor's performance. Monoclonal antibodies (mAbs) are more suited for use in diagnostics where specificity is beneficial, as they, by definition, bind to one particular epitope on the target. They are also repeatable, which is crucial for the production of specific immunosensors. Conversely, polyclonal antibodies (pAbs) are collections of multiple antibodies produced against different epitopes of the pathogen, therefore offering a greater degree of coverage but simultaneously increasing the possibility of spotting more than one target. The two most significant factors in the selection of antibodies are sensitivity and specificity. In this scenario, suitable high-affinity antibodies are utilised, as they can identify *E. coli* O157:H7 even at very minute concentrations in the material under examination. Specificity is crucial, as there is no binding to other bacteria or food components that can provide erroneous positive findings (Pebdeni et al. 2022).

Attaching the chosen antibodies to the surface of the immobilised sensor allows one to capture *E. coli* O157:H7. Physical adsorption, covalent attachment, affinity-based immobilisation, and self-assembled monolayers (SAMs) are just a few of the several ways in which antibodies can be immobilised. Many times, covalent binding is more desirable, as it enables stronger and stable binding of the antibodies to the surface than physical adsorption, which is simple but may take stable binding difficult to obtain. More accuracy and control come from affinity-based methods such as biotin-streptavidin interactions (Rani et al. 2021). In particular, it is crucial to choose a specific immobilisation technique very carefully so that antibodies retain their functional activity and therefore avoid affecting sensor performance during testing. In food matrix applications, where detection may be influenced by complex systems composed of fats, proteins, and sugars, optimisation strategies implementing cloudy and surface-active agents (e.g. bovine serum albumin) or improving sensors by surface-active nanoparticles are used to reduce nonspecific binding and enhance the sensor (Wu et al. 2015).

Many signal-enhancing methods are used to increase the sensitivity of the sensor. One such method that is most often used is the use of enzyme-linked antibodies, in which an antibody is coupled to an enzyme in charge of producing a discernible signal, colour, or fluorescence. Even with very low quantities of *E. coli* O157:H7, this method can provide an enhanced signal. Furthermore, gold colloids and magneto-nanoparticles, especially magnetic nanoparticles, are useful. Because of the phenomenon of surface plasmon resonance, the former is used to improve various optical signals; the latter serves to increase the sensitivity of detection by means of bacterial isolation from the sample. Moreover, new types of nanomaterials, known as quantum dots (QDs), have the ability to produce fluorescence and may be distinguished from one another on the basis of their unique ability to capture various targets at the same time. These are some of the nanoparticle-based techniques that show great potential when combined with enzyme amplification in the construction of immunosensors, where the detection of minimal amounts of *E. coli* O157:H7 is possible; thus, their use in food safety and the detection of pathogens is relevant. By choosing the correct antibodies and fixing methods, as well as by using strategic amplification, one may enhance such immunosensors and provide a fast, accurate food or environmental safety detection system (Razmi et al. 2020).

5 Application of Immunosensors in Different Food Matrices

Furthermore, the use of immunosensor technology to detect *E. coli* O157:H7 in various food matrices is complicated, as different food types have distinct compositions. Every food matrix might cause interferences that affect sensor performance; thus, particular actions should be taken to improve detection. Among vegetables and fruits, such materials include water, carbohydrates, and plant components, among others, which disrupt antibody–antigen binding, making the sensors less effective and sensitive. Techniques such as sample pretreatment (filtration or centrifugation) and the addition of blocking agents such as bovine serum albumin to the solution are used to overcome this problem and prevent the matrix from adhering to the sensor. Furthermore, depending on the degree of interference expected, microfluidic devices are coupled with the enhanced surface modification of nanoparticles to improve pathogen capture. For example, these techniques have been found to increase detection limits in the case of *E. coli* O157:H7 in raw fruits and vegetables, even at extremely low levels and for a shorter duration of the test (Ricci et al. 2007). The use of paper electrodes has made detection a very cost-effective approach (Panwar et al. 2022).

The high fat and protein contents of meat products aggravate the issue of immunosensor performance even more. For example, fatty acids and proteins might adsorb non-specifically onto the surface of a sensor and thereby interfere with the interaction of antibodies. Sensors are often altered with suitable coatings or more sophisticated biocompatible materials that help to lower the impacts of proteins and lipids to address this problem. Furthermore, too high a water content, typical of meat products, might influence the immunosensor's function; therefore, it is essential to use homogenising or drying techniques for sample preparation. The food sector has made use of immunosensor devices to successfully evaluate the presence of *E. coli* O157:H7 from contaminated beef and chicken. Some studies indicate a detection limit of 10 CFU/g, which is suitable for compliance with food safety measures in major meat manufacturing (Duffy and Moore 2017; Hosu et al. 2018).

Similar challenges as those associated with meat matrices arise from the naturally high-fat and high-protein concentrations of dairy products. These components might promote nonspecific adsorption onto the sensor, with either reduced sensitivity effects or possibly false positive influence. In response, milk-specific immunosensors usually utilise surface treatment along with functionalisation to make the antibodies accessible and reduce matrix interference. For example, while enzyme amplification techniques can be employed to increase the sensitivity of detection, hydrophilic coatings can help minimise the consequences of protein fouling (Moises and Schäferling 2009). This is especially challenging, as *E. coli* O157:H7 is likely present even at tiny concentrations in pasteurised dairy products, and the sensitivity criteria are generally high for dairy products. Studies have indicated that these sensors can offer farmers an efficient solution with high detection capabilities in the quickest time, allowing them to rapidly and efficiently evaluate milk, cheese, and yogurt contamination levels (Dzantiev et al. 2014).

Preservatives, thickeners, and processing by-products help prevent the interaction between antibodies and antigens, thereby generating extra critical dynamics in processed and packaged foods. Even some food colourings, such as sodium benzoate or sorbate, can attach non-specifically to the surfaces of sensors, which usually reduces the detection effectiveness. This immune problem while testing food goods with food-borne pathogens is generally resolved via enhanced signal amplification procedures or designs to concentrate on select markers such as the Shiga toxins or specific surface antigens of *E. coli* O157:H7. Moreover, the wear-and-tear dependability of the systems depends on the evaluation of how such systems might be used in industry, for instance, batch operations and point-of-use testing. Studies have shown that immunosensors can also be used for the detection of *E. coli* O157:H7 in processed meats, ready-made foods, and packaged vegetables, indicating that these systems can provide soft regulatory solutions for manufacturers in a very short period of time while being compliant with the laws. Because they can rapidly and specifically identify any pathogen that is necessary to prevent food-borne illnesses, immunosensors generally have great potential for enhancing food safety procedures in different food matrices (Jayan et al. 2020; Pan et al. 2023).

6 Challenges and Limitations of Immunosensors in Food Matrices

In particular, for the identification of pathogens such as *Escherichia coli* O157:H7, immunosensor devices are under development to improve pathogen detection in food systems. Nevertheless, several difficulties and restrictions exist that limit their usefulness in useful contexts. The so-called matrix effect, resulting from food matrices composed of lipids, proteins, fibrous material, and other components negatively affecting sensor performance, presents one of the key difficulties in analysis. Moreover, nonspecific binding or cross-reactivity are problems wherein antibodies attach themselves to components other than the target microorganism, thereby either lowering the general sensor sensitivity or producing false positive readings in the data. For example, some food ingredients, including grease or dairy powders, might coat sensor surfaces or bind to antibody active sites, thereby reducing the sensitivity and accuracy of the immunosensor. By either avoiding or interfering with antibody binding to *E. coli* O157:H7, other dietary compounds can essentially block antibody-based detection of the bacterium. Furthermore, these additional interferences caused by matrix compositions and elements in highly complex food types, such as raw vegetables, meat, or even dairy products, are especially noticeable because of the need for sensor surfaces and sample processing to be optimised to reduce such effects.

The ability to achieve strong sensitivity and low detection limits in complicated food matrices is another important restricting influence. Immunosensors are meant to be able to detect every particular pathogenic agent present even at the lowest quantity. Food matrix components might cause the sensor to lose sensitivity; hence, identifying infections at acceptable levels becomes challenging. Certain food products, especially those containing multiple interfering ingredients, comply with regulations for detection limits. For example, the 1–10 CFU/g required by legislation might be impossible. Conversely, although new methods are based on culture media, such as PCR, ELISA promises to yield findings with great sensitivity within a short period of time that are somewhat arduous and lengthy. Conversely, immunosensors provide rapid findings, but the complex sample matrix can influence the sensitivity of the technique. Therefore, even if immunosensors are faster than more conventional methods are used, the latter might be able to offer a higher degree of sensitivity and dependability than is attainable under some situations when immunosensors are used.

In particular, the use of immunosensors in the field still raises serious questions about their stability, repeatability, and storage. Most immunosensors are quite sensitive to several elements, including the use of biological components such as antibodies that cannot remain for long, therefore influencing the operation of the sensor. Furthermore, the lifetime of the sensor is rather crucial, especially in field use under different climatic conditions; specifically, the temperature and even the humidity affect the sensors. Maintaining the repeatability of the data acquired with reference to sensors is especially challenging, as any result may be altered depending on minor changes in sample handling, matrix effects, or sensor degradation. Combining these approaches requires much study and the introduction of new materials, synthetic antibodies, and nanomaterials, for example, as well as new techniques of storage, such as freeze drying or the use of stabilisers. Nevertheless, surface chemistry, good sensor designs, and amplification methods have been developed in an attempt to improve the dependability and applicability of food safety immunosensors.

7 Future Perspectives and Emerging Trends

For food safety, development trends, especially nanotechnology, mobility, and multiplexing, will be beneficial for immunosensors. Among the most important advances aimed at increasing the sensitivity of immunosensors is nanotechnology. By expanding the area for antibody-binding surfaces and signal augmentation via plasmonic or emissive processes such as resonance and fluorescence, gold, silver, and magnetic nanoparticles help the sensor overcome the impediment provided by any pathogen at low concentrations. One advantage over more traditional methods is that these types of nanomaterials can also enable fast pathogen detection in various applications (Srivastava et al. 2021; Joshi et al. 2022; Pramanik et al. 2023; Yadav et al. 2023). Because they are highly useful for in-field and real-time sensor applications, the addition of nanoparticles improves the functional properties of immunosensors, hence increasing their sensitivity and helping to reduce the sensor size. Nanotechnology addresses some of the fundamental issues associated with immunosensors, which concern complex food samples, by providing signal augmentation and enhanced detection limits.

The development and widespread adoption of portable and point-of-care immunosensors that enable rapid pathogen detection without being limited to laboratories is another area of valuable investigation (Mohanty et al. 2023). These sensors, which are typically coupled with smartphones and other wearable devices, are altering the field of food product safety assessment and thereby constitute a revolution. Mobile apps have allowed immunosensors to provide real-time data via a detection platform shaped like a smartphone, thereby improving decisions in food processing and customer safety. For both far-off and outdoor sites and even manufacturing and assembling lines, the inclusion also offers a clear assessment of the scope of field testing, accelerating the management of contamination hazards and avoiding the complexity of employing laboratories. The mobility of these devices also helps small-scale food producers who can scarcely afford the necessary basic testing facilities, especially in poor nations.

Multiplexed high-throughput immunosensors are yet another developing trend projected to improve pathogen detection efficiency and increase its capabilities. These sensors enable one test, which may simultaneously identify numerous infections or foodborne toxins (Srivastava et al. 2021; Mitra et al. 2022). Such benefits simplify the food safety consumer product testing process. Multiplexed sensors reduce testing time and expense by employing various antibodies or probes for different diseases (Mohammadzadeh et al. 2022; Yeganeh et al. 2022; Kumar et al. 2022; Sabarees et al. 2023; Pramanik et al. 2023; Manjunathan et al. 2024). These robotic sensors enable the performance of what is known as industrial testing, whereby large quantities of samples may be handled without much person involvement. For food manufacturers, government agencies, and research facilities, such systems are perfect, as they offer a solution for the difficulty of performing extensive tests quickly for reasons of quality assurance, control of disease outbreaks, and regulatory compliance. The developments in the technologies of multiplexed sensors also enable improved food safety procedures, where different pollutants may be concurrently watched, thereby improving food safety management and preventing the spread of foodborne diseases. These developments also provide a glimpse of the future for immunosensors, with speed, sensitivity, and scalability present to revolutionise food testing worldwide.

8 Conclusion

Immunosensors are increasingly indispensable in the management of *E. coli* O157:H7 in food items because of their fast, sensitive, and specific approaches for detecting this disease. Food safety and management of *E. coli* outbreaks depend on the detection of *E. coli* O157:H7 at quite low levels in certain food matrices. These sensors use the affinity of antibodies to recognise bacterial antigens or toxins, therefore removing the sluggish and demanding requirement for culture. Nevertheless, there are additional issues, particularly related to dietary components that might have multiple effects because proteins and lipids, among other elements, can compromise the sensors. To improve the sensitivity and dependability of immunosensors in field applications, aspects such as cross-reactivity, nonspecific binding, and matrix inhibition need to be considered.

With respect to technological foresight, the evolution of immunosensors for pathogen detection will include multiplex testing, biomolecular devices, and miniaturisation of diagnostic instruments. By producing more signals and resulting in less distortion, nanoparticles and other nanostructures can be employed to increase the detection speed and accuracy of surface detection. Such portable point-of-care immunosensors functioning with smartphones and other wearable devices will enable testing at the place of interest and simultaneous generation of findings for both high-volume and retail end users. Furthermore, the use of multiplexed immunosensors for the simultaneous detection of many pathogens would help consolidate testing and thereby enable considerably more effective and less expensive monitoring of food safety than in past years. Although several difficulties still exist, such as how to stabilise the sensors and achieve low detection limits under complex matrices, the developed technologies show that immunosensors will be rather helpful for ensuring food safety.

Overall, the application of immunosensors in food safety systems has significant potential for enhancing present health measures by means of quick, simple, wide-area pathogen detection. Evaluating the presence of bacteria such as *E. coli* O157:H7 is crucial for prompt intervention in food safety, thereby reducing the likelihood of outbreak development and reducing the burden on the health care system. Indeed, as they have developed, there is every reason to assume that immunosensors will become a necessary tool with the goal of improving food safety worldwide, the scope of which ensures the absence of infections in food items and thus promotes healthy food systems worldwide.

References

Addis M, Sisay D (2015) A review on major food borne bacteria l illnesses. J Trop Dis 3(4):1–7

Aydin M, Aydin EB, Sezgintürk MK (2021) Advances in immunosensor technology. Adv Clin Chem 102:1–62. https://doi.org/10.1016/bs.acc.2020.08.001

Begum SJP, Pratibha S, Rawat JM et al (2022) Recent advances in green synthesis, characterisation, and applications of bioactive metallic nanoparticles. Pharmaceuticals 15:455. https://doi.org/10.3390/ph15040455

Besser RE (1993) An outbreak of diarrhea and hemolytic uremic syndrome from *Escherichia coli* O157:H7 in fresh-pressed apple cider. JAMA J Am Med Assoc 269:2217. https://doi.org/10.1001/jama.1993.03500170047032

Delaquis P, Bach S, Dinu L-D (2007) Behavior of *Escherichia coli* O157:H7 in leafy vegetables. J Food Prot 70:1966–1974. https://doi.org/10.4315/0362-028X-70.8.1966

Duan M, Xiao X, Huang Y et al (2021) Immuno-HCR based on contact quenching and fluorescence resonance energy transfer for sensitive and low background detection of *Escherichia coli* O157:H7. Food Chem 334:127568. https://doi.org/10.1016/j.foodchem.2020.127568

Duffy GF, Moore EJ (2017) Electrochemical immunosensors for food analysis: a review of recent developments. Anal Lett 50:1–32. https://doi.org/10.1080/00032719.2016.1167900

Dzantiev BB, Byzova NA, Urusov AE et al (2014) Immunochromatographic methods in food analysis. TrAC Trends Anal Chem 55:81–93. https://doi.org/10.1016/j.trac.2013.11.007

Eyi A, Arslan S (2012) Prevalence of Escherichia coli in retail poultry meat, ground beef and beef. Med Weter 68:237–240. medycynawet.edu.pl

Gopinath SCB, Tang T-H, Citartan M et al (2014) Current aspects in immunosensors. Biosens Bioelectron 57:292–302. https://doi.org/10.1016/j.bios.2014.02.029

Güner A, Çevik E, Şenel M et al (2017) An electrochemical immunosensor for sensitive detection of *Escherichia coli* O157:H7 by using chitosan, MWCNT, polypyrrole with gold nanoparticles hybrid sensing platform. Food Chem 229:358–365. https://doi.org/10.1016/j.foodchem.2017.02.083

Hosu O, Selvolini G, Marrazza G (2018) Recent advances of immunosensors for detecting food allergens. Curr Opin Electrochem 10:149–156. https://doi.org/10.1016/j.coelec.2018.05.022

Jayan H, Pu H, Sun D-W (2020) Recent development in rapid detection techniques for microorganism activities in food matrices using biorecognition: a review. Trends Food Sci Technol 95:233–246. https://doi.org/10.1016/j.tifs.2019.11.007

Joshi NC, Rawat BS, Bisht H et al (2022) Synthesis and supercapacitive behaviour of SnO2/r-GO nanocomposite. Synth Met 289:117132. https://doi.org/10.1016/j.synthmet.2022.117132

Kumar A, Devi M, Kumar M et al (2022) Silicon nanostructures and nanocomposites for antibacterial and theranostic applications. Sensors Actuators A Phys 347:113912. https://doi.org/10.1016/j.sna.2022.113912

Lv X, Wang L, Zhang J et al (2020) Rapid and sensitive detection of VBNC *Escherichia coli* O157:H7 in beef by PMAxx and real-time LAMP. Food Control 115:107292. https://doi.org/10.1016/j.foodcont.2020.107292

Manjunathan J, Revathi M, Sowmya H et al (2024) Recent advancements in nanotechnological approaches for pollution monitoring and environmental sustainability. Clean Techn Environ Policy 26:3667–3683. https://doi.org/10.1007/s10098-023-02676-z

Mendes Silva D, Domingues L (2015) On the track for an efficient detection of *Escherichia coli* in water: a review on PCR-based methods. Ecotoxicol Environ Saf 113:400–411. https://doi.org/10.1016/j.ecoenv.2014.12.015

Mitra D, Mondal R, Khoshru B et al (2022) Actinobacteria-enhanced plant growth, nutrient acquisition, and crop protection: advances in soil, plant, and microbial multifactorial interactions. Pedosphere 32:149–170. https://doi.org/10.1016/S1002-0160(21)60042-5

Mohammadzadeh V, Rahiman N, Hosseinikhah SM et al (2022) Novel EPR-enhanced strategies for targeted drug delivery in pancreatic cancer: an update. J Drug Deliv Sci Technol 73:103459. https://doi.org/10.1016/j.jddst.2022.103459

Mohanty S, Panwar V, Anoop G et al (2023) Eco-friendly phyllanthus emblica-based ionic polymer composite for enhanced mechanical, electrical, and wearable sensing performance. Mater Chem Phys 303:127791. https://doi.org/10.1016/j.matchemphys.2023.127791

Moises SS, Schäferling M (2009) Toxin immunosensors and sensor arrays for food quality control. Bioanal Rev 1:73–104. https://doi.org/10.1007/s12566-009-0006-x

Nangan S, Kanagaraj K, Kaarthikeyan G et al (2023) Sustainable preparation of luminescent carbon dots from syringe waste and hyaluronic acid for cellular imaging and antimicrobial applications. Environ Res 237:116990. https://doi.org/10.1016/j.envres.2023.116990

Nesakumar N, Lakshmanakumar M, Srinivasan S et al (2021) Principles and recent advances in biosensors for pathogens detection. ChemistrySelect 6:10063–10091. https://doi.org/10.1002/slct.202101062

Pan M, Li H, Yang J et al (2023) Review: synthesis of metal organic framework-based composites for application as immunosensors in food safety. Anal Chim Acta 1266:341331. https://doi.org/10.1016/j.aca.2023.341331

Panwar V, Anoop G, Gaur SS et al (2022) Enhanced sensing and electrical performance of hierarchical porous ionic polymer-metal nanocomposite by minimising cracks in electrode. J Colloid Interface Sci 606:837–847. https://doi.org/10.1016/j.jcis.2021.08.074

Pebdeni AB, Roshani A, Mirsadoughi E et al (2022) Recent advances in optical biosensors for specific detection of *E. coli* bacteria in food and water. Food Control 135:108822. https://doi.org/10.1016/j.foodcont.2022.108822

Pramanik B, Sar P, Bharti R et al (2023) Multifactorial role of nanoparticles in alleviating environmental stresses for sustainable crop production and protection. Plant Physiol Biochem 201:107831. https://doi.org/10.1016/j.plaphy.2023.107831

Prasad B, Alarifi IM, Gill FS et al (2022) Development of conductive thin films as piezoresistive strain sensor. Mater Chem Phys 276:125371. https://doi.org/10.1016/j.matchemphys.2021.125371

Rani A, Ravindran VB, Surapaneni A et al (2021) Review: trends in point-of-care diagnosis for *Escherichia coli* O157:H7 in food and water. Int J Food Microbiol 349:109233. https://doi.org/10.1016/j.ijfoodmicro.2021.109233

Razmi N, Hasanzadeh M, Willander M et al (2020) Recent progress on the electrochemical biosensing of *Escherichia coli* O157:H7: material and methods overview. Biosensors 10:54. https://doi.org/10.3390/bios10050054

Ricci F, Volpe G, Micheli L et al (2007) A review on novel developments and applications of immunosensors in food analysis. Anal Chim Acta 605:111–129. https://doi.org/10.1016/j.aca.2007.10.046

Sabarees G, Tamilarasi GP, Velmurugan V et al (2023) Emerging trends in silk fibroin based nanofibers for impaired wound healing. J Drug Deliv Sci Technol 79:103994. https://doi.org/10.1016/j.jddst.2022.103994

Scallan E, Hoekstra RM, Angulo FJ et al (2011) Foodborne illness acquired in the United States—major pathogens. Emerg Infect Dis 17:7–15. https://doi.org/10.3201/eid1701.P11101

Shah K, Maghsoudlou P (2016) Enzyme-linked immunosorbent assay (ELISA): the basics. Br J Hosp Med 77:C98–C101. https://doi.org/10.12968/hmed.2016.77.7.C98

Sidari R, Caridi A (2011) Methods for detecting enterohaemorrhagic Escherichia coli in food. Food Rev Int 27:134–153. https://doi.org/10.1080/87559129.2010.535232

Srivastava S, Singh L, Kaushik V et al (2021) Electrically controlled nanophotonic slot structure based on photocatalytic nanocomposite for optical detection of foodborne pathogens. J Light Technol 39:6670–6677. https://doi.org/10.1109/JLT.2021.3104409

Trache D, Tarchoun AF, Abdelaziz A et al (2022) Cellulose nanofibrils–graphene hybrids: recent advances in fabrication, properties, and applications. Nanoscale 14:12515–12546. https://doi.org/10.1039/D2NR01967A

Wang Z, Cai R, Gao Z et al (2020) Immunomagnetic separation: an effective pretreatment technology for isolation and enrichment in food microorganisms detection. Compr Rev Food Sci Food Saf 19:3802–3824. https://doi.org/10.1111/1541-4337.12656

Worku LA, Tadesse MG, Bachheti A et al (2024) Experimental investigations on PVA/chitosan and PVA/chitin films for active food packaging using Oxytenanthera abyssinica lignin nanoparticles and its UV-shielding, antimicrobial, and antiradical effects. Int J Biol Macromol 254:127644. https://doi.org/10.1016/j.ijbiomac.2023.127644

Wu S-Y, Hulme J, An SSA (2015) Recent trends in the detection of pathogenic *Escherichia coli* O157: H7. Biochip J 9:173–181. https://doi.org/10.1007/s13206-015-9208-9

Wu Y, Liu Y, Zhan Q et al (2017) Rapid detection of *Escherichia coli* O157:H7 using tunneling magnetoresistance biosensor. AIP Adv 7:056658. https://doi.org/10.1063/1.4977017

Yadav S, Chamoli S, Kumar P et al (2023) Structural and functional insights in polysaccharides coated cerium oxide nanoparticles and their potential biomedical applications: a review. Int J Biol Macromol 246:125673. https://doi.org/10.1016/j.ijbiomac.2023.125673

Yeganeh FE, Yeganeh AE, Far BF et al (2022) Synthesis and characterisation of tetracycline loaded methionine-coated NiFe2O4 nanoparticles for anticancer and antibacterial applications. Nanomaterials 12:2286. https://doi.org/10.3390/nano12132286

Zhao Y-W, Wang H-X, Jia G-C et al (2018) Application of Aptamer-based biosensor for rapid detection of pathogenic *Escherichia coli*. Sensors 18:2518. https://doi.org/10.3390/s18082518

Zhong M, Yang L, Yang H et al (2019) An electrochemical immunobiosensor for ultrasensitive detection of *Escherichia coli* O157:H7 using CdS quantum dots-encapsulated metal-organic frameworks as signal-amplifying tags. Biosens Bioelectron 126:493–500. https://doi.org/10.1016/j.bios.2018.11.001

Detection of Foodborne Pathogens in Meat and Meat-Based Products

Harshita Ranjan, Nidhi Bhardwaj, Lukas Peter, Junaid Aman, Habiba Iqbal, and Kanika Dulta

Abstract

Recent advances have been made in the food industry regarding preservation and food safety technology, but nevertheless, there are still disease outbreaks that are linked to the food pathogens such as bacteria, fungi, and viruses, showing that these pathogens are still responsible for causing outbreaks worldwide, posing a threat to the consumer's health, particularly in meat and meat-based products, which are very often vulnerable to various types of contamination by various means—for example, during processing and handling of the food. This chapter focuses on a comprehensive overview of different detection methods used for foodborne pathogens such as *Salmonella, Escherichia coli, Listeria monocytogenes, and Campylobacter jejuni*, starting from traditional culture-based techniques to modern molecular approaches, including immunological assays and emerging technologies like biosensors. Among these biosensors, immunosensors have shown to be a very promising technology. And above all these advanced techniques, there is still room for improvement; for example, false

H. Ranjan
School of Bioengineering & Food Technology, Shoolini University of Biotechnology and Management Sciences, Solan, Himachal Pradesh, India

N. Bhardwaj · L. Peter · K. Dulta (✉)
Center of Advanced Innovation Technologies, VSB-Technical University of Ostrava, Ostrava, Czech Republic
e-mail: lukas.peter@vsb.cz

J. Aman
School of Applied and Life Sciences, Uttaranchal University, Dehradun, India

H. Iqbal
Centre of Excellence in Molecular Biology, University of the Punjab, Lahore, Pakistan

K. Kuča et al. (eds.), *Ensuring Wellness: Immunosensors Monitoring Food Quality and Pathogenicity Risks*, https://doi.org/10.1007/978-3-031-97477-9_6

positives/negatives and regulatory hurdles remain prevalent. The chapter has concluded with a call for combining available approaches of multiple detection methods to enhance accuracy and assure food security toward public health.

Keywords

Food · Meat · Meat products · Foodborne pathogens · Biosensors · Food safety

1 Introduction

In the food industry, microorganisms are key players. From the production line to the safety and the spoilage, the interactions occurring between these species can lead to the spoilage of food and hence become unfit for human consumption. Apart from causing the spoilage, these foodborne pathogens represent a threat toward the food safety, not only to the people who produce the food but also to the people who consume the food (Aladhadh 2023). According to the Centre for disease control and prevention, global statistics for foodborne outbreaks published for the years 2011 to 2012, 29,112 affected patients, 1750 hospitalizations, and this is not the end, also causing 68 deaths (Stephen Inbaraj and Chen 2016). The World Health Organization (WHO) states that these pathogens lead to 23 million foodborne illnesses, with 5000 reported deaths annually across Europe. Foodborne illnesses are frequently linked to pathogenic microorganisms, including bacteria, viruses, fungi, yeast, and parasites. Human foodborne illness is primarily caused by *Salmonella, Campylobacter species, Listeria monocytogenes, and Escherichia coli.* During their growth, face harvesting, processing, storage, or subsequent transportation and preservation phase, fresh food—such as fish, meat, poultry, eggs, and dairy—can become contaminated (Kabiraz et al. 2023). It is therefore very important to have efficiency and accuracy in these methods of detection in order to ensure healthy food and prevent any health outbreaks. Meat is considered to be one of the nutritious foods as it has the ability to provide essential nutrients, high-quality proteins, along with fatty acid content and micronutrients. Meat products being highly perishable, as a result, can be spoiled very easily and thus lose their suitability for human consumption. Microbial proliferation, chemical alterations, and endogenous enzyme breakdown are the causes (Armany et al. 2016). Since meat production and processing is quite complex, resulting in the risk of contamination by these pathogens (Bonaldo et al. 2023; Lianou et al. 2017). Several factors could contribute to the pathogen survival in meat and meat products, which may include improper handling of carcasses, not adequate cooking temperatures, cross-contamination during processing, and poor hygiene conditions. For example, *Salmonella* can survive in various environmental niches, which include animal intestines and processing facilities, making it a continuous threat in the meat supply chain (Lianou et al. 2017). Similarly, *E. coli O157:H7* is concerning due to its low infectious dose and potential for causing severe health complications, including hemolytic uremic syndrome (HUS) (Rahal et al. 2012). The potential ability of these pathogens to adapt to various environmental conditions further complicates detection and control efforts (Ferrer-Bustins et al. 2024). Considering all the facts, while assessing the safety of meat and meat derivatives, it is critical to take these pathogens' potential for transmission into account. As a result, having strong, quick, and trustworthy techniques to identify infections in various food products is essential (Kumar et al. 2020).

This chapter has discussed both traditional and cutting-edge techniques for identifying pathogens in meat and meat-based products in the field. We have also focused on the use of biosensors, and among those biosensors, immunosensors, utilizing antibodies as biorecognition elements, have gained a lot of attention due to their high specificity. We have also focused on the challenges faced in the industry and future advancements.

2 Foodborne Pathogens in Meat

2.1 Common Pathogens and Their Health Consequences

Salmonella species, *Escherichia coli (O157:H7)*, *Listeria monocytogenes*, and *Campylobacter jejuni* are among the most prevalent foodborne pathogens identified in meat and meat-based products. Furthermore, some opportunistic bacteria like *Cronobacter sakazakii* pose a growing risk because of their potential ability to contaminate a variety of items and cause a number of diseases that have a 40–80% death rate in babies and immunocompromised individuals (Bonaldo et al. 2023). Since each of these infections has unique traits and reservoirs, therefore it becomes difficult to detect them in the food matrix. The second most common gastrointestinal illness in the European Union (EU) is due to consuming foods contaminated with *Salmonella* is salmonellosis. Abdominal cramps, bloody diarrhea, fever, myalgia, headache, nausea, and vomiting are typical signs of gastroenteritis. More than half of the foodborne outbreak cases reported in the EU in 2018 were attributed to *Salmonella*, according to reports (Ehuwa et al. 2021). There are many serotypes from this genus, and most of them are considered a threat to humans. *Salmonella* can infect meat and meat products at various stages during processing and is present in animal intestines, especially poultry and livestock (Meyer et al. 2010).

A well-known harmful strain of *Escherichia coli* O157:H7 causes illnesses when contaminated food or water is consumed and is primarily found in the intestine of cattle, thus making it a suitable source of contamination in the carcass. For example, *E. coli* contamination in chicken proves the inadequate hygiene practices in slaughterhouse and trading areas (Kumar et al. 2020). This strain has the potential to cause a wide range of illnesses, starting from mild diarrhea and hemorrhagic colitis to hemolytic uremic syndrome (HUS), which can be fatal, and hemorrhagic colitis, which can cause kidney failure, especially in young children and the elderly. Due to the minimal infectious dosage required to produce infection, it has posed a significant concern to public health (Rahal et al. 2012).

Because of its unusual capacity to thrive at refrigerator temperatures, a pathogenic bacterium *Listeria monocytogenes*, causing *listeriosis*, poses a risk to meat items that are ready to consume. Immunocompromised people, the elderly, babies, and pregnant women are the most vulnerable ones (Kurpas et al. 2018). From minor flu-like symptoms to serious illnesses like meningitis, symptoms might vary. The ability to thrive in harsh environments and to create biofilms on a variety of surfaces makes it especially worrisome when it comes to meat and meat-based products. It is also common in ready-to-eat (RTE) meals and can grow at refrigerator temperatures. This pathogen is often present in the intestines of animals, soil, water, and plants, making it a significant risk factor in the meat supply chain (Lianou et al. 2017).

Campylobacter jejuni in meat endangers the general public's health, as it is a globally concerned bacterial food-borne pathogen which is frequently found in raw or undercooked poultry. It affects roughly 1% of people in Europe annually, and 13 out of every 100,000 people in the US get infected with it (Epps et al. 2013). It causes gastrointestinal disorders that include fever, cramping in the abdomen, and diarrhea. Cross-contamination during food preparation can result in infection. Food-borne pathogens have severe health complications worldwide and hence being a major threat to society. For example, gastrointestinal issues, especially affecting the small intestine tissue, are frequently the cause of meat-borne illnesses, which might include diarrhea (Ali and Alsayeqh 2022). Common symptoms may include fever, vomiting, diarrhea (frequently bloody), and abdominal pain. After eating contaminated meat, these symptoms usually appear hours to days later (WHO 2020). Apart from their immediate health problems, food-borne pathogens could create long-term health disorders and complications such as reactive arthritis or chronic kidney disease, which can ultimately follow hemolytic uremic syndrome (HUS) (Batz et al. 2013). Due to the healthcare and hospital costs, the economic impact, as well as lost productivity, further underscores the utmost need for effective and accurate detection and prevention.

3 Detection Methods

The detection of these pathogens is a crucial aspect of the regulatory compliance in the food industry chain and processing, with high recurrence and sensitivity, which is essential to assure food safety and consumer health (Aladhadh 2023; Kabiraz et al. 2023). Several detection methods, including from traditional methods to modern techniques, have been developed to identify foodborne pathogens. Figure 1 shows several methods for the detection of foodborne pathogens, from traditional to modern ones. In this chapter, we will discuss both methods along with their principles, advantages, and disadvantages.

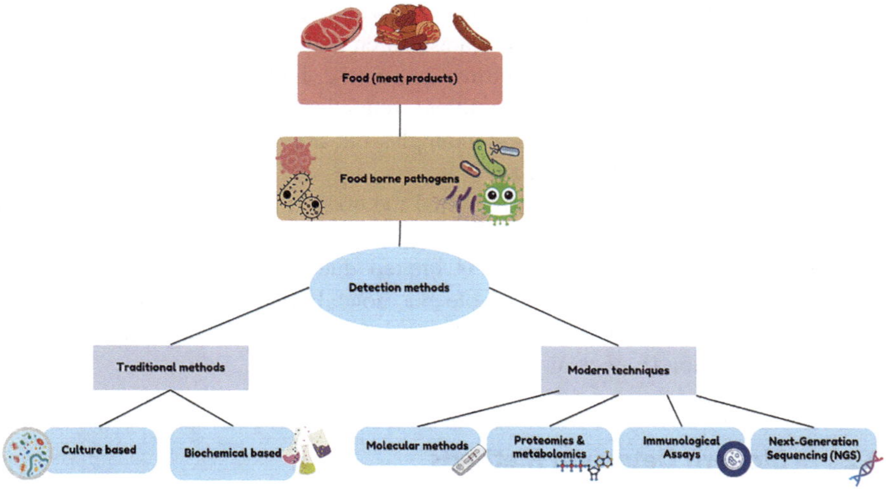

Fig. 1 Schematic representation of the methods for the detection of pathogens

3.1 Traditional Detection Methods

3.1.1 Culture-Based Methods

Conventional methods for identifying foodborne pathogens in food matrices typically involve culturing the microorganisms on agar plates, following the standard biochemical identification processes (Law et al. 2015). The culture-based is an established approach and a promising technique for identifying foodborne pathogens and is also economical. This approach has proven to be very selective and unique since it can eliminate the growth of unwanted microbes, and a differential medium is utilized to detect specific pathogenic microbes (Kabiraz et al. 2023) . For example, a specific type of culture medium, *Rogosa*, is created to differentiate between the *Lactobacillus* genus and nonlactic acid bacteria. It contains acetic acid, sodium acetate, and specific inhibitors such as nitrite, polymyxin B, and actidione, which hinder the development of nonlactic bacteria. This culture-based method, which is based on culture, enables distinct differentiation through color diversity. The appearance of the blue or purple colonies that appear on *Cefsulodin-Irgasan-Novobiocin* (CIN) agar indicates *Yersinia enterocolitica* presence (Stephen Inbaraj and Chen 2016).

3.1.2 Biochemical Tests

A well-known method which promotes growth and involves the utilization of compounds as signals that indicate the presence of pathogens. It involves the inhibition of the growth of competing microbes by decreasing competition. It frequently entails a culture technique in which the pathogen is incubated in a solid or liquid culture medium (Kabiraz et al. 2023). But it can involve a variety of biochemical tests that help identify pathogens based on their metabolic characteristics. Catalase, indole production, methyl red, Voges-Proskauer, triple sugar iron agar, oxidase, blood agar plates, starch hydrolysis, galactose hydrolysis, carbohydrate fermentation, citrate utilization, urease, hydrogen sulfide production, motility agar, mannitol salt agar, nitrate reduction, optochin sensitivity, and bacitracin sensitivity tests are a few examples but are often time-consuming and less specific (Saravanan et al. 2021).

The major disadvantage of conventional methods is that they are laborious, as they need culture media preparation, along with plate inoculation and colony counting. Additionally, this method is limited due to its low sensitivity. False negatives can appear because of viable but nonculturable or VBNC pathogens. These inadequacies in identifying foodborne infections would raise the danger of disease transmission (Law et al. 2015).

3.2 Modern Detection Methods

3.2.1 Molecular Methods

New advances have been made over time in the field of science, resulting in several lightning-fast methods with excellent sensitivity and specificity designed to overcome the shortcomings of traditional conventional methods to detect foodborne pathogens. And among those, molecular techniques have been shown to revolutionize pathogen detection because of their speed and specificity.

3.2.1.1 Polymerase Chain Reaction (PCR)

PCR is often used for detecting bacterial pathogens by selecting a specific targeted DNA, allowing for rapid identification. After the denaturation of double-stranded DNA into single strands at a high temperature, two synthetic oligonucleotides known as primers, one forward and one reverse, anneal to the single-stranded DNA. Then, the next step is the polymerization phase, where the complementary primers are extended to single-stranded DNA in the presence of deoxyribonucleotides and a thermostable DNA polymerase. The results, in the form of PCR amplification products after staining with ethidium bromide, are visualized as bands on an electrophoresis gel. PCR has been employed to detect various pathogens that are foodborne, including *Listeria monocytogenes, Escherichia coli O157:H7, Staphylococcus aureus, Campylobacter jejuni, Salmonella* spp., and *Shigella* spp. (Law et al. 2015; Zhao et al. 2014).

3.2.1.2 Multiplex-PCR

Compared to simple PCR, a multiplex PCR is even more rapid detection by continuous multiple gene target amplification. The principle is similar to the conventional PCR. The only difference is that there are specific primers used in mPCR assay, but when it comes to the conventional one, only a set of specific primers is being used. The design of the primer plays a very important role when it comes to developing mPCR because a successful mPCR experiment requires the primer set to have the same annealing temperature (Zhao et al. 2014). Also, the concentration of primer plays an important role in mPCR because of the potential interactions that may occur between the several primer sets in mPCR, which could produce primer dimers. Therefore, in order to ensure that dependable PCR products are produced, the primer concentration must be changed. Previously, mPCR could be used to detect two to three pathogens only at one time, but now it is more advanced because of the ability in detecting up to five or more pathogens at the same time (Law et al. 2015).

3.2.2 Proteomics and Metabolomics

3.2.2.1 Matrix-Assisted Laser Desorption Ionization–Time of Flight Mass Spectrometry

A well-known analytical technique, which is used for rapidly identifying and characterizing the microorganisms in proto mix, to attain a structural identification and characterization of proteins and peptides, which include several pro forms as there are posttranslational modifications (PTMs) or protein–protein interactions (PPIs). This technique can be used to study both macromolecules and micromolecules and offers the sensitivity of picomole and femtomole; also offering the direct analysis of biological samples, for example, biofluids, solid tissue/ cell homogenates, and cell culture lysates, with minimal sample preparation. The method operates by the ionization of microbial cells by a short laser pulse, causing acceleration of ions within the vacuum system. It will further create a molecular fingerprint specific to each microbe. The resulting mass spectra are then compared against established databases for identification (Darie-Ion et al. 2022; Zhao et al. 2014). The method proves to be effective because of the minimal sample preparation and its speed and simplicity. For instance, a study showed that this method could effectively differentiate between various types of raw meat, including beef, lamb, pork, and chicken, by analyzing their protein patterns (Nairn 2022).

3.2.2.2 Immunological Assays

This method of detecting pathogens in food is an interaction between antibody and antigen. Here, a particular antigen will bind to target antibody (monoclonal or polyclonal). However, monoclonal antibodies have been demonstrated to be more appropriate than polyclonal ones because of their selective detection of pathogens. This is because of their specificity, reproducibility, ability, sensitivity, and reliability. The antibody epitope position binding to specific antigen is responsible for the sensitivity as well as specificity of this method. It is widely shown to be a rapid method compared to culture-based methods for detecting these pathogens and their toxins (i.e., mycotoxin) (Kabiraz et al. 2023).

3.2.2.2.1 Enzyme-Linked Immunosorbent Assay (ELISA)

ELISA is a well-known immunological method that uses the affectability of an enzymatic reaction to target antigen-antibody complexes in tests. It includes pipet tips, paddles, membranes, dipsticks, solid matrices, and a microtiter that serves as a sturdy support. β-galactosidase, peroxidase, and alkaline phosphatase are the most used enzymes. It attaches with the samples by covalent bond microtiter plates coated with antigen in an alkaline buffer system, using antibodies against pathogen antigen with affinity which is quite high, and based on ultraviolet light, the fluorescent antigen/antibody complexes are detected. A particular protein or toxicant will then bind in this assay to the corresponding antigen and will generate colors visible with the naked eye, and hence the wavelength of the colors can be measured by an ELISA reader. This technique has the potential to identify pathogens from a food matrix quickly, with minimum labor work, with high specificity, and with a large number of samples. But also, there are some drawbacks of this method, which may include the challenges in determining enzyme activity, antigen variability, and the level of sensitivity limit. The assay needs to have the minimum sensitivity of a traditional medium to produce cell surface antigens and to detect them (Kabiraz et al. 2023).

3.2.2.2.2 Next-Generation Sequencing (NGS)

Currently, next-generation sequencing (NGS) technology has been proven to be more accurate for detection of pathogens in food. NGS approaches, when combined with bioinformatics, prove to be a powerful tool in securing food security. NGS can be primarily used in several ways, for example, in the field of metagenomics, to detect the sequences of many microorganisms present in a sample, and in whole genome sequencing, also known as WGS, to determine the full genome sequence of an isolate. Figure 2 shows the workflow for analyzing microbial communities in food samples using various NGS-based methods, including metatranscriptomics, shotgun genomics, metabarcoding, and metagenomics, highlighting the sequencing and bioinformatic steps leading to microbial identification and characterization. The process integrates molecular biology techniques with advanced sequencing technologies and bioinformatic analyses. In the latter use, this method can target specific microbiological groups (viruses, fungi, or bacteria) utilizing 16S rRNA or any other biological markers.

There are excellent remarks on NGS principles, types, and advantages (Aladhadh 2023). One of the biggest advantages over other ones is that it can detect several

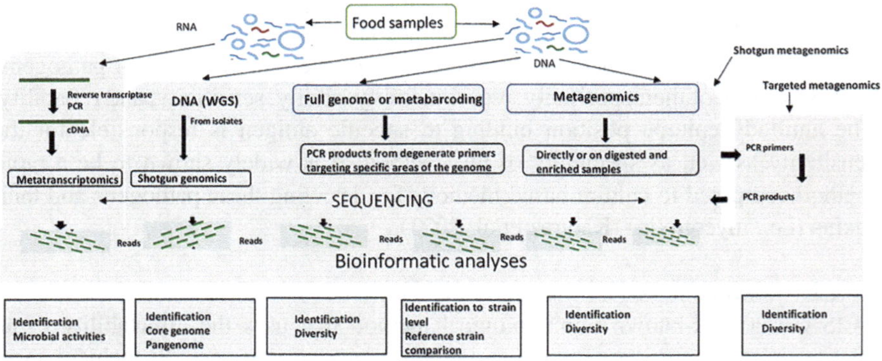

Fig. 2 Next-generation sequencing approach for detecting and identifying foodborne pathogens by (Aladhadh 2023), Creative Commons Attribution (CC BY) license

potential agents that cause spoilage and diseases, and without any target analysis, has the ability to determine the culturable and nonculturable ones. A study by (Calo-Mata et al. 2016) demonstrated the possibility of detecting nontargeted bacterial species using liquid chromatography-mass spectrometry. But the restriction was caused by the lack of standardization and public databases. Food microbiology investigations are now using NGS, majorly emphasizing on whole genome sequencing (WGS) of isolates of the bacteria and identifying microbiomes linked to specific commodities (Elbehiry et al. 2023).

3.3 Emerging Technologies

3.3.1 Biosensor and Its Types

An analytical tool known as a biosensor combines a biologically motivated molecular recognition molecule with a feasible physicochemical transducing mechanism to transform the biological response into an electrical signal (Janik-Karpinska et al. 2022). Figure 3 shows the fundamental components and step process of a biosensor, where a bioreceptor detects a target analyte, a transducer converts the recognition event into a measurable signal, and a monitoring system amplifies and displays the result for rapid pathogen detection. Biosensors are the latest emerging real-time detection technology that offers high sensitivity and specificity, being a promising advancement in pathogen monitoring. Biosensor-based technologies have caught attention in the field due to their analytical ability to incorporate a biological entity (e.g., tissue, microorganisms, nucleic acids, natural products, organelles, cell receptors, enzymes, antibodies, etc.), or a derived (e.g., recombinant antibodies, engineered proteins, aptamers, etc.), or a biomimic (e.g., synthetic catalysts, combinatorial ligands, and imprinted polymers) that can be integrated or associated within a physicochemical transducer or transducing microsystem (Zhao et al. 2014). It is basically operated by using three elements: a biological capture molecule (probes and antibodies), a method for converting capture molecule–target interactions into a signal, and an output data (Kabiraz et al. 2023). There are several types of biosensors

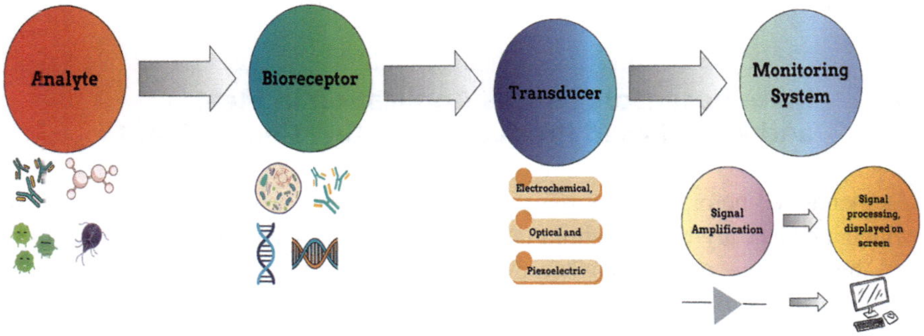

Fig. 3 Basic principle of biosensor and its components

that are available, which may include enzyme-based, magnetic, DNA biosensors, and antibody-based or immunosensors.

The biorecognition process of elements includes various types of bioreceptors, which may include enzymes, combining a diverse range of transducers which generate a signal. The very first biosensor of enzyme was introduced by Leland C. Clark in 1962. After this, enzymes were proven to be highly efficient and specific to be used widely in the development of biosensors (Wijayanti et al. 2023). The formulation of enzyme biosensor is based on mobilization method, which is the absorption of the enzymes by using vandal walls forces, ionic bonding, and covalent bonding. Oxidoreductases, polyphenol oxidases, peroxidase, and amino oxidases are the commonly used enzymes for formulation methods (Mehrotra 2016).

Magnetic biosensors are one of the types of biosensors that are miniaturized and used for the detection of magnetic micro- and nanoparticles, which are present in microfluid channels, by the use of magnetoresistance effect and have great potential in terms of their sensitivity and size (Mehrotra 2016).

Aptamer biosensors are basically oligonucleic acids, which may include either DNA or RNA, and are synthesized against a diverse target range and a light which has very high binding affinities. The use of ace can be selected against all the types of targets, which may include low immunogenicity or high toxicity. They are the analogues of amino acids of antibodies and have a very high affinity for the targets, with a dissociation constant quite close to nanopicomolar. The amine selection process is basically based on the successive steps of target binding, repetition, and the removal of oligonucleotides, which are unbound, followed by the elution, amplification, and purification of the selected oligonucleotides (Tyśkiewicz et al. 2023). The formulation of DNA-based biosensors is based on the principle that a single-strand nucleic acid molecule is able to recognize and, as a result, it binds to the complementary strand of it in the sample. And due to the formation of the hydrogen bond, which is stable, there is an interaction occurring between the two nucleic acid strands (Mehrotra 2016).

Antibody-based biosensors, also called as immunosensors, are a special type of biosensor that are based on the property of the antibody showing a strong affinity to their respective antigens, which means the antibodies will particularly bind to the pathogens only or their toxins, or there might be an interaction that can happen with the host's immune system components (Mehrotra 2016).

4 Immunosensors and Its Application in Meat and Meat-Based Products to Detect Contaminants

As mentioned before, immunosensors are known to be a special type of biosensor which are basically based on immunochemical reactions, and comprise the antibody or the antigen on a transducer's surface. Its biorecognition elements are immobilized, which results in stable complex formation (Tyśkiewicz et al. 2023). It has an affinity for solid-state-based biosensors, meaning that antigen (Ag) and antibody (Ab) form a stable complex and act as a capturing agent. Thereafter, a

Table 1 Immunosensors for the detection of contaminants in meat and meat products (Nanda et al. 2022, under the creative common attribution license)

Analyte	Biorecognition element	Electrode	Immobilization technique and detection	Target food	Limit of detection/ sensitivity or correlation
Salmonella enterica, Listeria monocytogenes, and Escherichia coli O157: H7	Alexa Fluor 647-labeled monoclonal antibodies	Optical waveguides (Streptavidin coated)	Biotinylated polyclonal antibodies	Ready-to-eat beef, chicken, and Turkey breast meat	LOD: 10^3 CFU/ ml
Salmonella pullorum	Anti-*Salmonella* polyclonal antibodies	Amperometric sensor with screen-printed electrode (MWCNT-chitosanperoxidase modified)	Cellulose nitrate membrane	Chicken	LOD: 10^2 CFU/ ml; time: 1.5–2 h
Staphylococcal enterotoxin B	Antibody	Electro-optical biosensor	Carboxymethyldextra	Potted meat (Hormel)	Sensitivity: 1–10 ng/ml
E coli K-12	Anti-*E. coli* antibody	Electrochemical impedance spectroscopy & SPR imaging	Gold surface	Frozen chicken	LOD: 3 log CFU/ml
S. Typhimurium	Antibodies	Quartz-crystal microbalance	Functionalized nanoparticles	Chicken	LOD: 1 CFU/ml
Salmonella gallinarum and S. pullorum	Antibodies	Screen-printed carbon electrode	Gold nanoparticles	Chicken	LOD: 3 log CFU/ml
E. coli O157:H7	Antibody	Screen-printed microelectrodes (Glucose oxidase-polydopamine nanocomposites & Prussian blue)	Gold nanoparticles	Ground beef	LOD: 2.05×10^3 CFU/ g

(continued)

Table 1 (continued)

Analyte	Biorecognition element	Electrode	Immobilization technique and detection	Target food	Limit of detection/ sensitivity or correlation
Campylobacter jejuni	Antibodies	Optical biosensor based on SPR	Immunoaffinity	Broiler	Good sensitivity: 10^3 CFU/ml against *C. jejuni*
Salmonella spp.	*Salmonella* spp. monoclonal antibodies	GCE	Grafted ethylene diamine and self-assembled gold nanoparticle monolayer	Pork	LOD: 2 log CFU/ml detection time: 40 min

LOD Limit of detection, *GCE* Glassy carbon electrode

measurable signal is produced and captured by the transducer. There are mainly three classifications of immunosensors based on the transducer mode, namely: electrochemical, which includes potentiometric, amperometric, impedimetric, and conductometric; optical and piezoelectric devices (Janik-Karpinska et al. 2022) (Table 1).

Electrochemical immunosensors are biosensors often used and further categorized into labeled and label-free sensors. The concept of the immunosensor follows the analyte (Ag) selective identification by capturing the analyte (Ag) immobilized on the electrode surface. In simple terms, these types of immunosensors have the ability to detect various types of analytes, as if there is any change in current, potential, conductance, or impedance brought on by an immunological reaction may be measured by them. This immunosensor can be further classified as amperometric, potentiometric, impedimetric, or conductometric based on the sort of signal that the transducer produces (Janik-Karpinska et al. 2022).

Optical immunosensors contain a light source and various components of optics in order to generate a light beam having special characteristics, directing the light in the direction of a modulating agent, a photodetector, and a modified sensing head (Jiang et al. 2008). Furthermore, they have the ability to detect the changes occurring in the field's optical properties of evanescent on the surface of an optical wave so that it can quantify the interactions happening between Ag and Ab (antigen and antibody) (Janik-Karpinska et al. 2022).

Piezoelectric immunosensors are a type of immunosensor which have crystals of quartz with Ag or Ab surface-mobilized, and can be utilized further by applying an external alternating electric field or a change in pH. Interestingly, the reaction happening between the antigen and antibody can be monitored in real time. The shape of the electrode, dielectric constant, effective viscosity, conductivity, density, and temperature of the liquid are some of the factors that may influence the oscillations' frequency (Janik-Karpinska et al. 2022; Jiang et al. 2008).

5 Future Directions

With the evolving food industry, the detection of foodborne pathogens must also evolve. Since meat processing is very critical as it involves various stages and exposure to the risk of contamination, meat safety becomes the most important concern for both the present and the future. An enhanced integration of conventional techniques with the molecular ones can improve the accuracy. Also, the emerging technologies such as biosensors have proven to be very effective. Further research should be done to have better results, as the biosensor technology's ability to detect pathogens in the meat processing industry at various stages as this technology is very promising for the detection of foodborne pathogens, along with their toxins and contaminants. However, there is a need for research to gain in-depth knowledge regarding the applications of biosensors, which have great potential to cure several dangerous diseases.

Data analysis framework plays a very important role in preventing or protecting the food matrix. Therefore, the use of data analytics and machine learning algorithms could be useful to predict potential contamination events based on historical data, environmental factors, and processing practices. Predictive microbiology can also be useful to predict any future threats. Nevertheless, the consumers are the utmost factor while producing any food or setting up an industry. Therefore, it becomes crucial to create awareness among consumers regarding the potential risks and to practice safe food handling. This can be achieved by organizing campaigns about proper cooking and storage temperatures, hygiene practices, and awareness of potential contaminants. It is also the responsibility of the governing authorities to strengthen the regulations and standardize the processes to ensure food safety.

6 Conclusion

The presence of pathogens in food, especially meat and products that are meat-based, is a big threat in both; the food industry and for the consumers. Therefore, it's crucial to have accurate, reliable, and efficient methods of detection in order to ensure public health. The current research in the detection technologies will further help us to improve our ability to monitor and respond to potential threats effectively. As mentioned in this chapter, foodborne pathogens, which may include *Listeria monocytogenes, E. coli O157:H7, Salmonella,* and *Campylobacter jejuni*, are prevalent in meat products and can cause severe health complications and long-term illness. Since the beginning, the conventional methods have served their purpose. However, the advancements in the existing technologies like in molecular techniques, proteomics, metabolomics, and immunoassays are paving the way for more effective pathogen detection strategies. By embracing these innovations and focusing on strategies for future directions outlined above, researchers and industries can significantly avoid the potential risks associated with foodborne pathogens in meat products. There is a need for continuous efforts in the research field, which should aim to improve the existing methods of detection not only to ensure the safety of food but also the protection of public health by reducing any potential incidence of foodborne illnesses and outbreaks throughout the globe. Attributes such as accuracy, affordability, sustainability, validation, and commercial use are crucial for developing a dependable technique that guarantees the highest level of public health safety. There is a rapid growth in the field of biosensors, focused on the application and development of molecular recognition elements, which are naturally occurring. And immunosensor, which is a special type of biosensor, has shown to be of great potential in becoming a very effective measurement tool in the field of food industry, focusing on meat and meat industry. This technology not only allows low-cost technology, small sample consumption, convenient instrument, operation, and real-time quantification, but also makes it a unique and promising technology. It is predicted that immunosensors are going to play a key role in the future in the detection of pathogens and toxin sensor detection. As we progress, there is a need for a combined role of a collaborative approach involving technology developers, food producers, regulators, and consumers to create a safe feed for the existing and future generations.

References

Aladhadh M (2023) A review of modern methods for the detection of foodborne pathogens. Microorganisms 11:1111

Ali S, Alsayeqh AF (2022) Review of major meat-borne zoonotic bacterial pathogens. Front Public Health 10:1045599

Armany G, Ahmed H, Ibrahim H et al (2016) Detection of some foodborne pathogens in meat products by polymerase chain reaction. Benha Vet Med J 30:323–330

Batz MB, Henke et al (2013) Long-term consequences of foodborne infections. Infect Dis Clin N Am 27:599–616

Bonaldo F, Avot BJP et al (2023) Foodborne pathogen dynamics in meat and meat analogues analysed using traditional microbiology and metagenomic sequencing. Antibiotics 13:16

Calo-Mata P, Carrera M, Böhme K, Caamaño-Antelo S et al (2016) Novel peptide biomarker discovery for detection and identification of bacterial pathogens by LC-ESI-MS/MS. J Anal Bioanal Tech 7:1000296

Darie-Ion L, Whitham D, Jayathirtha M (2022) Applications of MALDI-MS/MS-based proteomics in biomedical research. Molecules 27:6196

Ehuwa O, Jaiswal AK, Jaiswal S (2021) Salmonella, food safety and food handling practices. Food Secur 10:907

Elbehiry A, Abalkhail A, Marzouk E (2023) An overview of the public health challenges in diagnosing and controlling human foodborne pathogens. Vaccine 11:725

Epps S, Harvey R, Hume M (2013) Foodborne campylobacter: infections, metabolism, pathogenesis and reservoirs. Int J Environ Res Public Health 10:6292–6304

Ferrer-Bustins N, Yvon C, Martín B et al (2024) Genomic insights of Salmonella isolated from dry fermented sausage production chains in Spain and France. Sci Rep 14:11660. https://www.who.int/news-room/fact-sheets/detail/campylobacter

Janik-Karpinska E, Ceremuga M, Niemcewicz et al (2022) Immunosensors—the future of pathogen real-time detection. Sensors 22:9757

Jiang X, Li D, Xu X et al (2008) Immunosensors for detection of pesticide residues. Biosens Bioelectron 23:1577–1587

Kabiraz MP, Majumdar PR, Mahmud et al (2023) Conventional and advanced detection techniques of foodborne pathogens: a comprehensive review. Heliyon 9:e15482

Kumar H, Bhardwaj K, Kaur T, Nepovimova et al (2020) Detection of bacterial pathogens and antibiotic residues in chicken meat: a review. Food Secur 9:1504

Kurpas M, Wieczorek K, Osek J et al (2018) Ready-to-eat meat products as a source of *Listeria monocytogenes*. J Vet Res 62:49–55

Law JW-F, Ab Mutalib N-S, Chan et al (2015) Rapid methods for the detection of foodborne bacterial pathogens: principles, applications, advantages and limitations. Front Microbiol 5:770

Lianou A, Panagou EZ, Nychas G-JE et al (2017) Meat safety—I foodborne pathogens and other biological issues. In: Lawrie's meat science. Elsevier, pp 521–552

Mehrotra P (2016) Biosensors and their applications – a review. J Oral Biol Craniofacial Res 6:153–159

Meyer C, Thiel S, Ullrich et al (2010) Salmonella in raw meat and by-products from pork and beef. J Food Prot 73:1780–1784

Nairn MD (2022) Differentiation of raw meat using MALDI-TOF and PCA statistical analysis. https://doi.org/10.13140/RG.2.2.15376.07683

Nanda PK, Bhattacharya D, Das JK, Bandyopadhyay et al (2022) Emerging role of biosensors and chemical indicators to monitor the quality and safety of meat and meat products. Chemosensors 10:322

Rahal EA, Kazzi N, Nassar FJ et al (2012) Escherichia coli O157:H7—clinical aspects and novel treatment approaches. Front Cell Infect Microbiol 2:138

Saravanan A, Kumar PS, Hemavathy RV et al (2021) Methods of detection of food-borne pathogens: a review. Environ Chem Lett 19:189–207

Stephen Inbaraj B, Chen BH (2016) Nanomaterial-based sensors for detection of foodborne bacterial pathogens and toxins as well as pork adulteration in meat products. J Food Drug Anal 24:15–28

Tyśkiewicz R, Fedorowicz M, Nakonieczna A et al (2023) Electrochemical, optical and mass-based immunosensors: a comprehensive review of Bacillus anthracis detection methods. Anal Biochem 675:115215

Wijayanti SD, Tsvik L, Haltrich D (2023) Recent advances in electrochemical enzyme-based biosensors for food and beverage analysis. Food Secur 12:3355

Zhao X, Lin C-W, Wang J et al (2014) Advances in rapid detection methods for foodborne pathogens. J Microbiol Biotechnol 24:297–312

Detection of Gram-Positive Bacteria Using Electrochemical Immunosensors in Different Foods

Lorico D. S. Lapitan, Jr. and Cristina E. Tiangco

Abstract

Bacteria significantly influence food safety and quality, ranging from beneficial roles in fermentation to causing severe foodborne illnesses. Emphasising the importance of detecting pathogenic Gram-positive bacteria, rapid and reliable detection methods are crucial to mitigate health risks. This chapter discusses the development of electrochemical immunosensors for detecting Gram-positive bacteria in food and drinking water, focusing on nanomaterial enhancements. These immunosensors present an excellent alternative to conventional laboratory methods (e.g. culture methods), molecular methods like Polymerase Chain Reaction (PCR), and immunological-based methods including Enzyme-Linked Immunosorbent Assay (ELISA). This chapter delves into current trends and biosensor designs in recent electrochemical immunosensors. Additionally, this chapter highlights the advantages of using nanomaterials to improve sensitivity and specificity, providing insights into how these advanced technologies can be integrated into food safety practices for better monitoring and control. Specifically, immunosensors for direct detection of *S. aureus*, *L. monocytogenes*, and *B. cereus* that are recently developed are discussed.

L. D. S. Lapitan, Jr. (✉)
Department of Medical Diagnostics, Center for Advanced Materials and Technologies (CEZAMAT), Warsaw University of Technology, Warsaw, Poland

Department of Chemical Engineering, Faculty of Engineering, University of Santo Tomas, Manila, Philippines
e-mail: ldlapitan@ust.edu.ph; lapitan.santos@pw.edu.pl

C. E. Tiangco
Department of Chemical Engineering, Faculty of Engineering, University of Santo Tomas, Manila, Philippines

Research Center for the Natural and Applied Sciences, University of Santo Tomas, Manila, Philippines

K. Kuča et al. (eds.), *Ensuring Wellness: Immunosensors Monitoring Food Quality and Pathogenicity Risks*, https://doi.org/10.1007/978-3-031-97477-9_7

Keywords

Gram-positive bacteria · Immunosensors · Optical · Electrochemical · Food samples

1 Introduction

The Centers for Disease Control and Prevention (CDC) estimates that foodborne diseases in the United States affect approximately 48 million people annually, leading to 128,000 hospitalisations and 3000 deaths (2023). Several gram-positive bacteria are among the identified pathogens associated with fatalities resulting from foodborne illnesses. Relevant gram-positive pathogens and their toxins include the emetic and diarrhoeal enterotoxins from *Bacillus cereus* (*B. cereus*), neurotoxins from *Clostridium botulinum* (*C. botulinum*), enterotoxin from *Clostridium perfringens* (*C. perfringens*), a group of enterotoxins produced by *Staphylococcus aureus* (*S. aureus*) and certain other staphylococci, and *Listeria monocytogenes* (*L. monocytogenes*), which is the primary cause of listeriosis (Rajkovic et al. 2020). Serious food poisoning outbreaks have a severe impact on public health. In addition, bacterial contamination in food frequently results in product recalls, extended investigations, and legal repercussions for manufacturers.

The conventional method for detecting bacteria is based on culture and colony counting methods, microscopy and staining, and biochemical tests (Gracias and McKillip 2004; Vasavada et al. 2020). A comprehensive set of analyses is frequently required to identify the pathogen accurately. Plate culture involves inoculating a sample (e.g. food, water, clinical specimen) onto a selective agar plate that contains nutrients and growth-promoting factors. The plate is incubated at an appropriate temperature, typically 37 °C, for a designated period to enable microorganisms to grow and form visible colonies (Yousef and Carlstrom 2003; Ogodo et al. 2022). Bacterial colonies form on the surface of the plate, which can then be counted or identified using various biochemical tests, staining techniques, or morphology. Once a bacterial colony is isolated from a culture, it undergoes biochemical tests to determine the organism based on its metabolic activities (e.g. sugar fermentation and enzyme production like catalase, oxidase, and coagulase) (Yousef and Carlstrom 2003; Aryal 2022). Biochemical tests identify the bacteria at the species or genus level (Aryal 2022). Microscopy examines bacteria directly under a microscope after staining the sample with crystal violet and safranin dyes in a Gram stain. This method differentiates between Gram-positive and Gram-negative bacteria. The staining provides initial information about bacterial cell morphology (e.g. rods, cocci) and their Gram classification (Ogodo et al. 2022). These conventional bacterial detection methods remain widely used due to their reliability, especially in resource-limited settings. However, these methods are generally slow, requiring days to complete, labour-intensive, and require large bacterial concentrations for reliable identification. Moreover, some bacteria cannot be cultured, which limits the sensitivity and comprehensiveness of these techniques. Biochemical tests of closely related bacterial species may also produce similar biochemical profiles, which can lead to misidentification or inconclusive results. More importantly, conventional methods do not allow real-time monitoring or rapid decision-making, especially in time-critical scenarios like food safety or clinical diagnostics.

To address these challenges, researchers have developed various rapid bacterial detection techniques, including PCR-based methods and immunology-based methods (Lazcka et al. 2007; Foddai and Grant 2020; Canciu et al. 2021). These methods are faster and have higher sensitivity and specificity than the plate culture method. PCR-based assays are popular because PCR can detect as few as 1 to 10 copies of bacterial DNA per reaction (Lazcka et al. 2007). This sensitivity makes it possible to identify very low levels of bacteria that are undetectable by traditional methods (Canciu et al. 2021). Moreover, PCR can detect bacterial DNA that is viable but in a non-culturable state, which is a significant advantage over culture methods. PCR is highly specific due to targeted primers binding to unique bacterial genome sequences. This allows it to differentiate between bacterial species and strains with high accuracy. PCR can detect bacterial concentrations as low as 10^2 to 10^3 CFU/mL (colony-forming units per millilitre) without enrichment (Lazcka et al. 2007; Canciu et al. 2021). On the other hand, immunological methods for bacterial detection rely on the interaction between antibodies and specific antigens (proteins, polysaccharides, or other molecules on the bacterial surface). An example is the enzyme-linked immunosorbent assay (ELISA). This method is widely used to detect bacterial antigens or antibodies against bacteria in a sample. It uses enzyme-conjugated antibodies that produce a colour change in the presence of the target antigen (Lazcka et al. 2007). These methods allow for earlier detection and identification of bacteria, including viable but non-culturable organisms. Commercial ELISA kits utilise monoclonal and polyclonal antibodies to detect entire pathogens or their specific proteins. In many modern laboratories, molecular methods and immunological assays frequently supplement or replace conventional methods. While these techniques are powerful, they also have limitations. For instance, PCR is unable to distinguish both live and dead bacteria. Similarly, immunological methods can sometimes result in false positives due to antibody cross-reactivity with antigens from non-target bacteria. A further drawback is that these methods often require skilled personnel and specialised laboratory environments, which may pose financial challenges in resource-constrained settings (Canciu et al. 2021; Lopez-Tellez et al. 2022). Other analytical methods, such as matrix-assisted laser desorption ionisation time-of-flight (MALDI-TOF) mass spectrometry and fluorescence in situ hybridisation, provide quicker and more accurate pathogen identification from culture-positive samples (Singhal et al. 2015; Shah and Ramasamy 2022). However, their widespread adoption in laboratories is limited by the high cost and maintenance of the equipment, the requirement for skilled operators, and the complexity of sample preparation needed to isolate and concentrate bacteria before analysis.

Antibodies are the most utilised biological recognition elements in developing electrochemical biosensors. Electrochemical immunosensors represent a cutting-edge technology developed for faster and more specific detection of bacteria in various fields, such as food safety, clinical diagnostics, and environmental monitoring (Wen et al. 2017; Harshavardhan et al. 2019). These immunosensors integrate the specificity of immunoassays with the sensitivity of electrochemical detection methods, offering a powerful tool for pathogen detection. An electrochemical immunosensor typically consists of an electrode surface functionalised with antibodies that specifically bind to antigens (e.g. bacteria or bacterial markers) from the target pathogen. The binding of the antigen to the antibody generates an electrochemical signal that can be measured, which corresponds to the bacterial concentration (Cancelliere et al. 2024). Electrochemical immunosensors offer numerous advantages for bacterial detection. Electrochemical immunosensors are highly sensitive and provide excellent specificity by using antibodies to target specific bacterial antigens. Another benefit is the speed, as these methods can deliver results in minutes to a few hours, surpassing the time required by conventional methods. Electrochemical methods can offer label-free detection, thus simplifying the detection process by eliminating the need for additional reagents and secondary biomolecule tags, making them cost-effective. Electrochemical methods usually require only small sample volumes, making them ideal for situations with limited samples (Cancelliere et al. 2024). Electrochemical immunosensors are highly portable and enable on-site real-time testing, which is especially useful in fieldwork and point-of-care diagnostics. Electrochemical immunosensors can also be designed for multiplexed detection, which allows the simultaneous detection of multiple bacteria (Yáñez-Sedeño et al. 2017). They can be integrated with microfluidics for automated and high-throughput analysis (Akhtarian et al. 2024). Research has also demonstrated that immunosensors can be reused after successful regeneration of the biorecognition layer, which can further reduce costs (Lin et al. 2015; Mo et al. 2019). These features make electrochemical immunosensors an attractive, efficient, and versatile tool for food safety and monitoring (Table 1).

In direct analysis, bacteria bind to the modified electrode surface, which causes a change in the signal corresponding to the pathogen concentration. Electrochemical immunosensors operate through antibody–antigen interactions, making the stability of immobilised antibodies crucial to their performance. Factors such as temperature and pH can influence the sensor's durability (Wen et al. 2017). As with ELISA and PCR, complex food or biological samples can compromise electrochemical immunosensor performance. However, sample pre-treatment procedures can help minimise interference. Developing electrochemical immunosensors is often costly and requires significant investment, especially in producing specific antibodies and robust electrochemical sensing platforms (Kim and Park 2021). However, these biosensors provide numerous benefits once established and become highly effective for bacterial detection.

Table 1 Comparison of key features of traditional, molecular, immunological, and electrochemical immunosensors for detection of bacteria

Method	Analysis time	Sensitivity	Specificity	Cost-effectiveness	Portability	References
Plate culture	24 to 48+ h	Moderate	Moderate to high	Moderate	Low (laboratory-based)	Gracias and McKillip (2004) and Vasavada et al. (2020)
PCR	Hours	High (DNA level)	Very specific (based on primers)	High	Low (laboratory-based)	Lazcka et al. (2007) and Canciu et al. (2021)
ELISA	Hours	High	Moderate to high	High (requires secondary antibodies)	Low (laboratory-based)	Lopez-Tellez et al. (2022)
Electrochemical immunosensors	Minutes to hours	High (detects low CFU/mL)	Moderate to high	Low (once developed)	High	Harshavardhan et al. (2019) and Wen et al. (2017)

2 Electrochemical Transduction

Electrochemical biosensors typically consist of three primary electrodes: a working electrode, a reference electrode, and a counter electrode. Gold is extensively employed in biosensor development due to its ease of surface modification, chemical stability, and excellent conductivity. However, carbon-based electrodes like graphite and glassy carbon electrodes are also commonly utilised. The working electrode is the core component that facilitates the detection and quantification of the analyte in an electrochemical biosensor. The working electrode surface serves as the site where the target analyte interacts, undergoes oxidation or reduction, and produces an electrical signal (Kim and Park 2021). Hence, the material and surface characteristics of the working electrode are critical to the biosensor's analytical performance. For instance, modification of the working electrode surface with biorecognition molecules (e.g. antibodies) and nanoparticles can improve its sensitivity and specificity for target bacteria. The reference electrode maintains a stable potential to ensure accurate measurement of the working electrode's potential, while the counter electrode facilitates current flow to complete the electrical circuit. These elements work together to ensure precise and reliable detection of target analytes. Electrochemical sensors identify and measure chemical substances through electrochemical reactions, converting chemical events into electrical signals for analysis. The core principle behind these sensors involves interactions between the analyte and the working electrode surface, resulting in detectable changes in electrical parameters such as current, potential, or conductivity (Kim and Park 2021).

Electrochemical sensors include various types based on their transduction mechanisms, such as potentiometric, impedimetric, amperometric, and voltammetric (Kim and Park 2021). Potentiometric methods measure the electric potential between the indicator and reference electrodes with zero electric currents (Hernández et al. 2014). Impedimetric biosensors measure the electrode surface's electrical impedance (resistance to alternating current) in detecting biomolecular interactions (Lin et al. 2019). These biosensors monitor changes in the impedance of the electrode surface when a biological recognition event occurs, such as the binding of an antigen to an antibody. Impedimetric biosensors are frequently employed to detect bacterial pathogens or their metabolites (Leva-Bueno et al. 2020). An impedimetric biosensor applies a small-amplitude alternating sinusoidal voltage (around 5 mV) across a range of frequencies, typically from 0.01 Hz to 100 kHz, to the working electrode (Bhand and Bacher 2018). The resulting current response, which may be phase-shifted relative to the applied voltage, is measured. Impedance (Z) in an AC circuit is calculated as the ratio of the applied voltage to the resulting alternating current, according to Ohm's law. Importantly, impedance accounts for the system's resistive (real) and reactive (imaginary) components. Impedance data are typically displayed as Nyquist plots, illustrating the negative imaginary component of impedance against the real component. In impedimetric biosensors, the working electrode is typically modified with a biological recognition component, such as antibodies, enzymes, or nucleic acid strands, designed to selectively bind the target analyte. Upon binding, the surface properties of the electrode are altered, resulting in detectable changes in impedance. These biosensors are usually label-free, eliminating the need for extra markers or tags (such as fluorescent or radioactive labels) to identify the interaction, thereby simplifying and streamlining the detection process (Bekir et al. 2015).

Amperometric methods measure the current produced by the redox reaction of electroactive species with respect to time. This method applies a constant potential to an electrode, monitoring the resulting current. The current is directly proportional to the analyte concentration, which undergoes oxidation or reduction on the electrode surface. This makes amperometry a sensitive and simple technique for quantifying electroactive species. As an example, a disposable amperometric immunosensor using functionalised magnetic beads (MBs) and gold screen-printed electrodes (Au/SPEs) was developed for the specific detection and quantification of Staphylococcal protein A (ProtA) and *S. aureus*. The biosensor was constructed by immobilising anti-ProtA antibodies onto ProtA-modified MBs. Detection was carried out using a competitive immunoassay format, where the ProtA antigen was labelled with HRP. An external magnetic field captured the resulting modified MBs on the surface of tetrathiafulvalene-modified Au/SPEs. The response obtained at − 0.15 V versus the Ag pseudo-reference electrode of the Au/SPEs after adding H_2O_2 was employed and used as the signal. The developed immunosensor has a limit of detection of 1 CFU/mL *S. aureus* in raw milk samples. Moreover, the immunosensor showed excellent selectivity against common foodborne pathogens found in milk (Esteban-Fernández de Ávila et al. 2012).

Voltammetric methods measure current as a function of applied potential to detect the presence of bacteria. In these sensors, cyclic voltammetry (CV), square-wave voltammetry (SWV), and differential pulse voltammetry (DPV) are three common voltammetric techniques used to study redox reactions and quantify analytes (Lopez-Tellez et al. 2022). Cyclic voltammetry involves the potential being swept linearly from an initial potential value to a known potential (forward scan) value and then reversed back (reverse scan) to the initial potential value. The process is repeated in several cycles, and the current is continuously measured during the potential sweep (Elgrishi et al. 2018). The resulting current versus potential plot (voltammogram) shows distinct redox peaks appearing at the characteristic potentials of the analyte. Cyclic voltammetry is widely used to study the electrochemical behaviour of a system, such as oxidation-reduction processes, electron transfer kinetics, and reaction mechanisms (Batchelor-McAuley et al. 2015). Cyclic voltammetry is useful for exploratory studies because it provides information on oxidation and reduction processes. Wang et al. (2021) employed cyclic voltammetry to investigate an electrochemical detection system that evaluates the viability of *S. aureus* by analysing the oxidation signals of guanine and adenine secreted by the bacteria. The working electrode consists of a multi-walled carbon nanotube-modified glassy carbon electrode (MWCNTs/GCE). The electrochemical detection successfully identified two distinct electrochemical signals at +0.693 V and + 0.973 V versus Ag/AgCl, which are attributed to guanine and adenine, respectively. These electrochemical signals provide insights into bacterial proliferation and purine metabolism, thus offering an earlier indication of viability changes compared to traditional plate counting methods. Interestingly, the study further demonstrated that the guanine signal closely aligns with plate counting results, suggesting its potential as a reliable indicator. The approach was validated through experiments showing a strong linear relationship between electrochemical signals and bacterial concentration in various growth phases. The study also successfully demonstrated the inhibitory effects of levofloxacin on bacterial viability using guanine signals, providing results consistent with traditional techniques (Wang et al. 2021). This study highlights the advantages of electrochemical methods (i.e. cyclic voltammetry) in assessing microbial viability, including simplicity, rapid detection, and the ability to capture metabolic changes.

Differential pulse voltammetry (DPV) involves the potential being incrementally increased in small steps, and a pulse is applied at each step. The current is recorded both before and after each pulse, and the difference between these two readings is recorded. This provides better resolution by minimising the non-faradaic current. Hence, DPV is highly sensitive and is mainly used for detecting and quantifying low concentrations of analytes compared to cyclic voltammetry. In SWV, a square wave modulation to the potential is applied. The potential alternates between two levels (forward and reverse), with a symmetrical wave applied at each step. The frequency and amplitude of the wave can be adjusted. The current is measured at both the forward and reverse potential levels. The difference between these currents is plotted as a function of potential. SWV is used for sensitive and fast quantitative analysis of electroactive species (Chen and Shah 2013) .

3 Integration of Nanomaterials in Electrochemical Biosensors

Integrating various nanomaterials in constructing electrochemical sensors has paved the way for improved performance, particularly in sensitivity, response time, stability, and selectivity. Specifically, nanomaterials, including carbon nanotubes, graphene, and metal nanoparticles, have enhanced these sensors' sensitivity and selectivity. Carbon-based nanomaterials, including graphene and carbon nanotubes (CNTs), are well known for their high conductivity, excellent mechanical properties, and large surface area (Bhardwaj et al. 2017; Feng et al. 2022). Metallic nanomaterials such as gold (Au) and silver (Ag) are used for their large surface area and conductive properties (Han et al. 2020). Platinum (Pt) and palladium (Pd) nanoparticles are often used for their catalytic and conductive properties. Semiconductor nanocrystals (quantum dots), with size-dependent optical and electronic properties, are often used as signal amplifiers in electrochemical detection.

Nanomaterials on the surface of the working electrode enhance electron transfer between the analyte and the electrode, thus enhancing the sensor's electrical signal (Rusling 2012). Their large surface area allows more target molecules to interact with the sensor, improving detection limits and enabling the creation of miniaturised sensors (Zhang et al. 2020). In addition, catalytic nanomaterials can accelerate reactions, like oxidation or reduction, on the electrode surface, further enhancing sensor performance. By controlling the size and morphology of nanomaterials—such as nanotubes, nanowires, and nanoparticles—reactivity, sensitivity, and response times can be optimised for specific sensing applications. Extensive research has shown that nanomaterial-enhanced sensors enable fast, highly sensitive detection of bacterial pathogens in food and water samples, along with broader environmental monitoring and medical diagnostics applications (Kokkinos et al. 2016; Wen et al. 2017).

Nanomaterials are typically incorporated into the working electrode using various techniques that enhance the sensor's performance by improving sensitivity, selectivity, and conductivity. One of the simplest and most used approaches is the direct deposition of nanomaterials onto the electrode surface, which can be achieved through drop-casting, electrodeposition, and spin-coating (Ahmad et al. 2018). Drop-casting involves applying a nanomaterial solution to the electrode and drying it to create a thin layer. Electrodeposition involves depositing a material onto a conductive surface (i.e. electrode) by applying an electric current in an electrochemical cell. For example, electrodeposition can reduce metal salts from a solution directly onto the electrode surface to form a layer of metallic nanoparticles. Electroactive monomers can be electrodeposited to create electrodes modified with a semi-conducting polymer layer. Svalova et al. (2021) electrodeposited a 3,4-ethylendioxythiophene-3-ethynylthiophene composite film on the surface of the Pt electrode. The thin layer of polymer film was used for the covalent immobilisation of bovine serum albumin and antibodies (Svalova et al. 2021). The immunosensor was applied for label-free detection of *E. coli* and *S. aureus*. The immunosensors exhibited a detection limit of 7.2 CFU/mL for *E. coli* and 15.9 CFU/mL for *S. aureus*. The reported linear range was 10^2–10^6 CFU/mL and has great potential for environmental and food monitoring and control (Svalova et al. 2021).

Spin-coating involves a nanomaterial solution being applied to the electrode, which is then rotated at high speed to spread the material evenly. On the other hand, layer-by-layer assembly and self-assembly use electrostatic or non-covalent (e.g. hydrogen bonding, van der Waals) forces to organise nanomaterials into structured layers. Screen printing and physical vapour deposition (PVD) create uniform coatings for large-scale production. Nanomaterials can also be incorporated into polymer matrices, enhancing the electrode's mechanical stability and biocompatibility. Each method enables precise control over nanomaterial placement while optimising the sensor's electrochemical properties for various applications.

4 Recent Electrochemical Immunosensors for Selected Gram-Positive Bacteria

Immunosensors are affinity-based biosensors that interact with a target analyte, typically an antigen (Ag) and an antibody (Ab), to form a stable Ag–Ab immunocomplex. Immunological assays and electrochemical immunosensors differ primarily in their detection mechanisms and applications. Immunological assays, such as ELISA, rely on colourimetric or fluorescent methods to measure antigen–

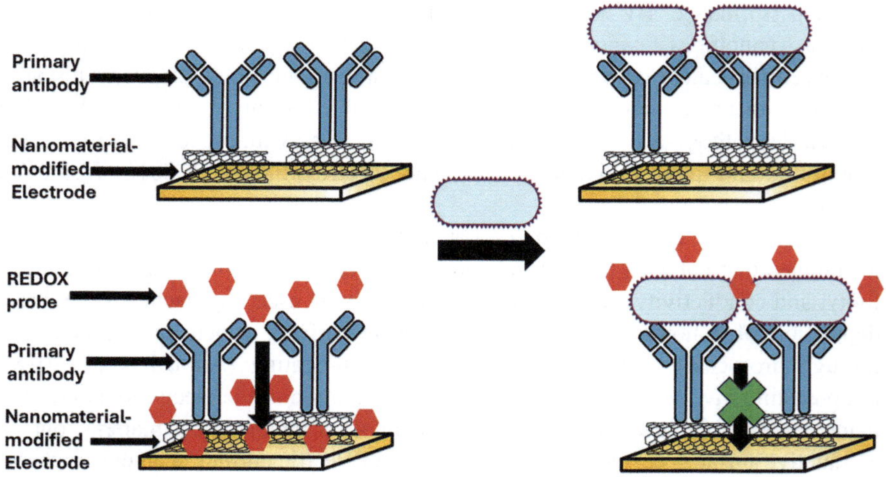

Fig. 1 Schematic rendering of the label-free biosensor of the selective binding of a bacteria to immobilised antibodies on the nanomaterial-modified electrode surface of the impedimetric immunosensor. A mechanism showing the detection of bacteria using a redox probe pair, typically $K_3[Fe(CN)_6]/K_4[Fe(CN)_6]$, but other electroactive iron redox probes can be used. The detection consists of capturing the bacteria on the electrode surface through interactions with a specific immobilised antibody. The formation of the antibody–bacteria complex causes a decrease in the electron transfer rate on the biosensor's surface with the redox probe

antibody interactions, often producing a visual readout. These assays typically require laboratory-based equipment, such as plate readers, and involve multiple steps, like washing and incubation, which can take several hours. While highly reproducible in controlled environments, immunological assays are more expensive due to the need for reagents and secondary antibodies. In contrast, electrochemical immunosensors detect antigen–antibody interactions by measuring electrical signals, such as current, voltage, or impedance (Harshavardhan et al. 2019). These sensors offer faster, real-time results and are often more sensitive because of direct electron transfer and the use of nanomaterials for signal amplification. Several parameters influence the effectiveness of an immunosensor, such as the efficiency of surface functionalisation, antibody orientation, surface density of the antibody, surface regeneration, and the overall design of the immunosensor. Thoughtful optimisation of these factors is essential to achieve sensitive and reliable results with any immunosensor technology.

Signal generation in electrochemical immunosensors can be accomplished in several ways, as shown in Fig. 1. Depending on the electrochemical technique employed, the target analyte can be detected using immobilised antibodies in either label-free or labelled immunosensors. Label-free detection of pathogens is schematically shown in Fig. 1. In this case, the antibody captures the pathogen, causing a change in the electrical impedance. The change in impedance is correlated with the bacterial loading. Redox probes can also be used for label-free immunosensors, as shown in Fig. 1. Typically, a 1–5 mM solution of K_3[Fe

$(CN)_6]/K_4[Fe(CN)_6]$ is used as a redox probe in a buffer solution (Wang et al. 2019). The biosensing principle involves the absence of the analyte (pathogen), where the redox probe will diffuse near the electrode surface, where it can undergo a redox reaction. However, in the presence of a pathogen bound to the antibody, the access of the redox probe to the surface is limited, causing reduced electron transfer with the surface of the electrode. For example, Bekir et al. (2015) used polyclonal anti-*S. aureus* immobilised onto the gold electrode (Bekir et al. 2015). The developed immunosensor monitored the charge transfer resistance at the solid/liquid interface using $[K_3[Fe(CN)_6]/K_4[Fe(CN)_6]$. The charge transfer resistance exhibited an increase when the concentration of bacteria increased due to the antibody–bacteria interaction. The immunosensor exhibited a detection limit of 10 CFU/mL for *S. aureus* and a linear detection range from 10 to 10^7 CFU/mL (Bekir et al. 2015).

Label-free electrochemical immunosensors offer several advantages, including reduced preparation time and lower analysis costs, since they do not require additional labels or markers (Cancelliere et al. 2024). Additionally, label-free detection allows for the direct monitoring of analyte binding in real time, making the process faster and more straightforward than labelled techniques (Templier et al. 2016; Han et al. 2020). Label-free electrochemical immunosensors have several disadvantages that can impact their performance and reliability. One key drawback is their generally lower sensitivity compared to labelled immunosensors, as they rely solely on the inherent properties of the analyte and the antibody without any amplification from labels (Cancelliere et al. 2024). Additionally, these sensors are more susceptible to non-specific binding, leading to increased background noise and reduced accuracy in detecting the target analyte. Non-specific binding can be reduced with proper surface blocking using polyethylene glycol (PEG) self-assembled monolayers (SAM), milk proteins (casein), and bovine serum albumin surface (BSA) passivation (Riquelme et al. 2016). Label-free immunosensors tend to have a limited dynamic range for quantifying analyte concentrations, making it challenging to measure very low or very high concentrations accurately (Templier et al. 2016). Reproducibility can also be an issue, as variations in the immobilisation of antibodies or surface characteristics can lead to inconsistent results. Some analytes may only produce strong enough electrochemical signals for reliable detection with labels, mainly if they are small or have low electrochemical activity. Lastly, while label-free detection may reduce preparation time, the assay can still take longer if the binding kinetics between the analyte and antibody are slow, potentially affecting real-time detection capabilities (Wang et al. 2019; Cancelliere et al. 2024).

Labelled electrochemical immunosensors typically utilise the sandwich format found in ELISA, which uses primary and secondary antibodies as capture probes and signal-amplifier probes (Kokkinos et al. 2016). Labelled antibodies rely on horseradish peroxidase, the most widely used agent for indirect detection. In this setup, an enzymatic reaction forms redox-active products that can be detected through electrochemical methods, as illustrated in Fig. 2. Nanomaterials can be used as carriers containing multiple copies of enzymes for powerful signal amplifiers (Rusling 2012). Labelled electrochemical immunosensors offer several notable advantages, including high sensitivity due to signal amplification. This enables the detection of

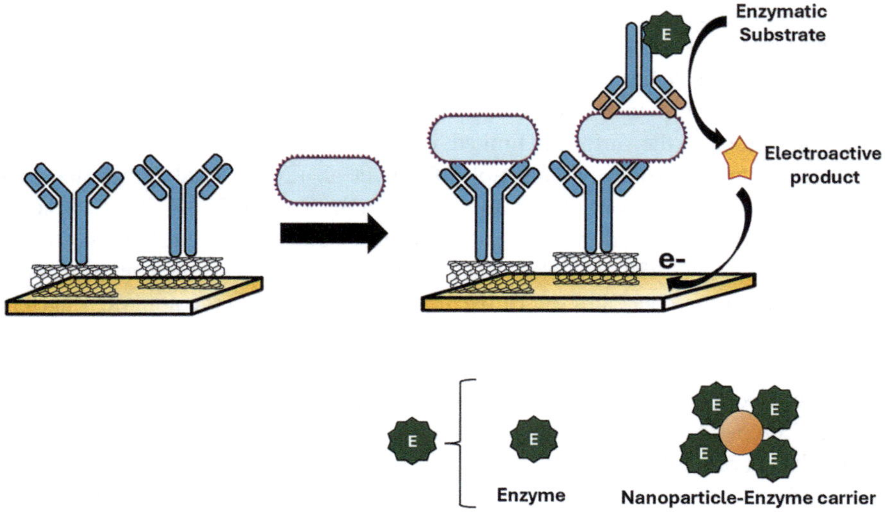

Fig. 2 A schematic representation of labelled electrochemical immunosensors. The labelled antibody specifically binds to bacterial antigens on the electrode surface. Upon addition of an enzymatic substrate, the enzyme catalyses a reaction that produces an electrochemically active product, enabling electron transfer and detection. Nanoparticles can be used as support materials for conjugating multiple enzymes to improve the detection limit

low analyte concentrations. The specificity of the sandwich format minimises interference from other substances (Kokkinos et al. 2016). On the other hand, labelled sensors have disadvantages. The preparation process can be complex and time-consuming, often involving multiple steps for conjugating antibodies with labels. This complexity can lead to higher costs due to the need for various reagents. The stability of labels is another concern, as environmental factors can affect their activity, potentially compromising sensor performance. Furthermore, the operational lifespan of enzymatic labels may be limited, leading to reduced effectiveness over time and necessitating frequent recalibration, along with the risk of signal saturation at high analyte concentrations.

4.1 *Staphylococcus aureus*

Staphylococcus aureus (*S. aureus*) is a spherical, gram-positive bacterium that commonly resides on the skin, in the nasal passages, and in the throat as a commensal organism (Bencardino et al. 2021). It is also a common human pathogen responsible for a range of infections, including bacteremia, infective endocarditis, osteoarticular infections, and infections of the skin and soft tissues (Tong et al. 2015). Some strains of S. *aureus* produce enterotoxins that withstand high temperatures, which can contaminate foods such as dairy products, milk, meat products, etc. (Bencardino et al. 2021). Hence, research on *S. aureus* and

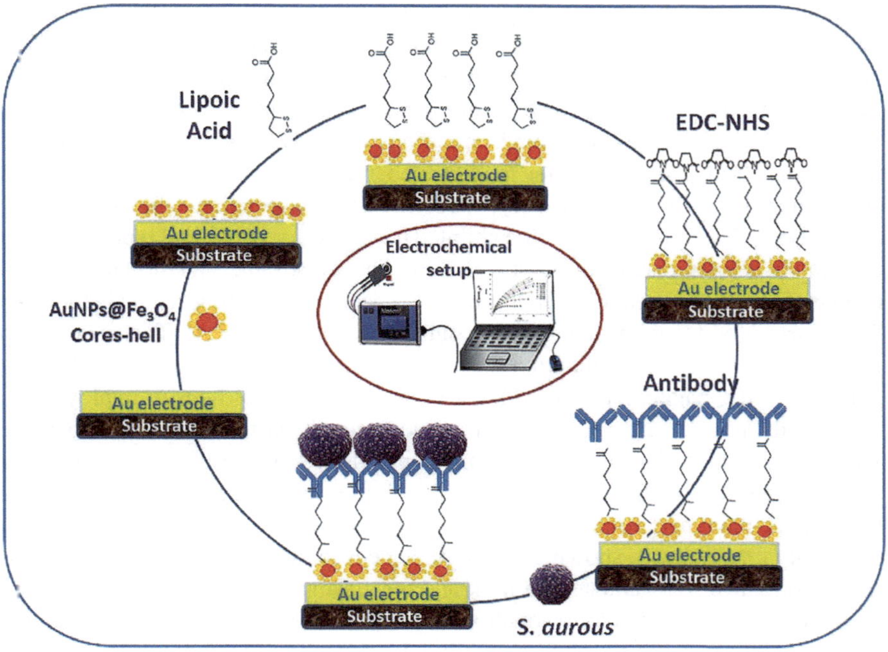

Fig. 3 A schematic representation of a rapid immunosensor for detecting *S. aureus* in milk and beef meat samples based on AuNPs@Fe_3O_4 nanocomposites. The procedure begins with the deposition of a thin nanocomposite layer onto screen-printed gold electrodes, followed by the formation of a lipoic acid in a self-assembled monolayer (SAM) and the attachment of monoclonal anti-*S. aureus* antibodies onto the surface. The process concludes with the sensor being exposed to the target pathogen. The immunosensor is designed to capture whole *S. aureus* cells through a specific immune reaction, resulting in a significant increase in the impedance signal. (Reprinted with permission from Magar et al. 2023, License Number 5924231010492)

Staphylococcal Enterotoxins (SE) remains essential. Specifically, efforts to develop new detection methods are crucial to ensure the safety of foods, reduce contamination risk, minimise industrial losses, and protect public health.

Several label-free biosensors have been recently developed for *S. aureus*. Gold nanoparticles (Au NPs) are frequently employed as stable anchoring sites for immobilising antibodies using various covalent coupling techniques to achieve strong and specific binding (Jazayeri et al. 2016). Au NPs provide high surface area, biocompatibility, and chemical stability. The nano-sized dimensions of Au nanomaterials (e.g. spheres, rods) are excellent platforms for antibody attachment to enhance the sensitivity and specificity of biosensors. As shown in Fig. 3, Magar et al. (2024) recently introduced an innovative biosensor system designed for the rapid and selective detection of *S. aureus* in food and water samples (Magar et al. 2024). This system utilises gold nanoparticles (Au NPs) combined with a copper metal-organic framework (AuNPs@Cu-MOF) on screen-printed electrodes to immobilise antibodies. The biosensor exhibited a broad detection range from 1×10^1 to 1×10^7 CFU/mL and an impressively low detection limit of 0.132 CFU/mL. The immunosensor also exhibited strong selectivity, responding specifically to *S. aureus* while remaining unaffected by other common pathogens such as *E. coli* and *Salmonella*. The high selectivity of the biosensor is an essential feature for accurate results in environments with mixed microbial presence. The biosensor's effectiveness was validated through real-world testing on various food and water samples, including meat, tuna, fruit juice, tap water, and bottled water, where it detected *S. aureus* contamination with high accuracy, underscoring its potential for food safety applications. The biosensor system exhibited robust performance and reproducibility and could deliver results in about 20 min. The disposable and portable design enables field use and easy operation by non-specialists, making it a valuable tool for ensuring food hygiene and quality control (Magar et al. 2024). The same group also developed a rapid immunosensor for detecting *S. aureus* in food using AuNPs@Fe_3O_4 nanocomposites (Magar et al. 2023). The biosensor employs self-assembled lipoic acid monolayers for stable antibody immobilisation. Label-free sensing was achieved using electrochemical impedance spectroscopy. The biosensor demonstrated remarkable performance with a detection limit of 0.10 CFU/mL and a broad detection range from 10^1 to 10^7 CFU/mL. Notably, the sensor maintains high specificity for *S. aureus*, showing no interference from other pathogens. The biosensor performance was validated on milk and meat samples, and the sensor provided accurate results comparable to traditional methods, even without requiring sample pre-treatment. The biosensor also exhibited strong reproducibility and stability, retaining effectiveness over several weeks at 4 °C, making it suitable for practical, on-site food safety applications (Magar et al. 2023).

In a separate study, Roushani et al. (2020) developed a simple and highly sensitive electrochemical immunosensor for detecting *S. aureus* by employing IgY antibodies immobilised on a gold nanoparticle-modified glassy carbon electrode (Roushani et al. 2020). The immunosensor leverages the stable and efficient antibody attachment provided by spherical gold nanoparticles (Au NPs). The detection was achieved through impedance spectroscopy and voltammetry. The biosensor exhibited a broad detection range from 10 to 10^7 CFU/mL and a low detection limit of 3.3 CFU/mL. The sensor also showed high specificity for *S. aureus* while remaining unaffected by interference from other bacterial species. The biosensor was validated in complex matrices like milk and serum. The sensor recovery rates achieved 97–103%, proving its reliability for real-world applications. The immunosensor maintains over 96% of its activity after 1 week at 4 °C and shows reproducible performance with a 3% relative standard deviation (Roushani et al. 2020). In an additional study, Han et al. (2020) developed a label-free electrochemical immunosensor for sensitively detecting *S. aureus* (Han et al. 2020). The sensors utilised self-assembled gold nanorods (Au NRs) on a glassy carbon electrode. The Au NRs are immobilised on the electrode surface using PDDA and PSS polyelectrolyte linkers, providing a stable platform for antibody attachment. The detection was carried out through electrochemical impedance spectroscopy. The sensor demonstrated impressive performance with a linear detection range from 1.8×10^3 to 1.8×10^7 CFU/mL and a detection limit of 2.4×10^2 CFU/mL. The biosensor exhibited strong selectivity for *S. aureus* and achieved recovery rates between 96.3%

and 104.1% in milk samples. The sensor also retained 92.6% of its response after 2 weeks of storage at 4 °C, showing reproducibility with low assay variability. This immunosensor offers a reliable and efficient tool for monitoring food safety and potential clinical diagnostics (Han et al. 2020).

Cost-effective biosensors present a promising solution for easy, on-site detection of *S. aureus*. A paper-based electrochemical immunosensor created for rapidly detecting *S. aureus* in food samples provides a practical and cost-effective approach to pathogen detection. Bhardwaj et al. (2017) successfully prepared antibody-modified single-walled carbon nanotubes (SWCNTs) on a paper substrate (Bhardwaj et al. 2017). This biosensor exhibited a wide linear detection range from 10 to 10^7 CFU/mL and a limit of detection of 13 CFU/mL. The biosensor can provide results in 30 min and is highly specific to *S. aureus*. The biosensor demonstrates high stability and retains functionality for up to 5 weeks when stored at a temperature of 4 °C. Overall, the biosensor provides a reliable, low-cost, and efficient method for monitoring food safety for real-world applications in food and environmental safety (Bhardwaj et al. 2017).

4.2 *Listeria monocytogenes*

Listeria monocytogenes (*L. monocytogenes*) is a rod-shaped, gram-positive bacterium that can cause listeriosis, a severe infection particularly harmful to pregnant women, newborns, elderly individuals, and those with weakened immune systems (Thakur et al. 2018). Widely present in soil, water, and contaminated food, it can survive and proliferate at refrigeration temperatures, posing a major concern for food safety (Lecuit 2020). This pathogen can be present during food processing, distribution, and retail. *L. monocytogenes* has been detected in various foods, including raw milk, cheese, uncooked meat products, and salads (Todd and Notermans 2011). *L. monocytogenes* is ubiquitous in the farm environment. Hence, contamination often stems from inadequate hygiene practices during milking and slaughter processes (Rodriguez et al. 2021). Additionally, cross-contamination on food-processing equipment contributes to the proliferation of *L. monocytogenes* in processed foods (Todd and Notermans 2011; Lecuit 2020).

Several electrochemical biosensors developed for *L. monocytogenes* have been reported in the literature (Silva et al. 2020). As a recent example, Lee et al. (2022) developed a single-walled carbon nanotube (SWCNT)-based electrochemical impedance immunosensor for on-site detection of *L. monocytogenes* (Lee et al. 2022). The immunosensor is constructed with a gold-plated wire functionalised using polyethyleneimine, SWCNTs, streptavidin, and biotinylated antibodies specific to *L. monocytogenes*. The detection is achieved by measuring changes in electron transfer resistance as antibody–antigen binding occurs through electrochemical impedance spectroscopy. The biosensor showed a broad linear detection range of 10^3 to 10^8 CFU/mL and a limit of detection of 10^3 CFU/mL. It can detect *L. monocytogenes* within 10 min and demonstrates high selectivity. The biosensor was validated in lettuce homogenate and showed high recovery rates (90–95%) for *L. monocytogenes*, underscoring its potential for practical food safety applications.

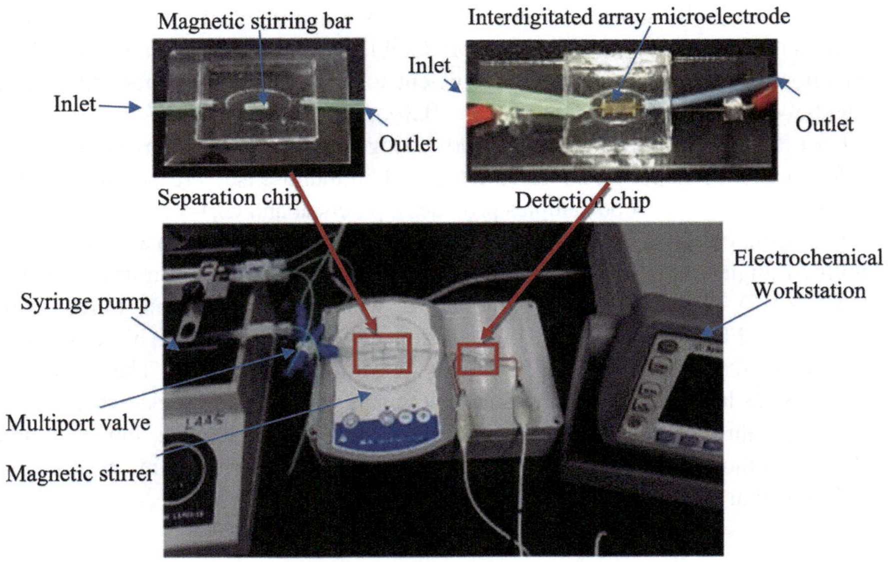

Fig. 4 A microfluidic biosensor set-up for fast and sensitive detection of food-borne pathogens, including *L. monocytogenes*. The impedance integrates immunomagnetic separation with urease catalysis to detect *L. monocytogenes* using an immobilisation-free interdigitated array microelectrode. (Reprinted with permission from Chen et al. 2016, License Number 5924231350584)

The biosensor was also integrated into a smartphone-controlled portable platform and reported consistent results compared to a standard laboratory potentiostat. Thus, making it a valuable tool for real-time, on-site food pathogen monitoring (Lee et al. 2022). As illustrated in Fig. 4, Chen et al. (2016) fabricated a microfluidic biosensor to rapidly detect *L. monocytogenes*. The biosensor design integrates electrochemical impedance analysis, urease catalysis, and immunomagnetic separation. The biosensor uses magnetic nanoparticles (MNPs) coated with anti-*Listeria* monoclonal antibodies and gold nanoparticles (AuNPs) coated with polyclonal antibodies and urease to create complexes with *Listeria* cells. These complexes are captured in a high-gradient magnetic field, and urea catalysis is then used to generate ions that increase the ionic strength of the solution. This increase in ionic strength is measured through impedance on a microfluidic chip with interdigitated microelectrodes. The system demonstrated efficient capture, with approximately 93% of *Listeria* cells captured within 30 min, and achieved a detection limit as low as 1.6×10^2 CFU/mL. The entire analysis time lasts about 1 h. The biosensor also displays high specificity while showing minimal interference from non-target bacteria such as *Escherichia coli*. The biosensor was validated on lettuce samples spiked with *Listeria*. The system achieved recovery rates between 82.1% and 89.6%, showcasing its effectiveness in real-world food safety applications. This microfluidic biosensor offers a fast, sensitive, and specific method for on-site pathogen detection, with significant potential for integration into automated food safety monitoring systems (Chen et al. 2016).

4.3 *Bacillus cereus*

Bacillus cereus (*B. cereus*) is a gram-positive, rod-shaped bacterium commonly found in soil and food (Lindbäck and Granum 2013). *B. cereus* is known for its capability to produce toxins. It can cause foodborne illnesses, including two types of food poisoning: an emetic (vomiting) form often linked to rice and a diarrhoeal form associated with various foods (Schoeni and Lee Wong 2005). *B. cereus* spores are heat-resistant, enabling the bacteria to survive cooking processes, making it a frequent concern in food safety (den Besten et al. 2018). Few examples of immunosensors have been developed for *B. cereus*. For instance, Yahsini et al. (2022) reported an electrochemical biosensor based on a chitosan/gelatin-modified electrode for detecting *Bacillus cereus* in cereal-based food products (Yashini et al. 2022). The biosensor utilises antibodies specific to *B. cereus*, immobilised on a glassy carbon electrode modified with chitosan and gelatin via glutaraldehyde cross-linking. The biosensor construction and antibody attachment were thoroughly characterised using a scanning electron microscope, Fourier Transform Infrared Spectroscopy, contact angle measurements, and cyclic voltammetry, confirming the integrity of the fabrication process. The biosensor exhibited a linear detection range from 10^1 to 10^8 CFU/mL and a limit of detection of 10 CFU/mL. The biosensor can perform rapid detection in under 50 min. It recovered between 85% and 107% when applied to infant food and rice flour samples spiked with *B. cereus*. The immunosensor also maintains 70% of its activity after 3 weeks at 4 °C and has a reproducibility rate of 10.63% RSD. The biosensor offered high specificity for *B. cereus* and minimal sensor response from common bacteria like *E. coli* and *Staphylococcus aureus*. The chitosan/gelatin-based sensor is a sensitive, reliable, and practical tool for rapid pathogen detection in food safety applications (Yashini et al. 2022).

As highlighted in Fig. 5, Kang et al. (2013) reported an electrochemical immunosensor for detecting *Bacillus cereus* in food, utilising a double layer of gold nanoparticles (GNPs) and chitosan to enhance sensitivity and stability. The sensor is fabricated by immobilising *B. cereus* antibodies onto a chitosan-modified glassy carbon electrode, with the GNPs providing a stable platform that improves antibody binding and overall performance. The biosensor developed showed a linear detection range of 5×10^1 to 5×10^4 CFU/mL and a low detection limit of 10 CFU/mL. The immunosensor also shows high specificity for *B. cereus* with minimal cross-reactivity from other bacteria such as *Bacillus subtilis* and *Escherichia coli*. The biosensor was tested on milk samples, and recovery rates between 90% and 106% were achieved in spiked samples. The developed biosensor also maintained 93% of its initial response after 20 days, making it a robust and practical tool for food safety applications (Kang et al. 2013).

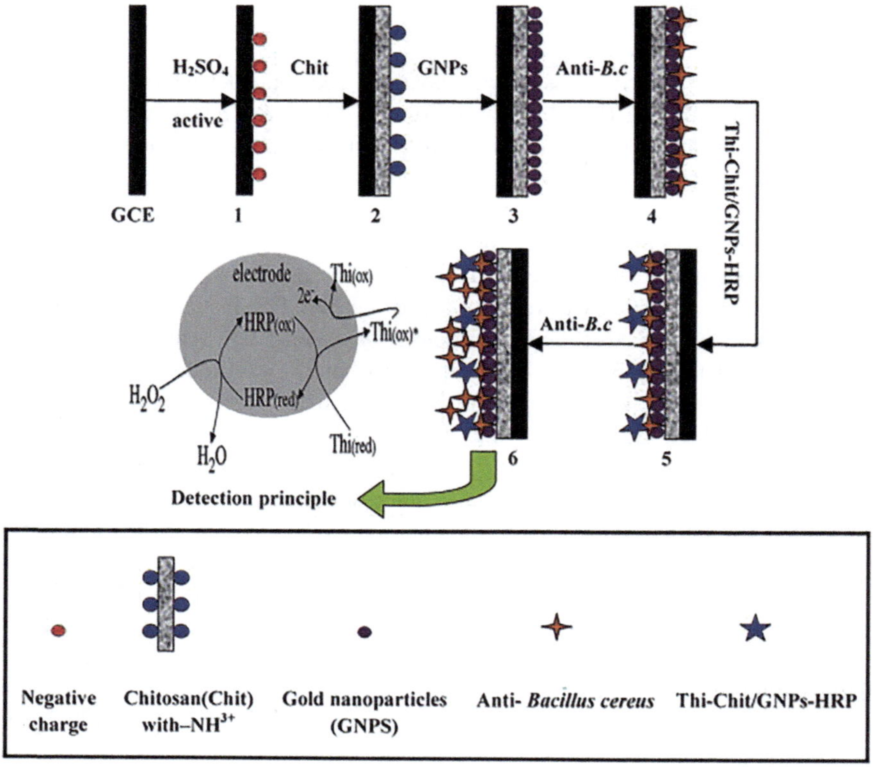

Fig. 5 Schematic representation of fabrication procedures for *Bacillus cereus* electrochemical immunosensors utilising double-layer gold nanoparticles and chitosan as the bridge cross-linker. Pre-treatment of the GCE was followed by fabrication of the *B. cereus* immunosensor from chitosan acetate solution, followed by preparation of the GNPs-Chit-GCE electrode dipped into an antibody solution containing *B. cereus*. (Reprinted with permission from Kang et al. 2013, License Number 5924240825349)

5 Integrating Electrochemical Immunosensors for Food Safety

Integrating immunosensors for detecting gram-positive bacteria in food safety requires a systematic approach to ensure accuracy and usability in real-world environments. Selecting and immobilising antibodies that specifically target gram-positive pathogens like *S. aureus, L. monocytogenes,* or *B. cereus* is essential. These antibodies can be anchored onto sensor surfaces, such as gold nanoparticles or conductive polymers, to create a sensor capable of selective pathogen capture and detection. The antibodies in developing immunosensors are limited by their production and purification steps (Rathore et al. 2021). Producing high-quality antibodies is essential for achieving the sensitivity required to meet regulatory detection limits in sensor development. This can be a time-consuming task and expensive in the fabrication of immunosensors. The limited availability of highly specific antibodies has been a major challenge in developing antibody-based methods for detecting *L. monocytogenes and B. cereus.* As a result, despite their significant threat to public health, the number of studies reported is relatively low compared to other foodborne pathogens. The limited availability of highly specific antibodies has likely been a significant obstacle in creating antibody-based methods for detecting *L. monocytogenes and B. cereus.* As a result, fewer studies have been conducted than those for other foodborne bacteria (e.g. *S. aureus*), despite these pathogens' serious public health risks. On the other hand, recombinant antibody fragments are being used in medicine, biotechnology, and therapeutics because they offer several advantages over traditional monoclonal antibodies (mAbs). These fragments can be created by cloning or amplifying variable genes from antibody-producing cells, preserving the core functionality of conventional mAbs while allowing greater optimisation through in vitro production (Karunakaran et al. 2015). They provide benefits such as improved tissue penetration, consistent antigen recognition, reduced size, and simplified production. Recent advancements in bacterial expression and phage display techniques have also facilitated the selection of specific antibody fragments (André et al. 2022). The antibody fragments are often derived from genetically engineered phage display libraries, thus broadening the applications of recombinant antibodies in immunological research and practice.

Designing the electrochemical immunosensor platform for portability and ease of use allows it to function effectively in food processing settings. Options like paper-based sensors, microfluidic chips, or portable electrochemical devices enable on-site testing, which is particularly valuable for routine monitoring without laboratory access. These devices are user-friendly and require minimal training, allowing them to be integrated into various points along the food production line. Furthermore, these portable sensors can facilitate real-time decision-making by quickly identifying contamination events. Timely and effective decision-making can reduce potential downtime and prevent widespread contamination. Moreover, portable immunosensors allow cost-effective monitoring at multiple production, storage, and distribution stages (Jeon et al. 2014). This further enhances food safety by detecting pathogenic threats early and efficiently. It is also crucial to validate immunosensors against traditional detection methods like culture-based assays or PCR to confirm reliability and accuracy. Regular calibration ensures consistent immunosensor performance.

Effective integration of immunosensors also involves sample preparation tailored to handle complex food matrices, such as dairy, meat, or produce, ensuring that food residues or particles do not compromise the sensor's performance. Real-time monitoring systems can be established by embedding immunosensors within production lines, enabling continuous contamination checks and immediate feedback on bacterial levels. This setup helps food safety teams respond quickly to potential hazards. Connecting immunosensors to data analysis platforms facilitates instant result interpretation, centralised tracking, and trend analysis, which is invaluable for maintaining compliance and ensuring timely intervention. The monitoring of food contamination via the Internet of Things (IoT) could provide connectivity for real-time access to contamination data across different production sites, creating a comprehensive view of safety practices and potential hazards (Ataei Kachouei et al. 2023). For example, by leveraging data analysis and cloud-based storage, food safety managers can remotely monitor trends, detect emerging risks, and trace contamination sources throughout the supply chain (Seo et al. 2016). Automated alerts and data visualisation tools further enhance responsiveness, empowering teams to make proactive adjustments, optimise cleaning schedules, and implement targeted quality control measures (Chen et al. 2024). This integrated approach not only supports regulatory adherence but also enhances operational efficiency, reduces costs associated with contamination, and builds consumer confidence by upholding the highest standards of food safety.

6 Conclusion

Gram-positive pathogens such as *S. aureus*, *L. monocytogenes*, and *B. cereus* pose significant public health risks due to their prevalence in contaminated food and their ability to produce toxins. Although reliable, traditional bacterial detection methods, like culture-based assays, are time-consuming and labour-intensive. Modern molecular and immunological techniques, such as PCR and ELISA, provide improved sensitivity and specificity but face limitations like high costs, complex sample preparation, and the need for skilled operators. On the other hand, electrochemical immunosensors have emerged as a rapid, sensitive, and cost-effective alternative for pathogen detection in food safety monitoring to address these gaps. These sensors offer a powerful combination of speed, sensitivity, and adaptability, addressing the limitations of conventional methods. Electrochemical immunosensors offer label-free and labelled detection options, using impedance spectroscopy, voltammetry, and amperometry to measure electrical changes caused by antigen–antibody interactions. Nanomaterial advances, such as gold nanoparticles and carbon nanotubes, have significantly enhanced sensor sensitivity, specificity, and stability. Label-free methods simplify detection processes and reduce costs, while labelled systems provide amplified signals for higher sensitivity. Despite some challenges, like potential non-specific binding and complex manufacturing, electrochemical immunosensors stand out for their rapid detection capabilities, portability, and ability to monitor multiple pathogens simultaneously. Recent electrochemical immunosensors have been developed for *S. aureus* and *L. monocytogenes*. However, very few electrochemical immunosensors have been reported for *B. cereus*. In general, these developments demonstrate the potential of electrochemical immunosensors to enhance food safety monitoring across diverse environments. Integrating electrochemical immunosensors into food safety practices requires attention to antibody specificity, sample preparation, and real-time data connectivity. High-quality antibodies are crucial for achieving regulatory detection limits, and recent advances in recombinant antibody fragments have improved their availability and functionality. Portable and user-friendly sensor designs enable on-site testing at multiple food production and distribution stages. Embedding these sensors into production lines facilitates continuous monitoring, while connecting them to data analysis platforms supports centralised tracking, trend analysis, and regulatory compliance. Integrating Internet of Things (IoT) technology further enhances these systems by enabling real-time contamination monitoring across production sites. Electrochemical immunosensors have immense potential for revolutionising food safety monitoring and providing a safer food supply chain by bridging the gap between laboratory accuracy and field usability.

References

Ahmad R, Wolfbeis OS, Hahn Y-B et al (2018) Deposition of nanomaterials: a crucial step in biosensor fabrication. Mater Today Commun 17:289–321. https://doi.org/10.1016/j.mtcomm.2018.09.024

Akhtarian S, Kaur Brar S, Rezai P (2024) Electrochemical impedance spectroscopy-based microfluidic biosensor using cell-imprinted polymers for bacteria detection. Biosensors 14:445. https://doi.org/10.3390/bios14090445

André AS, Moutinho I, Dias JNR, Aires-da-Silva F (2022) In vivo phage display: a promising selection strategy for the improvement of antibody targeting and drug delivery properties. Front Microbiol 13. https://doi.org/10.3389/fmicb.2022.962124

Aryal S (2022) Biochemical test and identification of bacteria. https://microbenotes.com/biochemical-test-of-bacteria/. Accessed 12 Nov 2024

Ataei Kachouei M, Kaushik A, Ali MA (2023) Internet of things-enabled food and plant sensors to empower sustainability. Adv Intell Syst 5:2300321. https://doi.org/10.1002/aisy.202300321

Batchelor-McAuley C, Kätelhön E, Barnes EO et al (2015) Recent advances in voltammetry. ChemistryOpen 4:224–260. https://doi.org/10.1002/open.201500042

Bekir K, Barhoumi H, Braiek M et al (2015) Electrochemical impedance immunosensor for rapid detection of stressed pathogenic Staphylococcus aureus bacteria. Environ Sci Pollut Res 22:15796–15803. https://doi.org/10.1007/s11356-015-4761-7

Bencardino D, Amagliani G, Brandi G (2021) Carriage of Staphylococcus aureus among food handlers: an ongoing challenge in public health. Food Control 130:108362. https://doi.org/10.1016/j.foodcont.2021.108362

Bhand S, Bacher G (2018) Impedimetric sensors in environmental analysis: an overview. In: Bhattacharya S, Agarwal AK, Chanda N et al (eds) Environmental, chemical and medical sensors. Springer, Singapore, pp 67–85

Bhardwaj J, Devarakonda S, Kumar S, Jang J (2017) Development of a paper-based electrochemical immunosensor using an antibody-single walled carbon nanotubes bio-conjugate modified electrode for label-free detection of foodborne pathogens. Sens Actuators B Chem 253:115–123. https://doi.org/10.1016/j.snb.2017.06.108

Burden of Foodborne Illness: Findings | Estimates of Foodborne Illness | CDC (2023). https://www.cdc.gov/foodborneburden/2011-foodborne-estimates.html. Accessed 28 Sept 2024

Cancelliere R, Paialunga E, Grattagliano A, Micheli L (2024) Label-free electrochemical immunosensors: a practical guide. Trends Anal Chem 180:117949. https://doi.org/10.1016/j.trac.2024.117949

Canciu A, Tertis M, Hosu O et al (2021) Modern analytical techniques for detection of bacteria in surface and wastewaters. Sustain For 13:7229. https://doi.org/10.3390/su13137229

Chen A, Shah B (2013) Electrochemical sensing and biosensing based on square wave voltammetry. Anal Methods 5:2158–2173. https://doi.org/10.1039/C3AY40155C

Chen Q, Wang D, Cai G et al (2016) Fast and sensitive detection of foodborne pathogen using electrochemical impedance analysis, urease catalysis and microfluidics. Biosens Bioelectron 86:770–776. https://doi.org/10.1016/j.bios.2016.07.071

Chen Y, Wang Y, Zhang Y et al (2024) Intelligent biosensors promise smarter solutions in food safety 4.0. Food Secur 13:235. https://doi.org/10.3390/foods13020235

den Besten HMW, Wells-Bennik MHJ, Zwietering MH (2018) Natural diversity in heat resistance of bacteria and bacterial spores: impact on food safety and quality. Annu Rev Food Sci Technol 9:383–410. https://doi.org/10.1146/annurev-food-030117-012808

Elgrishi N, Rountree KJ, McCarthy BD et al (2018) A practical beginner's guide to cyclic voltammetry. J Chem Educ 95:197–206. https://doi.org/10.1021/acs.jchemed.7b00361

Esteban-Fernández de Ávila B, Pedrero M, Campuzano S et al (2012) Sensitive and rapid amperometric magnetoimmunosensor for the determination of Staphylococcus aureus. Anal Bioanal Chem 403:917–925. https://doi.org/10.1007/s00216-012-5738-8

Feng K, Li T, Ye C et al (2022) A novel electrochemical immunosensor based on Fe3O4@graphene nanocomposite modified glassy carbon electrode for rapid detection of Salmonella in milk. J Dairy Sci 105:2108–2118. https://doi.org/10.3168/jds.2021-21121

Foddai ACG, Grant IR (2020) Methods for detection of viable foodborne pathogens: current state-of-art and future prospects. Appl Microbiol Biotechnol 104:4281–4288. https://doi.org/10.1007/s00253-020-10542-x

Gracias KS, McKillip JL (2004) A review of conventional detection and enumeration methods for pathogenic bacteria in food. Can J Microbiol 50:883–890. https://doi.org/10.1139/w04-080

Han E, Li X, Zhang Y et al (2020) Electrochemical immunosensor based on self-assembled gold nanorods for label-free and sensitive determination of Staphylococcus aureus. Anal Biochem 611:113982. https://doi.org/10.1016/j.ab.2020.113982

Harshavardhan S, Rajadas SE, Vijayakumar KK et al (2019) Electrochemical immunosensors. In: Bioelectrochemical Interface engineering. John Wiley & Sons, Ltd, pp 343–369

Hernández R, Vallés C, Benito AM et al (2014) Graphene-based potentiometric biosensor for the immediate detection of living bacteria. Biosens Bioelectron 54:553–557. https://doi.org/10.1016/j.bios.2013.11.053

Jazayeri MH, Amani H, Pourfatollah AA et al (2016) Various methods of gold nanoparticles (GNPs) conjugation to antibodies. Sens Bio-Sens Res 9:17–22. https://doi.org/10.1016/j.sbsr.2016.04.002

Jeon J-W, Kim J-H, Lee J-M et al (2014) Rapid immuno-analytical system physically integrated with lens-free CMOS image sensor for food-borne pathogens. Biosens Bioelectron 52:384–390. https://doi.org/10.1016/j.bios.2013.09.004

Kang X, Pang G, Chen Q, Liang X (2013) Fabrication of Bacillus cereus electrochemical immunosensor based on double-layer gold nanoparticles and chitosan. Sens Actuators B Chem 177:1010–1016. https://doi.org/10.1016/j.snb.2012.12.018

Karunakaran C, Pandiaraj M, Santharaman P (2015) Chapter 4 – Immunosensors. In: Karunakaran C, Bhargava K, Benjamin R (eds) Biosens bioelectron. Elsevier, pp 205–245

Kim J, Park M (2021) Recent progress in electrochemical immunosensors. Biosensors 11:360. https://doi.org/10.3390/bios11100360

Kokkinos C, Economou A, Prodromidis MI (2016) Electrochemical immunosensors: critical survey of different architectures and transduction strategies. Trends Anal Chem 79:88–105. https://doi.org/10.1016/j.trac.2015.11.020

Lazcka O, Campo FJD, Muñoz FX (2007) Pathogen detection: a perspective of traditional methods and biosensors. Biosens Bioelectron 22:1205–1217. https://doi.org/10.1016/j.bios.2006.06.036

Lecuit M (2020) Listeria monocytogenes, a model in infection biology. Cell Microbiol 22:e13186. https://doi.org/10.1111/cmi.13186

Lee BE, Kang T, Jenkins D et al (2022) A single-walled carbon nanotubes-based electrochemical impedance immunosensor for on-site detection of Listeria monocytogenes. J Food Sci 87:280–288. https://doi.org/10.1111/1750-3841.15996

Leva-Bueno J, Peyman SA, Millner PA (2020) A review on impedimetric immunosensors for pathogen and biomarker detection. Med Microbiol Immunol 209:343–362. https://doi.org/10.1007/s00430-020-00668-0

Lin D, Tang T, Jed Harrison D et al (2015) A regenerating ultrasensitive electrochemical impedance immunosensor for the detection of adenovirus. Biosens Bioelectron 68:129–134. https://doi.org/10.1016/j.bios.2014.12.032

Lin D, Pillai RG, Lee WE, Jemere AB (2019) An impedimetric biosensor for E. coli O157:H7 based on the use of self-assembled gold nanoparticles and protein G. Microchim Acta 186:169. https://doi.org/10.1007/s00604-019-3282-3

Lindbäck T, Granum PE (2013) Bacillus cereus. In: Guide to foodborne pathogens. John Wiley & Sons, Ltd, pp 75–81

Lopez-Tellez J, Ramirez-Montes S, Ferreira TA et al (2022) Application of voltammetric sensors for pathogen bacteria detection: a review. Chemosensors 10:424. https://doi.org/10.3390/chemosensors10100424

Magar HS, Abdelghany H, Abbas MN et al (2023) Fast analysis of Staphylococcus aureus in food products using disposable label-free nano-electrochemical immunosensor chips. Microchem J 193:109097. https://doi.org/10.1016/j.microc.2023.109097

Magar HS, Hemdan BA, Rashdan HRM, Hassan RYA (2024) Rapid and selective detection of foodborne pathogens using a disposable bio-sensing system designed by stepwise antibody immobilization on AuNPs@Cu-MOF nanocomposite. J Anal Test 8:478. https://doi.org/10.1007/s41664-024-00319-w

Mo X, Wu Z, Huang J et al (2019) A sensitive and regenerative electrochemical immunosensor for quantitative detection of Escherichia coli O157:H7 based on stable polyaniline coated screen-printed carbon electrode and rGO-NR-Au@Pt. Anal Methods 11:1475–1482. https://doi.org/10.1039/C8AY02594K

Ogodo AC, Agwaranze DI, Daji M, Aso RE (2022) Chapter 13 – Microbial techniques and methods: basic techniques and microscopy. In: Egbuna C, Patrick-Iwuanyanwu KC, Shah MA et al (eds) Analytical techniques in biosciences. Academic Press, pp 201–220

Rajkovic A, Jovanovic J, Monteiro S et al (2020) Detection of toxins involved in foodborne diseases caused by gram-positive bacteria. Compr Rev Food Sci Food Saf 19:1605–1657. https://doi.org/10.1111/1541-4337.12571

Rathore AS, Nikita S, Thakur G, Deore N (2021) Challenges in process control for continuous processing for production of monoclonal antibody products. Curr Opin Chem Eng 31:100671. https://doi.org/10.1016/j.coche.2021.100671

Riquelme MV, Zhao H, Srinivasaraghavan V et al (2016) Optimizing blocking of nonspecific bacterial attachment to impedimetric biosensors. Sens Bio-Sens Res 8:47–54. https://doi.org/10.1016/j.sbsr.2016.04.003

Rodriguez C, Taminiau B, García-Fuentes E et al (2021) Listeria monocytogenes dissemination in farming and primary production: sources, shedding and control measures. Food Control 120:107540. https://doi.org/10.1016/j.foodcont.2020.107540

Roushani M, Rahmati Z, Golchin M et al (2020) Electrochemical immunosensor for determination of Staphylococcus aureus bacteria by IgY immobilized on glassy carbon electrode with electrodeposited gold nanoparticles. Microchim Acta 187:567. https://doi.org/10.1007/s00604-020-04547-6

Rusling JF (2012) Nanomaterials-based electrochemical immunosensors for proteins. Chem Rec 12:164–176. https://doi.org/10.1002/tcr.201100034

Schoeni JL, Lee Wong AC (2005) Bacillus cereus food poisoning and its toxins. J Food Prot 68:636–648. https://doi.org/10.4315/0362-028X-68.3.636

Seo S-M, Kim S-W, Jeon J-W et al (2016) Food contamination monitoring via internet of things, exemplified by using pocket-sized immunosensor as terminal unit. Sens Actuators B Chem 233:148–156. https://doi.org/10.1016/j.snb.2016.04.061

Shah JS, Ramasamy R (2022) Fluorescence in situ hybridization (FISH) tests for identifying protozoan and bacterial pathogens in infectious diseases. Diagnostics 12:1286. https://doi.org/10.3390/diagnostics12051286

Silva NFD, Neves MMPS, Magalhães JMCS et al (2020) Emerging electrochemical biosensing approaches for detection of Listeria monocytogenes in food samples: an overview. Trends Food Sci Technol 99:621–633. https://doi.org/10.1016/j.tifs.2020.03.031

Singhal N, Kumar M, Kanaujia PK, Virdi JS (2015) MALDI-TOF mass spectrometry: an emerging technology for microbial identification and diagnosis. Front Microbiol 6. https://doi.org/10.3389/fmicb.2015.00791

Svalova TS, Medvedeva MV, Kozitsina AN (2021) A "clickable" electrodeposited polymer films based on 3-Ethynylthiophene for the covalent immobilization of proteins. Application to a label-free electrochemical immunosensor for Escherichia coli and Staphylococcus aureus determination. Electroanalysis 33:2469–2475. https://doi.org/10.1002/elan.202100320

Templier V, Roux A, Roupioz Y, Livache T (2016) Ligands for label-free detection of whole bacteria on biosensors: a review. TrAC Trends Anal Chem 79:71–79. https://doi.org/10.1016/j.trac.2015.10.015

Thakur M, Asrani RK, Patial V (2018) Listeria monocytogenes: a food-borne pathogen. In: Holban AM, Grumezescu AM (eds) Foodborne diseases. Academic Press, pp 157–192

Todd ECD, Notermans S (2011) Surveillance of listeriosis and its causative pathogen, Listeria monocytogenes. Food Control 22:1484–1490. https://doi.org/10.1016/j.foodcont.2010.07.021

Tong SYC, Davis JS, Eichenberger E et al (2015) Staphylococcus aureus infections: epidemiology, pathophysiology, clinical manifestations, and management. Clin Microbiol Rev 28:603–661. https://doi.org/10.1128/cmr.00134-14

Vasavada PC, Lee A, Betts R (2020) Conventional and novel rapid methods for detection and enumeration of microorganisms. In: Demirci A, Feng H, Krishnamurthy K (eds) Food safety engineering. Springer International Publishing, Cham, pp 85–128

Wang H, Xiu Y, Chen Y et al (2019) Electrochemical immunosensor based on an antibody-hierarchical mesoporous SiO 2 for the detection of Staphylococcus aureus. RSC Adv 9:16278–16287. https://doi.org/10.1039/C9RA00907H

Wang Z, Gao H, Cui J et al (2021) Two-signal electrochemical detection system for evaluation viability of Staphylococcus aureus. Electrochim Acta 367:137258. https://doi.org/10.1016/j.electacta.2020.137258

Wen W, Yan X, Zhu C et al (2017) Recent advances in electrochemical immunosensors. Anal Chem 89:138–156. https://doi.org/10.1021/acs.analchem.6b04281

Yáñez-Sedeño P, Campuzano S, Pingarrón JM (2017) Multiplexed electrochemical immunosensors for clinical biomarkers. Sensors 17:965. https://doi.org/10.3390/s17050965

Yashini M, Auddy I, Shanmugasundaram S et al (2022) Characterization of antibody immobilization on chitosan/gelatin-modified electrode and its application to Bacillus cereus detection in cereal-based food. Food Anal Methods 15:2382–2393. https://doi.org/10.1007/s12161-022-02299-y

Yousef AE, Carlstrom C (2003) Food microbiology: a laboratory manual. John Wiley & Sons

Zhang R, Belwal T, Li L et al (2020) Nanomaterial-based biosensors for sensing key foodborne pathogens: advances from recent decades. Compr Rev Food Sci Food Saf 19:1465–1487. https://doi.org/10.1111/1541-4337.12576

Detection of *Salmonella* spp. in Different Foods Using Immunosensors

Kritika Saini, Anita Sharma, Parul Dhar, Shyna Bhalla, and Anjana Thakur

Abstract

Food-borne pathogens pose significant health risks to both humans and animals worldwide. Health experts highlight that food safety is a critical global health issue, with the estimated total cost of detecting food-borne illnesses ranging from $5 to $6 billion. *Salmonella* is the most common pathogen responsible for food-borne illnesses, leading to salmonellosis and spreading daily through raw and processed foods, which presents a serious threat to public health. Detection methods for *Salmonella*, particularly in both developed and developing countries, are outdated. The primary detection methods are labor-intensive, costly, and can take several hours to days to yield results, often with a low detection limit. These methods are not suitable for analyzing large numbers of samples. Therefore, rapid detection of *Salmonella* is of great importance. Immunosensors present several advantages over traditional methods as they incorporate nanoparticles and micro- and nano-structured materials. Various immunosensors have been developed and commercialized for the detection of *Salmonella*. This chapter aims to provide a comprehensive overview of these immunosensors by highlighting the electrochemical signal-transducing mechanisms and analyzing recent trends, especially in conjunction with nanomaterials.

Keywords

Food safety · Foodborne pathogen · *Salmonella* detection · Nanoparticles

K. Saini (✉) · A. Sharma · P. Dhar · S. Bhalla · A. Thakur
University School of Life Sciences, Rayat Bahra University Kharar, Mohali, India
e-mail: kritika.19210@rayatbahrauniversity.edu.in

K. Kuča et al. (eds.), *Ensuring Wellness: Immunosensors Monitoring Food Quality and Pathogenicity Risks*, https://doi.org/10.1007/978-3-031-97477-9_8

Abbreviations

AgNPs	Silver nanoparticles
ASV	Anodic stripping voltammetry
AuNPs	Gold nanoparticles
Cd/Se ZnS	Cadmium selenide core and a zinc sulphide
Cds	Carbon dots
CeO_2 NPs	Cerium oxide nanoparticles
CFU	Colony forming unit
CNTs	Carbon nanotubes
CoFe	Cobalt hexacyanoferrate
CQDs	Carbon quantum dots
CV	Cyclic voltammetry
EDC	1-Ethyl-3-(3-dimethylaminopropyl) carbodiimide
EIS	Electrochemical impedance spectroscopy
ELISA	Enzyme-linked immunosorbent-based techniques
Fe_3O_4	Ferric oxide
FTO	Fluorine doped tin oxide
H	Hour
HRP	Horseradish peroxidase
IMB	Immunomagnetic beads
ISE	Ion-selective electrodes
KDa	Kilo Dalton
LFIA	Lateral flow immunoassay
LOD	Limit of detection
Min	Minutes
mL	MilliLitre
mV	Millivolts
MWCNT	Multiwalled carbon nanotubes
Ng	Nanogram
NHS	*N*-hydroxy succinimide
Nm	Nanometer
PCR	Polymerase chain reaction
Pt NPs	Platinum nanoparticles
QDs	Quantum dots
rGO	Reduced graphene oxide
SERS	Surface-enhanced Raman spectroscopy
SPCE	Screen-printed carbon electrode
SPE	Screen-printed electrode
SWASV	Square wave anodic stripping voltammetry
μg	Microgram

1 Introduction

Over the past 30 years, human health globally has been impacted by the rise of food-borne diseases, which are linked to the increasing prevalence of food-borne pathogens (Todd 2020). Food-borne pathogens have infiltrated all aspects of life, being present in everything from packaged foods to street vendor offerings. Among the various pathogens, *Salmonella* is notable for being a common, widespread, rod-shaped, flagellated bacterium, measuring 0.2–1.5 × 2–5 μm, and is classified as facultative anaerobic and gram-negative, responsible for both water borne and food-borne illnesses that have high morbidity and mortality rates (Bibi et al. 2015). The genus *Salmonella* encompasses two species: *Salmonella. enterica* and *Salmonella. bongori*; the first is subdivided into six subspecies and numerous serovars (Cobo-Simón et al. 2023). This bacterium can thrive across a broad temperature range (7 to 45 °C), with an optimal growth temperature of 37 °C. It is recognized as a zoonotic pathogen that causes infections in humans, animals, and birds. Over 2500 serotypes have been identified from *Salmonella* sp. utilizing the Kaufmann-White typing scheme, and all are considered potential pathogens (Abdel Rahman et al. 2024). Typical symptoms of *Salmonella* infection include chills, abdominal discomfort, nausea, cramps, joint pain, weight loss, and headache, in addition to enteric fever, diarrhea, and gastroenteritis, which are specific manifestations (Ünüvar 2018).

Salmonella can survive for several weeks or even months in dry environments and is found everywhere in nature. The main sources of *Salmonella* infections include products from meat, milk, and poultry such as eggs and chicks. Recently, contaminated fruits and fresh vegetables have been associated with a rise in *Salmonella*-causing pathogen. The schematic representation is as shown in Fig. 1a. According to Cho et al. (2020), fecal matter has also been recognized as a potential

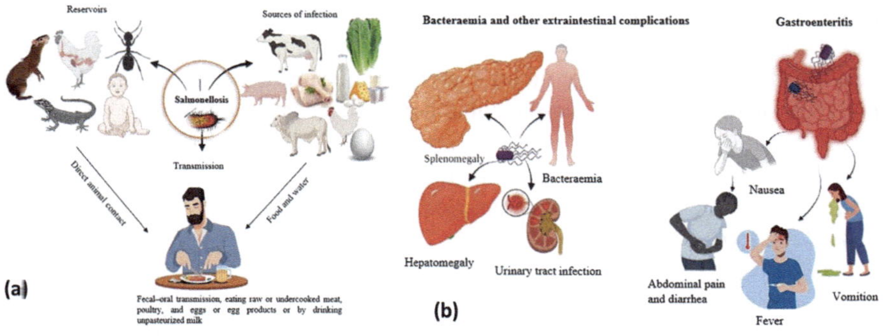

Fig. 1 Illustration ofthe main reservoir, sources of infection, and transmission routes for human salmonellosis. Insects, reptiles, humans, and animals can carry the bacteria. Common sources of infection include uncooked or contaminated foods such as poultry, beef, pork, eggs, milk, vegetables, flour, and nuts (**a**). Symptoms of human salmonellosis typically presents in two main ways. The first is gastroenteritis, characterized by headache, nausea, vomiting, abdominal pain, non-bloody diarrhea, and muscle aches. The second involves bacteremia and can lead to complications like splenomegaly, hepatomegaly, and urinary tract infections (**b**). (Teklemariam et al. 2023, under the terms and conditions of the Creative Commons Attribution (CC BY)

source of *Salmonella*. In the USA, *Salmonella* is the most common cause of food-borne infection-related hospitalizations and deaths, causing about 420 deaths and 26,500 hospitalizations annually (Han et al. 2024). According to the European Food Safety Authority (EFSA), human salmonellosis may have an annual economic impact of up to €3 billion (Galán-Relaño et al. 2023). Salmonella is thought to be the second most common cause of food-borne illnesses in the USA after norovirus, with an estimated 1.35 million infections annually (Subedi 2022). According to Kurtz et al. (2017), *Salmonella* infections are a major source of illness and mortality worldwide, particularly in developing countries. *Salmonella* can infect humans with as little as one colony forming unit per milliliter (CFU/mL) in a 25 g serving of ready-to-eat food according to the study of European Commission. Individuals with weakened immune systems, children under five, and the elderly are especially vulnerable to infection (Zizza et al. 2024).

Salmonella resides in the intestinal tracts of humans and animals and can easily spread in unsanitary conditions (Oludairo et al. 2023). It is a common cause of acute gastrointestinal infections among people around the world (Bhat et al. 2022). Gastroenteritis is an infection affecting the terminal ileum and colon and is the most common illness worldwide. Diarrhea typically begins within 6 to 48 hours after consuming contaminated food, and the introduction of *Salmonella* into the bloodstream can sometimes lead to pneumonia, gastroenteritis, bacteremia, meningitis, osteomyelitis, or other intestinal complications. Notably, *Salmonella* can also be found in pregnant women and is widely established in adults, children, and individuals with severely compromised immunity (Bula-Rudas et al. 2015). The schematic representation as shown in Fig. 1b.

Additionally, diagnosing *Salmonella* can be challenging because the symptoms of this infection are somewhat non-specific and resemble those of many other illnesses. There is currently no specific treatment available for this infection. Therefore, a timely and rapid diagnosis can aid in controlling and preventing the spread of this infection (Guadarrama et al. 2014). At this time, several methods have been developed to detect the pathogen, but many of them struggle with validation prior to commercialization due to a high likelihood of false negative outcomes (Valderrama et al. 2016). This chapter discusses various recently developed electrochemical immunosensors that provide rapid, reliable, and standardized methods and techniques to address the shortcomings of traditional methods for identifying *Salmonella* in food samples.

2 Conventional and Advanced Techniques for Detecting *Salmonella*

Conventional approaches (including those based on cultures, biochemical tests, immunological tests, and nucleic acid tests) and advanced techniques (such as hybridization-based, array-based, spectroscopy-based, and biosensor-based process) approaches are linked to the identification of food-borne pathogen. The schematic

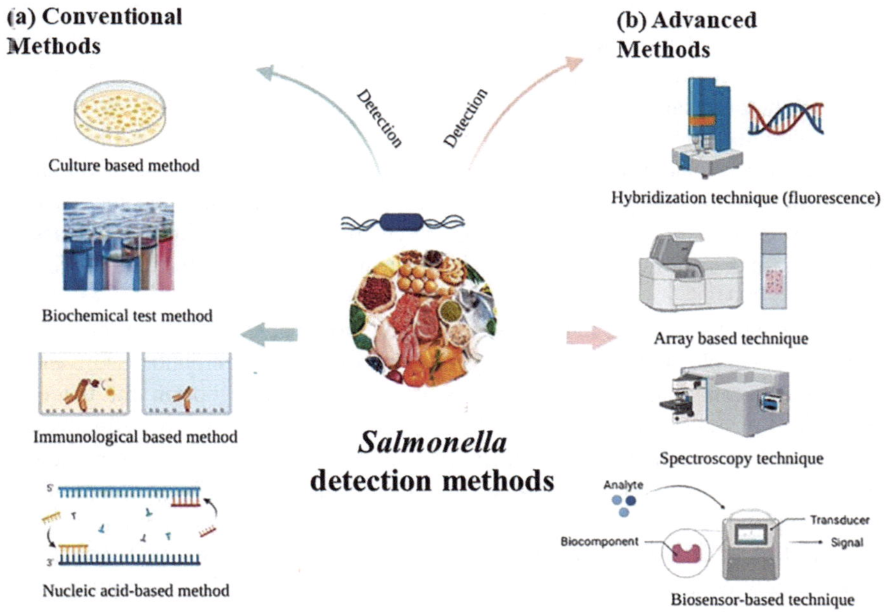

Fig. 2 Methods for detecting foodborne pathogen (*Salmonella*): (**a**) Conventional methods. (**b**) Advanced methods. (Kabiraz et al. 2023, under the CC BY-NC-ND license)

presentation for the detection of *Salmonella* utilizing both conventional and advanced methods is as shown in Fig. 2a, b.

The culture-based approach is considered the "gold standard" for *Salmonella* identification because of its sensitivity and selectivity (Awang et al. 2021). The cultural techniques for isolating *Salmonella* generally involve enriching a portion of the sample to recover damaged bacterial cells caused by heat, cold, acid, or osmotic stress (Fang et al. 2021). While conventional microbiological methods can identify single bacterium, they need signal amplification through colony growth. *Salmonella* can be isolated and identified using an array of techniques, which include bacterial identification, biochemical and serological testing, selective enrichment procedures, isolation on selective agar media, and an enrichment step on non-selective media (Rasamsetti et al. 2021). As a result, culture-based methods are not suited for detecting a high volume of samples at the industrial level (in food industries), necessitating a faster approach for analyzing food samples (Bell et al. 2016). The traditional plate culture method used for identifying bacteria typically involves evaluating the shape and growth patterns of microorganisms in different media and conditions. These methods are highly sensitive, cost-effective, and capable of providing both qualitative and quantitative data regarding the microorganisms present in a food sample (Fang et al. 2021).

Researchers have developed a number of advanced techniques for detecting foodborne pathogens, to overcome the time-consuming nature of conventional techniques. These techniques include the array method (Raspoet et al. 2014), lateral flow immunoassay (LFIA) (Gong et al. 2023), polymerase chain reaction (PCR) (Saini et al. 2019), enzyme-linked immunosorbent assay (ELISA) (Liao et al. 2024), and surface-enhanced Raman spectroscopy (SERS) (Zhang et al. 2015). While LFIA is a quick, easy, and affordable diagnostic method, it does have limitations, as its sensitivity needs to be enhanced (Liu et al. 2023; Wang et al. 2022). Although PCR is very sensitive and effective, it necessitates the nucleic acid extraction and requires sophisticated and costly equipment (Jiao et al. 2023). ELISA shortens the detection time by 2 to 3 hours by integrating enzymatic reactions with immunoassays; however, it has drawbacks such as low sensitivity and complicated procedures (Yu et al. 2021).

Consequently, there is a need for a quick method that is sensitive, user-friendly, non-toxic, simple, and fast. In recent years, electrochemical immunosensors and the use of different nanomaterials have become an attractive option for the quick detection of viable bacteria due to their operational simplicity, cost-effectiveness, and excellent specificity and sensitivity (Jiang et al. 2020). In electrochemical immunosensors, a potential is applied, and the electrical signal is measured. Therefore, amplifying the electrical signal is crucial for improving detection efficiency (Zhan et al. 2024). Furthermore, electrochemical immunosensor performance has been greatly enhanced by developments in nanomaterials and biotechnology (Lim and Ahmed 2016; Yang et al. 2024).

3 Biosensor

Novel biosensors have been developed for rapid and sensitive detection of food-borne pathogens and toxins in food samples, offering advantages over conventional approaches such as culture, immunological, and nucleic acid-based. These biosensors require minimal sample pre-processing. The principle behind a biosensor is the incorporation of a biological analyte (such as proteins, ions, microbes, toxins) within a physicochemical transducer. A transducer is an analytical device which converts output signals into input signals, functioning like an amplifier that transforms biological interactions into measurable electrical or optical signals. The intensity of the measurable signal is directly proportional to the concentration of the target analyte (Singh et al. 2014). Transducers come in various forms, including piezoelectric, electrochemical, thermal, ion-selective, and optical devices (Chen et al. 2020). Among these, electrochemical biosensors have shown significant advantages and characteristics, making them the preferred choice for the past 30 years (Gaudin 2017). The transducers used in these electrochemical biosensors are typically made of inert materials such as glassy carbon, gold, platinum, and graphite. These biosensors offer rapid, label-free, cost-effective, specific, and highly sensitive detection options (Ahmed Omar et al. 2014). Biosensors that use antibodies as bio-receptors are known as immunosensors. Biosensors don't require highly skilled workers like traditional methods do. Additionally, a biosensor that is extremely sensitive and selective can yield results more quickly than conventional techniques, which makes it perfect for field and practical applications (Perumal and Hashim 2014; Ahmad et al. 2018).

4 Immunosensors

Immunosensors, a kind of solid-state biosensor based on affinity, have gained prominence. It can easily identify the immunoreactions that form a stable immunological complex between an antibody and its target antigen. Here, the transducer produces a detectably strong signal once the antibody serves as a capturing agent (Patel et al. 2022). The premise behind immunosensors, is the specificity of molecular recognition between antigens and antibodies, which results in the formation of a stable, quantifiable complex. In order to identify food-borne pathogenic bacteria (*Salmonella*), immunosensors are further divided into labeled (indirect) and label-free (direct) approaches (Fang et al. 2020). Label-free detection allows for real-time target probe binding detection, which is generally not possible with label-based systems, and thereby reducing preparation time, analysis costs, and sample complexity (de Castro et al. 2020). However, the non-specific adsorption can be a limitation when using label-free immunosensors. Labeled immunosensors employ nanomaterials like carbon dots (Cds) and gold nanoparticles (AuNPs), quantum dots (QDs) as well as signal-generating labels, including metal ions, fluorescent dyes, and enzymes (catalase, glucose oxidase) (Fang et al. 2020). In labeled immunosensors, the labels can be attached to antigen or antibody, which causes the transfer of electrons and the amount of labels observed during measurement correlates with the quantity of target analytes (Mollarasouli et al. 2019). Because of the analytical properties of the label-free, labeled immunosensors have more adaptability and sensitivity and a lesser effect of non-specific signal adsorption.

4.1 Electrochemical Immunosensors

Electrochemical immunosensors are designed to detect specific sequences of antigens and antibodies. These sensors operate on the principle of coupling biological analytes with an electron transducer, which converts biological signals into electrical signals (Rotariu et al. 2016). The schematic representation of different elements of electrochemical immunosensor is as shown in Fig. 3a. These immunosensors are commercially available, and they exhibit high sensitivity and specificity. An antigen-antibody binding process takes place on the electrode of these immunosensors once the antibody is immobilized on the electrode surface, resulting in a voltage that can be measured.

Electrochemical immunosensors can be categorized into various groups based on the signal transduction mechanisms or transducers used (Metkar and Girigoswami 2019). They are particularly effective for detecting foodborne pathogens, offering high sensitivity and affinity, and are grouped among the immunosensors on the way they transmit the information (Mollarasouli et al. 2019; Wang et al. 2009). Several

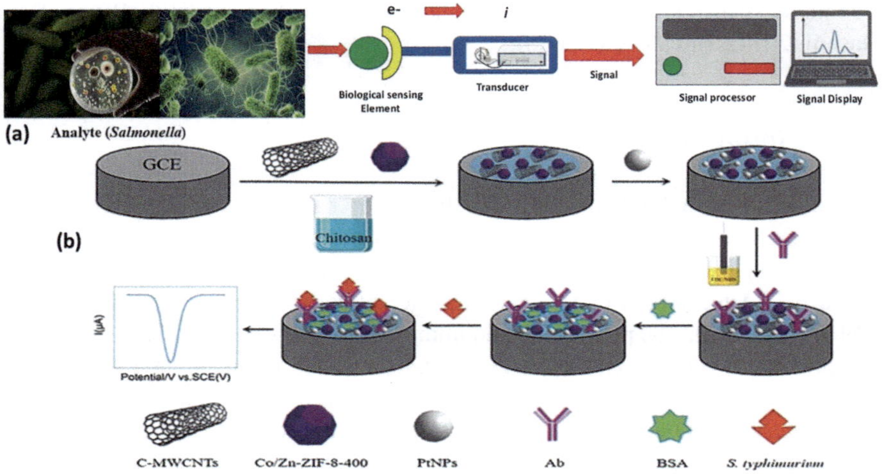

Fig. 3 (**a**) Schematic representation of different elements of electrochemical immunosensor for *Salmonella* detection. (**b**) Fabrication of an electrochemical immunosensor using immobilization of Pt nanoparticles and Co/Zn-MOFs@ carboxylic CNT on glassy carbon electrode, for the rapid detection of *Salmonella*. (Zhan et al. 2024, License number 5983721005181)

types of nanomaterials are used to detect *Salmonella*, further improving the application of electrochemical immunosensors (Martín et al. 2015; Soares et al. 2020). The use of QD's, AuNp's, ferric oxide (Fe_3O_4), Platinum nanoparticles (Pt NPs) in immunosensor building is becoming more and more relevant because of its benefits, which include highly specific and enhanced surface areas, superior biocompatibility, low toxicity, and easy to make (Li et al. 2024; Ding et al. 2022; Zamzami et al. 2024). Gold nanoparticles are often more used in immunosensing because of the simple synthesis, narrow in size and the electrochemical properties (Iglesias-Mayor et al. 2019). The use of nanoparticles in immunosensing provides better and more surface area for the immobilization. *Salmonella* in milk samples was detected using an electrochemical immunosensor. In Zhan et al. (2024), group of researchers immobilized (Pt NPs) and Co/Zn-MOFs@ carboxylic carbon nanotube on glassy carbon electrode for enhancing stability, sensitivity, and specificity. The sensor exhibited the linear range of 1.3×10^2–1.3×10^8 CFU/mL and an LOD of 9.4×10^1 CFU/mL. The schematic illustration of this electrochemical immunosensor is shown in Fig. 3b.

4.2 Amperometric Immunosensors

Amperometric immunosensors rely on measuring changes in current and operate using either a two-electrode or a three-electrode system. A two-electrode system consists of a reference electrode and a working electrode. However, a limitation of this system is the inability to control the potential on the surface of the working electrode at larger currents, which may limit the linear operating range. In the three-electrode arrangements, voltage is applied between the working and reference electrodes, whereas current travels between the working and auxiliary electrodes. Because of their stability and sensitivity, amperometric immunosensors are preferred over other kinds (Ghindilis et al. 1998). The working area of the screen-printed electrodes can be coated with some noble metals like gold and silver and biorecognition component (Wang 1999).

In the context of foodborne bacterial pathogen detection, amperometric immunosensors are utilized, which are based on antigen-antibody interactions (Jain et al. 2012). These are extensively used to identify *Salmonella* in dairy and food processing units (Melo et al. 2021). These immunosensors are less technically demanding, easy to operate and offer higher sensitivity than other immunosensors (Karunakaran et al. 2015). The working of these immunosensors is based on immobilizing specific anti-*Salmonella* antibodies on the surface of the electrode. When the antigen in the sample reacts with the antibody, it prompts an immunochemical reaction that serves as a recognition indicator. Additionally, this technology is very important for its miniaturization, which is beneficial when working with lower sample volumes (Leonard et al. 2003). Fei et al. (2015) created a sandwich-type amperometric immunosensor for the identification of *S. pullorum* and *S. gallinarum*. For strong signals and conjugation of the antibodies, a carbon-based screen-printed electrode (SPE) was deposited with AuNPs. A suitable milieu with respect to both activity and stability of the antibody was also designed on the surface of the electrode by modifying with an ionic liquid. Although it requires over 13 hours for developing the immunosensor, the authors have indicated that it is stable at 4 °C until used. Further, antibody conjugated with horseradish peroxidase (HRP) was introduced and incubated for another 40 min to create the sandwich complex after the 40-min interaction of the capture antibody with *Salmonella*. The reduction peak was found using cyclic voltammetry (CV) with the enzymatic product-thionine oxidized form and reduced thionine as the HRP substrates. The limit of detection (LOD) was 3×10^3 CFU/mL. This immunosensor recognized *S. pullorum* and *S. gallinarum* at the concentration between 10^4 and 10^9 CFU/mL.

In another study, immunomagnetic beads (IMB) and reduced graphene oxide (rGO) coated with AuNPs were used to successfully develop a fast and highly sensitive amperometric immunoassay for detecting *S. pullorum* on a screen-printed carbon electrode (SPCE), with a limit of detection (LOD) of 89 CFU/mL (Fei et al. 2016). Alexandre et al. (2018) reported an amperometric immunosensor for detecting *S. typhimurium* in skim milk without requiring a pre-enrichment step. In this study, a gold electrode was used for the biosensor assembly, which was modified with 10 mM cysteamine and left for 3 h at room temperature. Protein A (prA) was used to orient polyclonal antibodies, and 1-Ethyl-3-(3-dimethylaminopropyl) carbodiimide (EDC): N-hydroxy succinimide (NHS) crosslinking chemistry was used to immobilize them. The biosensor assembly was electrochemically analyzed by CV using a solution of 1 M KCl and 4 mM $K_3[Fe(CN)_6]$. An evaluation of the biosensor's performance showed that it had a detection time of 125 minutes and a detection limit of 10 CFU/mL for both whole and skim milk. For many years, label-free amperometric immunosensors have been employed to detect various illnesses. Using this technique, *salmonella* was found in milk samples. Because of its exceptional stability and strong sensitivity, the detection range for *Salmonella* in milk samples was 3.65×10^2–3.65×10^8 CFU mL^{-1}, and its LOD was 1.12×10^2 CFU mL^{-1} (Li et al. 2024).

4.3 Impedimetric Immunosensors

Impedimetric immunosensors (biosensors) operate on the principle of measuring changes in impedance. This device tracks either the overall impedance (Z) or its components: resistance (R) and capacitance (C); typically, inductance plays a minor role in standard electrochemical setups. The formula for impedance is expressed as follows: $Z^2 = R^2 + (1/2\pi FC)^2$, where F represents frequency. An impedance immunosensor consists of two electrodes that have an alternating voltage applied, with amplitude values ranging from a few millivolts (mV) to 100 mV (Davis and Higson 2015). This kind of immunosensor measures the sensor's impedance, which is influenced by the biological response (Yang and Bashir 2008). Electrochemical impedance spectroscopy (EIS) is a useful method for examining how compounds form between biomolecules on an electrode's surface by investigating the interfacial characteristics between the electrode and electrolyte. A little sinusoidal AC voltage perturbation signal is measured by EIS, which then calculates the corresponding AC current. The Randles equivalent-electric-circuit and impedimetric signal, which depend on the alteration of one of these equivalent-electric-circuit characteristics upon analyte binding (Hayat et al. 2012). The most popular substitute for microbial electrochemical analysis is impedimetric biosensors because of its lower detection limit and miniaturization potential (Silva et al. 2018).

The impedimetric immunosensor was used to employ to identify the *S. typhimurium* in spiked water samples and apple juices. In this study, anti-OmpD antibodies were generated which were employed as detector probes in immunoassay format on graphene-graphene oxide (G-GO)-modified screen-printed carbon electrodes. *S. typhimurium* was particularly detected by the devised immunosensor in spiked water with a limit of detection of 10^1 CFU/mL and in juice samples with a LOD of 1.04×10^1 CFU/mL and 1.07×10^1 CFU/mL, respectively. According to Mutreja et al. (2016), the developed immunosensor exhibits very minimal cross-reactivity with other strains and great selectivity. In order to identify *Salmonella* spp. in hen egg samples, an impedimetric immunosensor was developed using the lectin *Hechtia argentea* immobilized on screen-printed gold electrodes. *H. argentea*, a member of the *Bromeliaceae* family, contains an isolated lectin that was around 27.4 kDa in size. Through the use of the change in impedance versus concentration, the lectin's carbohydrate selectivity was estimated. As a result, bacterial binding to the biosensor surface increased as *Salmonella* spp. concentrations increased, reaching a linear range of detection of 15–2.57×10^7 CFU/mL with a limit of detection of 5 CFU/mL. Using Nyquist plots and electrochemical impedance spectroscopy, the rise in impedance was detected and examined (Lopez-tellez et al. 2020).

4.4 Potentiometric Immunosensors

Potentiometric methods biosensors measure the charge accumulation (potential) on the working electrode due to the interaction between the analyte and bioreceptor relative to the reference electrode under zero or negligible current flow. These biosensors utilize ion-selective electrodes (ISE) and ion-sensitive field-effect transistors to measure changes in voltage based on specific interactions with ions in the solution. The primary output signal results from ions accumulating at the interface of the ion-selective membrane (Pohanka and Skladal 2008). The potential difference is measured between the working electrode and the reference electrode. Even small changes in analyte concentration can lead to significant changes in potential at the working electrode (Ahmed Omar et al. 2014). These biosensors are among the least commonly used for pathogen detection. In one of the studies, a label-free potentiometric biosensor based on paper was developed to detect *S. typhimurium*. Two different immunosensing interfaces were designed and assembled using a paper-strip electrode. The first involved the direct conjugation of the antibody to the polymer membrane, whereas the second involved the use of an intermediate layer of a polyamidoamine dendrimer with an ethylenediamine. The blocking surface principle served as the basis for the detection. Following the evaluation of the various interface modification techniques using electrochemical impedance spectroscopy, the analytical performance of the two immunosensors was compared. With a detection limit of 5 cells/mL and a detection time of less than an hour, the developed potentiometric immunosensor successfully identified *S. typhimurium* in apple juice (Silva et al. 2019).

5 *Salmonella* Detection Electrochemical Immunosensors Based on Nanomaterials

In order to identify foodborne pathogens in food samples with precision, immunosensors have recently been employed as a substitute for conventional approaches. The diverse nanomaterials utilized in electrochemical immunosensors to detect *Salmonella* in various food samples are compiled in Table 1. There are many nanoparticles that are utilized as labels in immune-sensing, and are widely accessible on the market. Among the often-used nanoparticles are rGO, QDs, Pt NPs, AuNPs, cerium oxide nanoparticles (CeO_2 NPs) and silver nanoparticles (AgNPs), and so on.

Table 1 Nanomaterials for the *Salmonella* detection in food samples using electrochemical immunosensor

Source nanomaterial	Limit of detection CFU/mL	Source	Time duration for the detection	Detection type	Reference
Dendron molecules and QD	4 CFU/mL	Milk sample	2.5 h	SWASV	Murasova et al. (2020)
Gold electrode	10 CFU/mL	Milk sample	125 min	CV	Melo et al. (2016)
SiO2@MnO2	21 CFU/mL	Milk sample	50 min	EIS	Wang et al. (2020)
Graphene	13 ± 7 CFU/mL	Chicken sample	22 min	EIS	Soares et al. (2020)
CoFe-MOFs-graphene on AuNp's	1.2×10^2 CFU /mL	Milk sample	–	DPV	Feng et al. (2021)
rGO/FTO	218 and 173 cells	Meat sample	–	CV, EIS	Mahari and Gandhi (2022)
Fe_3O_4@graphene	2.4×10^2 CFU/mL	Milk sample	1.5 h	CV	Feng et al. (2022)
Cd/Se ZnS QDs	5 CFU/ML	Chicken sample	25 min	ASV	Angelopoulou et al. (2023)
Pt nanoparticles	9.4×10^1 CFU mL^{-1}	Milk sample	50 min	CV	Zhan et al. (2024)

6 Future Prospects

The future of *Salmonella* detection in food using immunosensors is poised for significant advancements, driven by the integration of nanotechnology, artificial intelligence (AI), and portable biosensing platforms. Emerging trends focus on improving the sensitivity, specificity, and real-time application of immunosensors while addressing challenges such as cost, stability, and large-scale implementation. The use of nanomaterials, such as gold nanoparticles, silver nanoparticles, carbon nanotubes, and graphene-based materials, enhances signal amplification, leading to ultra-sensitive detection of *Salmonella* spp. (Zarei 2017). Future research should explore the development of multifunctional nanocomposites to improve sensor stability and response time. AI-driven data analysis can enhance the accuracy of immunosensor readings by identifying patterns and minimizing false positives or negatives (Jiang et al. 2022). The combination of AI with biosensors can revolutionize *Salmonella* monitoring in food industries. Collaborative efforts between researchers, food industries, and regulatory bodies will be essential in realizing the full potential of immunosensors for *Salmonella* detection.

7 Conclusion

Food-borne illness is an emerging problem, leading to increased awareness about the impact of food on human health. Rapid food-borne pathogen detection methods are therefore becoming more and more in demand. Conventional methods for detecting *Salmonella* are considered the "gold standard," but they can be labor-intensive and time-consuming. Different rapid identification approaches have been developed in order to overcome the shortcomings of these conventional detection methods. Rapid techniques are especially crucial for detecting *Salmonella* in food samples, which helps stop the spread of infections and food-borne disease outbreaks. These techniques enable detection even at extremely low pathogen concentrations and are more.

sensitive, specific, labor-saving, and time-efficient. One promising approach is the development of various immunosensor types, including amperometric, impedimetric, and potentiometric immunosensors, which have been successfully developed and validated for detecting *Salmonella* in food samples like fruit juice, meat, dairy, eggs, and fresh produce. Despite their potential, challenges such as matrix interference, detection limits, and the need for cost-effective large-scale deployment remain. Future research should focus on improving sensor sensitivity, integration with smart detection platforms, and automation for field-based applications. The continuous advancement in immunosensor technology holds promise for enhancing food safety monitoring and reducing the global burden of *Salmonella*-related infections.

References

Abdel Rahman MA, Adel A, Samir A et al (2024) Application of PCR-based serogrouping of *Salmonella enterica* serotypes isolated from the poultry in Egypt. SVU-Int J Vet Sci 7(1):62–70

Ahmad R, Wolfbeis OS, Hahn Y-B et al (2018) Deposition of nanomaterials: a crucial step in biosensor fabrication. Mater Today Commun 17:289–321

Ahmed Omar B, Asghar AH, Ibrahim HA et al (2014) Detection of *Salmonella* in food samples by culture and polymerase chain reaction methods. J BacteriolParasitol 5(3):1

Alexandre DL, Melo AMA, Furtado RF et al (2018) A rapid and specific biosensor for *Salmonella Typhimurium* detection in Milk. Food Bioprocess Technol 11:748–756

Angelopoulou M, Kourti D, Mertiri M et al (2023) A 3D-printed electrochemical Immunosensor employing Cd/Se ZnS QDs as labels for the rapid and ultrasensitive detection of *Salmonella typhimurium* in poultry samples. Chem 11(9):475

Awang MS, Bustami Y, Hamzah HH et al (2021) Advancement in *Salmonella* detection methods: from conventional to electrochemical-based sensing detection. Biosensors 11(9):346

Bell RL, Jarvis KG, Ottesen AR et al (2016) Recent and emerging innovations in *Salmonella* detection: a food and environmental perspective. Microbiol Biotechnol 9(3):279–292

Bhat KA, Manzoor T, Dar MA et al (2022) *Salmonella* infection and pathogenesis. In: Bhardwaj SB (ed) Enterobacteria. https://doi.org/10.5772/intechopen.102061

Bibi F, Qaisrani SN, Ahmad AN et al (2015) Occurrence of *Salmonella* in freshwater fishes: a review. J Anim Plant Sci 25(3):303–310

Bula-Rudas FJ, Rathore MH, Maraqa NF (2015) *Salmonella* infections in childhood. Adv Pediatr Infect Dis 62(1):29–58

Chen Y, Qian C, Liu C et al (2020) Nucleic acid amplification free biosensors for pathogen detection. Biosens Bioelectron 153:112049

Cho S, Jackson CR, Frye JG (2020) The prevalence and antimicrobial resistance phenotypes of *Salmonella, Escherichia coli* and *Enterococcus* sp. in surface water. Lett Appl Microbiol 71(1): 3–25

Cobo-Simón M, Hart R, Ochman H (2023) Gene flow and species boundaries of the genus *Salmonella*. Msystems 8(4):e00292–e00223

Davis F, Higson SP (2015) Impedance Immunosensors. In: Electrochemical biosensors. Jenny Stanford Publishing, pp 304–323

de Castro ACH, Alves LM, Siquieroli ACS et al (2020) Label-free electrochemical Immunosensor for detection of Oncomarker Ca125 in serum. Microchemistry 155:104746

Ding S, Hu H, Yue X et al (2022) A fluorescent biosensor based on quantum dot–labeled streptavidin and poly-l-lysine for the rapid detection of *Salmonella* in milk. J Dairy Sci 105(4):2895–2907

Fang L, Liao X, Jia B et al (2020) Recent Progress in Immunosensors for pesticides. Biosens Bioelectron 164:112255

Fang T, Wu Y, Xie Y et al (2021) Inactivation and subsequent growth kinetics of *Listeria monocytogenes* after various mild bactericidal treatments. Front Microbiol 12:621

Fei J, Dou W, Zhao G (2015) A sandwich electrochemical immunosensor for *Salmonella pullorum* and *Salmonella gallinarum* based on a screen-printed carbon electrode modified with an ionic liquid and electrodeposited gold nanoparticles. Microchim Acta 182:2267–2275

Fei J, Dou W, Zhao G (2016) Amperometric immunoassay for the detection of *Salmonella pullorum* using a screen-printed carbon electrode modified with gold nanoparticle-coated reduced graphene oxide and immunomagnetic beads. Microchim Acta 183(2):757–764

Feng K, Li T, Ye C et al (2021) A label-free electrochemical immunosensor for rapid detection of *Salmonella* in milk by using CoFe-MOFs-graphene modified electrode. Food Control 130: 108357

Feng K, Li T, Ye C et al (2022) A novel electrochemical immunosensor based on Fe3O4@ graphene nanocomposite modified glassy carbon electrode for rapid detection of *Salmonella* in milk. J Dairy Sci 105(3):2108–2118

Galán-Relaño Á, Valero Díaz A, Huerta Lorenzo B et al (2023) *Salmonella* and salmonellosis: an update on public health implications and control strategies. Animals 13(23):3666

Gaudin V (2017) Advances in biosensor development for the screening of antibiotic residues in food products of animal origin-a comprehensive review. Biosens Bioelectron 90:363–377

Ghindilis AL, Atanasov P, Wilkins M et al (1998) Immunosensors: electrochemical sensing and other engineering approaches. Biosens Bioelectron 13:113–131

Gong L, Wang K, Liang J et al (2023) Enhanced sensitivity and accuracy via gold nanoparticles based multi-line lateral flow immunoassay strip for *Salmonella typhimurium* detection in milk and orange juice. Talanta 265:124929. https://doi.org/10.1016/j.talanta.2023.124929

Guadarrama C, Villaseñor T, Calva E (2014) The subtleties and contrasts of the LeuO regulator in *Salmonella typhi*: implications in the immune response. Front Immunol 5:581

Han J, Aljahdali N, Zhao S et al (2024) Infection biology of *Salmonella enterica*. EcoSal Plus eesp-0001

Hayat A, Barthelmebs L, Marty J-L (2012) Electrochemical Impedimetric Immunosensor for the detection of Okadaic acid in mussel sample. Sens Actuators B Chem 171:810–815

Iglesias-Mayor A, Amor-Gutiérrez O, Costa-García A (2019) Nanoparticles as emerging labels in electrochemical immunosensors. Sensors 19(23):5137

Jain S, Singh SR, Horn DW et al (2012) Development of an antibody functionalized carbon nanotube biosensor for foodborne bacterial pathogens. J Biosens Bioelectron 11:2

Jiang H, Yang J, Wan K et al (2020) Miniaturized paper-supported 3D cell-based electrochemical sensor for bacterial lipopolysaccharide detection. ACS Sensors 5(5):1325–1335. https://doi.org/10.1021/acssensors.9b02508

Jiang X, Wang Y, Zhang H et al (2022) Application of artificial intelligence in biosensors for pathogen detection. Trends Anal Chem 146:116509

Jiao Y, Zhang Z, Wang K et al (2023) Rapid detection of *Salmonella* in food matrices by photonic PCR based on the photothermal effect of Fe3O4. Food Chem X 19:100798. https://doi.org/10.1016/j.fochx.2023.100798

Kabiraz MP, Majumdar PR, Mahmud MC et al (2023) Conventional and advanced detection techniques of foodborne pathogens: a comprehensive review. Heliyon 9(4):e15482

Karunakaran C, Pandiaraj M, Santharaman P (2015) Immunosensors. Biosens Bioelectron:205–245

Kurtz JR, Goggins JA, McLachlan JB (2017) *Salmonella* infection: interplay between the bacteria and host immune system. Immunol Lett 190:42–50

Leonard P, Hearty S, Brennan J et al (2003) Advances in biosensors for detection of pathogens in food and water. Enzym Microb Technol 32:3–13

Li S, Zhang Y, Guo M et al (2024) Rapid detection of *Salmonella* in milk by labeling-free electrochemical immunosensor based on an Fe3O4–ionic liquid-modified electrode. Talanta 270:125576

Liao X, Zhang Y, Liang Y et al (2024) Enhanced sandwich immunoassay based on bivalent nanobody as an efficient immobilization approach for foodborne pathogens detection. Anal Chim Acta 1289:342209. https://doi.org/10.1016/j.aca.2024.342209

Lim SA, Ahmed MU (2016) Electrochemical immunosensors and their recent nanomaterial-based signal amplification strategies: a review. RSC Adv 6(30):24995–25014. https://doi.org/10.1039/c6ra00333h

Liu X, Chen Y, Bu T et al (2023) Nanosheet antibody mimics based label-free and dual-readout lateral flow immunoassay for *Salmonella enteritidis* rapid detection. Biosens Bioelectron 229:115239. https://doi.org/10.1016/j.bios.2023.115239

Lopez-Tellez J, Sanchez-Ortega I, Hornung-Leoni CT et al (2020) Impedimetric biosensor based on a *Hechtia argentea* lectin for the detection of *Salmonella* spp. Chem 8(4):115. https://doi.org/10.3390/chemosensors8040115

Mahari S, Gandhi S (2022) Electrochemical immunosensor for detection of avian Salmonellosis based on electroactive reduced graphene oxide (rGO) modified electrode. Bioelectrochemistry 144:108036

Martín M, Salazar P, Jiménez C et al (2015) Rapid *legionella pneumophila* determination based on a disposable core–shell Fe3O4@ poly (dopamine) magnetic nanoparticles immunoplatform. Anal Chim Acta 887:51–58

Melo AM, Alexandre DL, Furtado RF et al (2016) Electrochemical immunosensors for *Salmonella* detection in food. Appl Microbiol Biotechnol 100:5301–5312

Melo AMA, Furtado RF, de Fatima Borges M et al (2021) Performance of an Amperometric Immunosensor assembled on Carboxymethylated cashew gum for *Salmonella* detection. Microchem J 167:106268

Metkar SK, Girigoswami K (2019) Diagnostic biosensors in medicine–a review. Biocatal Agric Biotechnol 17:271–283

Mollarasouli F, Kurbanoglu S, Ozkan SA (2019) The role of electrochemical immunosensors in clinical analysis. Biosensors 9(3):86

Murasova P, Kovarova A, Kasparova J et al (2020) Direct culture-free electrochemical detection of *Salmonella* cells in milk based on quantum dots-modified nanostructured dendrons. J Electroanal Chem 863:114051

Mutreja R, Jariyal M, Pathania P et al (2016) Novel surface antigen based impedimetric immunosensor for detection of *Salmonella typhimurium* in water and juice samples. Biosens Bioelectron 85:707–713

Oludairo OO, Kwaga JK, Kabir J et al (2023) Transmission of *Salmonella* in humans and animals and its epidemiological factors. Zagazig Vet J 51(1):76–91

Patel M, Agrawal M, Srivastava A (2022) Signal amplification strategies in electrochemical biosensors via antibody immobilization and nanomaterial-based transducers. Mater Adv 3: 8864–8885

Perumal V, Hashim U (2014) Advances in biosensors: principle, architecture and applications. J Appl Biomed 12:1–15

Pohanka M, Skladal P (2008) Electrochemical biosensors—principles and applications. J Appl Biomed 6:57–64

Rasamsetti S, Berrang M, Cox NA et al (2021) Selective pre-enrichment method to lessen time needed to recover *Salmonella* from commercial poultry processing samples. Food Microbiol 99: 103818

Raspoet R, Appia-Ayme C, Shearer N et al (2014) Microarray-based detection of *Salmonella enterica* serovar Enteritidis genes involved in chicken reproductive tract colonization. Appl Environ Microbiol 80:7710–7716

Rotariu L, Lagarde F, Jaffrezic-Renault N et al (2016) Electrochemical biosensors for fast detection of food contaminants–trends and perspective. Trends Anal Chem 79:80–87

Saini K, Kaushal A, Gupta S et al (2019) Rapid detection of *Salmonella enterica* in raw milk samples using Stn gene-based biosensor. 3 Biotech 9(11):425

Silva NF, Magalhães JM, Freire C et al (2018) Electrochemical biosensors for *Salmonella*: state of the art and challenges in food safety assessment. Biosens Bioelectron 99:667–682

Silva NF, Almeida CM, Magalhães JM et al (2019) Development of a disposable paper-based potentiometric immunosensor for real-time detection of a foodborne pathogen. Biosens Bioelectron 141:111317

Singh R, Mukherjee MD, Sumana G et al (2014) Biosensors for pathogen detection: a smart approach towards clinical diagnosis. Sens Actuators B Chem 197:385–404

Soares RR, Hjort RG, Pola CC et al (2020) Laser-induced graphene electrochemical immunosensors for rapid and label-free monitoring of *Salmonella enterica* in chicken broth. ACS Sensors 5(7):1900–1911

Subedi Y (2022) Understanding the occurrence of *salmonella* in irrigation water used to grow fresh produce in the United States, Mexico, and Canada: a systematic review and meta-analysis (Doctoral dissertation)

Teklemariam AD, Al-Hindi RR, Albiheyri RS et al (2023) Human salmonellosis: a continuous global threat in the farm-to-fork food safety continuum. Foods 12(9):1756

Todd E (2020) Food-borne disease prevention and risk assessment. Int J Environ Res Public Health 17(14):5129

Ünüvar S (2018) Microbial foodborne diseases. In: Dodd ERC, Aldsworth T, Riemann HP (eds) Foodborne diseases, 3rd edn. Academic Press, pp 1–31

Valderrama WB, Dudley EG, Doores S et al (2016) Commercially available rapid methods for detection of selected food-borne pathogens. Crit Rev Food Sci Nutr 56(9):1519–1531

Wang J (1999) Amperometric biosensors for clinical and therapeutic drug monitoring: a review. J Pharm Biomed Anal 19:47–53

Wang L, Liu Q, Hu Z et al (2009) A novel electrochemical biosensor based on dynamic polymerase-extending hybridization for *Escherichia coli* O157: H7 DNA detection. Talanta 78(3):647–652

Wang S, Peng T, Meng Q et al (2020) Rapid and ultrasensitive detection of *Salmonella typhimurium* using a novel impedance biosensor based on SiO2@MnO2 nanocomposites and interdigitated array microelectrodes. Sens Actuators B Chem 324:128654

Wang H, Wu Q, Zhou M et al (2022) Development of a CRISPR/Cas9-integrated lateral flow strip for rapid and accurate detection of *Salmonella*. Food Control 142:109203. https://doi.org/10.1016/j.foodcont.2022.109203

Yang L, Bashir R (2008) Electrical/electrochemical impedance for rapid detection of foodborne pathogenic bacteria. Biotechnol Adv 26:135–150

Yang D, Li X, Li X et al (2024) Design and synthesis of nano-iron oxyhydroxide-based molecularly imprinted electrochemical sensors for trace-level carbendazim detection in actual samples. Microchim Acta 191(3):163. https://doi.org/10.1007/s00604-024-06236-0

Yu S, Xu Q, Huang J et al (2021) Rapid and sensitive detection of *Salmonella* in milk based on hybridization chain reaction and grapheneoxide fluorescence platform. J Dairy Sci 104(12): 12295–12302. https://doi.org/10.3168/jds.2021-20713

Zamzami M, Ahmad A, Alamoudi S et al (2024) A highly sensitive and specific gold electrode-based electrochemical immunosensor for rapid on-site detection of *Salmonella enterica*. Microchem J 199:110190

Zarei M (2017) Advances in nanomaterial-based electrochemical biosensors for food safety. Biosens Bioelectron 96:33–47

Zhan K, Chen L, Li S et al (2024) A novel metal–organic framework based electrochemical immunosensor for the rapid detection of *Salmonella typhimurium* detection in milk. Food Chem 444:138672

Zhang H, Ma X, Liu Y et al (2015) Gold nanoparticles enhanced SERS aptasensor for the simultaneous detection of *Salmonella typhimurium* and *Staphylococcus aureus*. Biosens Bioelectron 74:872–877

Zizza A, Fallucca A, Guido M et al (2024) Foodborne infections and *Salmonella*: current primary prevention tools and future perspectives. Vaccine 13(1):29

Detection of Antibiotics in Milk Using Immunosensors

Zahra Talebi Vandishi, Masood Dodangeh,
and Kimia Zarean Mousaabadi

Abstract

The presence of antibiotics in milk is a significant concern for food safety and public health due to potential risks of antibiotic resistance and adverse effects on consumer health. This chapter reviews the application of immunosensors for the detection of antibiotics in milk and provides an overview of the current methodologies and advancements in this area. Immunosensors demonstrate high specificity and sensitivity through antigen–antibody interactions in the detection of trace amounts of antibiotics. This chapter focuses on the principles of immunosensor technology, covering the types of immunosensors in pharmaceutical analysis, which include enzyme-linked immunosorbent assays (ELISA), surface plasmon resonance, and electrochemical immunosensors. We focus next on the challenges for antibiotic detection in complex milk matrices, considering possible interference by other milk components and requirements of sample preparation techniques. An update on nanomaterials and signal amplification strategies is also presented regarding their contribution to the higher performance of immunosensors. Moreover, this chapter contains some information about the prospects this immunosensor technology is facing in the view of food safety and regulatory requirements in terms of integration into routine quality control processes. This chapter gives the general overview of antibiotic residues monitoring systems development and implementation for effective milk quality control process.

Z. Talebi Vandishi · M. Dodangeh
Department of Chemistry, Isfahan University of Technology, Isfahan, Iran

K. Zarean Mousaabadi (✉)
Center for Electromicrobiology, Section for Microbiology, Department of Biology, Aarhus University, Aarhus C, Central Jutland, Denmark
e-mail: kzarean@bio.au.dk

K. Kuča et al. (eds.), *Ensuring Wellness: Immunosensors Monitoring Food Quality and Pathogenicity Risks*, https://doi.org/10.1007/978-3-031-97477-9_9

Keywords

Antibiotics · Detection techniques · Milk quality · Immunosensor · Dairy animals

1 Introduction

Milk is one of the most consumed foods in diets globally and plays a huge significance in human health and development. Besides being cheap, milk is a rich source of important nutrients like saturated fats, proteins, calcium, and several other related compounds (Samanidou and Nisyriou 2008). In order to maintain the good health and productivity of dairy animals, antibiotics are used as a precautionary measure to treat bacterial infections like mastitis, among others, and enhance feed efficiency (Singh et al. 2023). The antibiotics included in this context are β-lactams, sulphonamides, penicillin, tetracyclines, fluoroquinolones, lincosamides, macrolides, glycopeptides, amphenicols, and aminoglycosides. In the use of such drugs, the risk of their residuals staying in milk raises serious issues about human health and dairy product safety (Samanidou and Nisyriou 2008; Singh et al. 2023).

While most antibacterial agents currently in use are not toxic, even at higher concentrations, there are a few antibiotics that pose significant health risks. The residues from the antimicrobial additives have a certain risk of accumulation in the human organism itself or being metabolized into secondary metabolic products (Samanidou and Nisyriou 2008). This may lead to serious consequences: the development of drug resistance, allergic reactions, and multiple toxicological manifestations. In some cases, these can include carcinogenic and mutagenic impacts and even the development of severe diseases, such as optic neuropathy and brain abscesses. The severity and clinical manifestations of these health problems can vary greatly, depending on the type of antibiotic, its concentration, and the individual's susceptibility (Singh et al. 2023). Monitoring and detection of antibiotic residues in milk are essential to ensure milk safety and maintain consumer trust in dairy products. In order to avoid misuse of the antibiotics and consequently reduce their adverse effects, the European Union has urged milk producers to comply with the set maximum residual limit (MRL) for antibiotics. MRL is the maximum residue level of a veterinary drug that does not pose any risk to consumer health when the residue is present in milk. Strict adherence to such limits ensures safety for public health and upholds the integrity of the dairy industry (Jayalakshmi et al. 2017).

An effective and precise method is highly required to ensure that antibiotic residues in milk do not exceed the MRL. The current methods of detection are principally classified into two types, namely: (1) classical and (2) modern. The conventional techniques, which include immunological and microbiological methods, are simple but usually exhibit poor sensitivity and selectivity. In contrast, the techniques of current methodology based on HPLC, LC-MS, and GC-MS techniques offer considerably higher accuracy, specificity, and sensitivity. However, such advanced techniques also possess some disadvantages associated with highly trained personnel being required, time-consuming steps of sample preparation, and expensive equipment. Due to these limitations, in recent times, biosensors have come into the frame as a new technology and find extensive applications in the medical sector in terms of determination of drugs, including blood glucose testing. Biosensors are low-cost, small-sized, user-friendly, and rapid-response devices. Major classes of biosensors include: (1) immunosensors, (2) aptamers, (3) MIPs, and (4) techniques based on enzyme- and direct electrochemistry. This gives a promising alternative for rapid and reliable screening in monitoring residues of antibiotics in milk, ensuring safety compliance and protection of public health (Pollap and Kochana 2019; Singh et al. 2023).

Among the various types of biosensors, immunosensors are uniquely positioned through the offering of specific antigens, generally the target analyte, through interactions between antibodies and antigens on the surface of a transducer. These types of immunosensors are remarkably attractive for a huge range of target analytes because antigen and antibody interactions offer spectacular selectivity and a significant level of sensitivity. The extensive adoption of these devices can be ascribed to various benefits, such as user-friendliness, straightforwardness, dependability, adaptability in application, and the capacity for incorporation into multipurpose analytical instruments. Immunosensors have garnered considerable interest due to their prospective capability to identify antibiotic residues in milk, representing a promising methodology for safeguarding food safety.

This chapter will review the different types of immunosensors used for the detection of antibiotics in milk. Besides that, this chapter covers the limitations faced within this field, while describing recent developments in this technology (Pollap and Kochana 2019).

2 Principles of Immunosensor Technology

A biosensor is a small analytical device that integrates a bioreceptor with a physical or chemical detector to measure specific substances known as analytes. These sensors identify the presence of chemical, biological, or physical agents by converting biological interactions into measurable signals, which can be electrical, optical, or thermal. An immunosensor is a specialized type of biosensor that employs antibodies or antigens as the biological recognition elements. It functions by detecting the immune reaction between an antibody and its corresponding antigen, converting this interaction into a measurable signal (Aydin et al. 2021). The detailed principles of immunosensors were discussed in chapter "Introduction to Immunosensors: Principles and Applications in Food Safety", but here's a brief overview.

The design of an immunosensor depends on specific requirements, including the choice of materials. Ideal materials exhibit high electrical conductivity, large surface areas, and robust mechanical properties. Carbon-based materials, particularly graphene, a two-dimensional (2D) carbon structure made of a single layer of sp^2 carbon atoms, are commonly used. When functionalized, graphene effectively captures biomolecules through nucleophilic or Schiff base reactions. Moreover, polymers and other nanomaterials can be employed for modifications (Manasa et al. 2022).

To improve sensitivity, selectivity, and immobilization, a molecular recognition element (MRE) with high affinity is first immobilized onto the sensing interface (Kim and Kang 2008). Immunosensors can be classified based on detection mechanisms, biological recognition elements, signal transduction methods, and labeling techniques. Detection mechanisms include:

Direct immunosensors measure changes in physical properties, such as refractive index or electrical conductivity, during antigen–antibody interactions. In contrast, indirect immunosensors employ secondary reactions, utilizing labeled antibodies that bind to antigen–antibody complexes to amplify the detection signals. Competitive immunosensors function by allowing the target antigen to compete with a labeled antigen for binding to a limited number of antibodies, resulting in decreased signal intensity as the concentration of the target antigen increases. Together, these different types of immunosensors offer a range of approaches for effective detection of specific biomolecules (Harshavardhan et al. 2019).

Regarding biological recognition elements, antibodies are the most common in immunosensors, selectively binding to antigens for precise detection of pathogens, proteins, or toxins. Enzymes may also be used, either to amplify signals or as direct recognition components reacting with substrates to generate measurable outputs. Emerging alternatives like aptamers—short DNA or RNA molecules—offer high affinity and stability, making them suitable for various applications. Nucleic acids can further enhance sensitivity by binding to specific DNA or RNA sequences. Additionally, peptide-based immunosensors provide high-selectivity antigen detection, while enzyme-linked immunosensors rely on enzymes to produce measurable signals. Together, these elements enable the detection of a wide range of biological targets (Harshavardhan et al. 2019).

Immunological biosensors utilize various transduction mechanisms, focusing on signal generation, such as electrochemical or optical signals, or property changes, like mass or heat variations, resulting from antigen–antibody complex formation. Electrochemical immunosensors measure electrical changes during antigen–antibody interactions, making them effective for detecting small molecules. Optical immunosensors detect changes in light properties, with techniques like Surface Plasmon Resonance (SPR) being common (Harshavardhan et al. 2019). Piezoelectric immunosensors measure changes in mass or mechanical properties on the sensor surface, while thermal immunosensors assess heat changes during interactions, often using calorimetry. These diverse signal transduction methods enhance the detection capabilities of immunosensors. Lastly, immunosensors can be categorized by labeling methods. Label-free immunosensors operate without secondary labels, using techniques like SPR and certain electrochemical sensors for real-time detection. In contrast, labeled immunosensors rely on markers, such as fluorescent dyes or enzymes, to generate detectable signals when bound to antigen–antibody complexes, often enhancing sensitivity (Harshavardhan et al. 2019).

3 Immunosensors for Antibiotic Detection in Milk

In this section, we first start by discussing electrochemical immunosensors, which are among the most effective and widely used methods for detecting antibiotics in milk. Their high sensitivity, rapid response time, and ability to measure electrical changes during antigen–antibody interactions make them ideal for monitoring antibiotic residues in dairy products.

3.1 Electrochemical Immunosensors for Antibiotic Detection in Milk

Electrochemical sensors are an up-and-coming technology in the field of food safety, serving as detection systems for antibiotic residues in milk. These sensors convert chemical data into an electric pulse, which is the fastest, cheapest, and most sensitive way for measurement (Frigoli et al. 2024). Thus, although electrochemical sensors demonstrate significant potential in milk testing, they can be hard to use since milk is a complex matrix and the antibiotics are the analytes (de Faria et al. 2021).

3.1.1 Benefits of Electrochemical Sensors

Electrochemical sensors demonstrate high sensitivity with detection limits in the nanomolar range; they are appropriate for the usual small amounts of antibiotics that are present in milk (Wang et al. 2020). These sensors are designed to be selective to particular antibiotics, for example, penicillin, tetracycline, and streptomycin, with the use of aptamers, antibodies, or MIPs as recognition elements. The specificity of these recognition elements assures that small antibiotic concentrations are detectable under the condition of diverse milk components (Seth and Rathinasabapathi 2022).

One of the major benefits of electrochemical sensors is their very fast detection ability. To be precise, it takes electrochemical sensors a couple of minutes to send signals, thus also providing us with an ideal real-time or near real-time monitoring in dairy facilities (Novakovic et al. 2024). Also, quite a few electrochemical sensors are portable and can be tested on the spot at milk collection points or processing facilities (de Faria et al. 2021). This portability reduces the need for expensive laboratory equipment and skilled personnel, making the testing process more efficient.

The on-site testing capability means that there is less need for expensive laboratory equipment and qualified personnel. This enables the electrochemical testing process to be more efficient than the traditional methods such as liquid chromatography or mass spectrometry, which is a cost-effective solution in itself (Zhang et al. 2022). The equipment is of cheap cost, and the sensors can often be reused several times without the need to be replaced (Gaudin 2021). In some cases, there are disposable sensor strips which may also lead to the reduction of operational costs, and they also remove the step of cleaning or reconditioning the sensor surface (Abera et al. 2021).

Another advantage is that electrochemical sensors have the least possible sample preparation for it to work. Contrary to chromatography or spectrometry methods, electrochemical sensors can work with raw or minimally processed milk samples without the need for purification and dilution, which are normally done in the process (Singh et al. 2023). This issue is of significant relevance in the dairy industry, where timely and simple operations are crucial.

3.1.2 Limitations of Electrochemical Sensors

Matrix interference is the predominant obstacle of applying electrochemical sensors in milk, which arises from the intricate composition of milk, which contains proteins, fats, and carbohydrates. These components can adsorb on the sensor surface, and lead to non-specific binding and false-positive or false-negative (Raykova et al. 2021). For instance, the sensor's active sites might be covered with casein proteins, which could lead to its inability to interact with the target antibiotic (Wang et al. 2023).

In general, electrochemical sensors despite their selectivity can be sensitive to cross-reactivity, particularly when using antibodies or MIPs as recognition elements (Wang et al. 2021a). Structurally similar antibiotics or other substances in milk may bind to the sensor, leading to false positives. This issue becomes more problematic when it is necessary to detect a large number of drugs at the same time with a sensor because, as demand becomes higher, it is more doubtful to confirm that one specific antibiotic was detected.

The stability of electrochemical sensors, especially the ones that use biological recognition elements like enzymes or antibodies, can be a critical factor in the case of those sensors. In the long run, these biological parts could be damaged, resulting in the sensor's accuracy and shelf-life reduction as well. The equalization of these environmental issues, the temperature, humidity, and pH also play a significant part in a sensor's stability. Frequent calibration is necessary for upholding precision, while in a few cases, sensors must be replaced after a specified number of uses.

3.1.3 Factors Affecting the Performance of Electrochemical Sensors in Milk

The pH of the milk sample have a major impact on the electrochemical sensors (Wang et al. 2021a, b). The changes in pH can lead to the charge redistribution on the sensor surface, which will subsequently affect the binding affinity between the sensor and the antibiotic (Raykova et al. 2021). Electrochemical sensors also depend on temperature for their performance (Zhang et al. 2022). High temperatures can also denature elements like enzymes or antibodies, which, in turn, cause the sensor degradation. On the contrary, very low temperatures have the potential to slow down the reaction kinetics, which can be accompanied by longer detection periods (Singh et al. 2023). In addition, proper storage conditions, like keeping a stable temperature, will guarantee the performance of the sensor (Feizollahi et al. 2021).

Fouling is the deposition of proteins, lipids, or other components of milk on a sensor surface covering the active sites, and thus the antibiotics cannot be detected. This problem exists specifically in raw milk because the amount of these interfering substances is higher in non-pasteurized milk than in milk that has undergone processing. Proper and regular cleaning as well as the use of antifouling coatings play vital roles in preventing the mentioned problem. The following explains examples of electrochemical immunosensors used for detecting antibiotics in milk.

Recently, a research group developed an electrochemical immunosensor for the rapid detection of amoxicillin in raw milk. This sensor consists of anti-amoxicillin antibodies linked to amino-functionalized magnetic nanoparticles (MNP-TA/Ab_{Amx}) and incorporates a 3D-printed cell with a magnetic attachment and a smartphone-linked potentiostat. The magnetic field pulls the antibody–analyte complexes toward the surface of a screen-printed electrode, allowing for the concurrent application of SWV to achieve quick and sensitive electrochemical detection. The sensor can serve well for countries wherein very low amoxicillin residue of up to 0.44 μM is desired, as it has a linear range of 0.5–100 μM. Immobilization of anti-amoxicillin antibodies was carried out using glutaraldehyde as a crosslinker, with improved binding achieved through the addition of silica and amine modifications (TEOS and APTES). The first stage was the synthesis of sensors through the construction of magnetic nanoparticles using the coprecipitation technique. In terms of selectivity studies, lactose, glucose, riboflavin, and glutamic acid caused no interference. The sensor generated results that correlated well with other techniques used for the work in high performance liquid (Sanli 2024). Another paper describes development of a highly sensitive electrochemical immunosensor of ciprofloxacin (CPX) residues in milk using lanthanum oxide nanoparticles (La_2O_3 NPs) and reduced graphene oxide (rGO) nanocomposite. The nanocomposite was prepared by hydrothermal treatment and then it was modified in a way that CPX antibodies could be coupled to it. Because the La_2O_3 NPs increased rGO's conductive surface area, the prepared sensor can detect over the range of concentrations of CPX from 10^{-6} to 600 mg/mL with limit of detection (LOD) being at an impressive level of 0.055 mg/mL. The immunosensor immobilized on an indium tin oxide (ITO) electrode was analyzed by X-ray diffraction (XRD), Fourier transform infrared spectroscopy (FT-IR), scanning electron microscope (SEM), and electrochemical impedance spectroscopy (EIS) techniques. Differential pulse voltammetry (DPV) was used to test the sensor with the results indicating that the peak current and CPX concentration relation was linear indicating successful CPX binding and low CPX distribution. The sensor performed at a sensitivity of 6.52 mA mL mg^{-1} cm^{-2} with regression analysis carried out providing $R^2 = 0.992$ with dependable outcomes. The soliton also showed excellent stability being operational for 25 days. This new method permits the fast and inexpensive detection of milk food products which face safety concerns due to antibiotic residue (Chaudhary et al. 2024).

Besides, researchers designed a cost-effective electrochemical immunosensor for the detection of gentamicin (GEN) residues in milk. A nanocomposite ($rGO@Gd_2O_3$) was synthesized from rGO and gadolinium oxide (Gd_2O_3) nanoparticles through co-precipitation and immobilized on a paper-based screen-printed electrode (SPE). Such materials are expected to improve the sensor's conductivity, stability, and electrocatalytic activity, resulting in enhanced GEN detection sensitivity. The anti-GEN antibodies were attached to the $rGO@Gd_2O_3$

substrate for the sensor while bovine serum albumin (BSA) was used to saturate nonspecific binding sites. Under this configuration, only GEN molecules are allowed to attach to the GEN sensor and therefore minimization of interference from other solvents is guaranteed. The device possesses a remarkably wide detection range of 1 pM to 100 μM and a LOD of 0.424 pM with a linear sensitivity of 44.87 μA pM^{-1} cm^{-2}. Using spiked milk samples, it was confirmed that the immunosensor can perform as accurately as chromatography methods but with advantages of being cheaper, smaller, and easier to use. The selectivity tests indicated there is no interference from milk contaminants implying that the sensor is selective toward GEN presence even in complicated matrix. The sensor's performance remained consistent for up to 12 weeks, demonstrating a reproducibility of 2.02% RSD across multiple tests. This immunosensor makes a good sustainable option for the dairy industry in ensuring compliance with safety standards and protecting public health through rapid, on-site monitoring of antibiotics (Himanshu et al. 2024).

Another research describes the preparation of a paper-based optical analysis device (oPAD) for detection of antibiotic residues, in particular, Norfloxacin (Nor) and Enrofloxacin (Enr) antibiotics in milk. They are common in veterinary medicinal products used in animal husbandry, so there may be residues in the products that are dangerous to health, e.g., due to the development of antibiotic resistance. To address these health concerns, MRLs for these antibiotics are usually set by regulatory authorities in the range of 50–300 μg/kg. Antibody detection by ELISA tends to work well but is relatively costly and complex and hence, there is a requirement for cheaper and simpler methods. Nor and Enr are determined simultaneously with the help of Nor and Enr targeted oPAD that has been developed in this work. This involves three fundamental components: Two immobilization zones of antibodies directed against Nor and Enr molecules, compartments of the coating area with an electrocatalytic redox center thionine and ferrocenecarboxylic acid and places where a drop of buffering solution required for electrochemical analysis is placed. Aside from improving reliability, graphene ink-based electrodes provide additional sensitivity, where antibodies against Nor and Enr are attached firmly to the cellulose paper surface.

To analyze the milk samples, they are subjected to dilution with phosphate buffered saline (PBS) and allowed to stand on the oPAD for 60 minutes after which SWV is utilized to determine the electrochemical activity. The device presented a linear concentration detection from 0.01 to 10 μg/ml for anti-Nor (15 μg/ml) and 5 μg/ml for anti-Enr. It was inclusive of commercial milk samples hands on, where the oPAD could detect successfully antibiotic residue, showing its usefulness in the field of food safety. This portable oPAD which utilizes new technology is an antibiotic residue detection which makes it an ideal way to comply with food safety authorities in preventing agricultural product contamination (Chomthong et al. 2024). Here, in Table 1, a few studies around electrochemical immunosensors for antibiotic analysis in milk are introduced.

3.2 Enzyme-Linked Immunosensors for Antibiotic Detection in Milk

Enzyme-linked immunosensors are now utilized as instruments for identifying substances such as antibiotics in food items like milk. This is achieved by utilizing the nature of antibody interactions with antigens and the signal-enhancement feature of enzymes. These sensors provide sensitivity and accuracy in analysis applications (Mollarasouli et al. 2019). However, similar to any device, enzyme-linked immunosensors have their constraints, particularly when employed in matrices, like milk (Pollap and Kochana 2019). This part investigates the advantages and drawbacks of enzyme-linked immunosensors, in relation to detecting antibiotics in milk. It will also cover factors that may affect how well they work.

3.2.1 Benefits of ELISA in Detecting Antibiotics in Milk

ELISAs are well-known for their excellent sensitivity and specificity in the detection of antibiotics, even in complex matrices such as milk (Majdinasab et al. 2020). ELISA's ability to detect trace amounts of antibiotics makes it uniquely helpful in the field of food safety, especially in the dairy industry, where trace amounts of antibiotics may be present if veterinarians do not respect the drug withdrawal periods (Pollap and Kochana 2019). The enzyme amplification process in ELISA brings about a significant signal abundance which makes it able to check antibiotics like tetracycline, penicillin, and sulfonamides even at the lowest of concentration (ppb) (Talib et al. 2018).

One of the most notable benefits of ELISA is its adaptability. Antibody modification in ELISA is powerful enough to enable the assay to recognize different kinds of antibiotics. That is to say, the ELISA method can also be used in the detection of a wide range of antibiotics from the β-lactams and aminoglycosides to macrolides in a single test. This flexibility gives ELISA the capability to make scans that can meet the standards for the levels of milk residues left by the antibiotics, which is really the most challenging thing for the regulatory bodies (Adrian et al. 2009). Furthermore, one assay can account for several antibiotics altogether, which minimizes time and expense in scenarios such as high-throughput testing.

ELISA is more cost-effective and also requires less intricate devices compared to traditional methods like liquid chromatography or mass spectrometry (Sheng et al. 2011). This will be extremely beneficial to the dairy processing stations in particular, where regular checks are essential to avoid the contaminated milk from entering the food chain (Fernández et al. 2011). ELISA tests can usually provide results in a few hours and thus decision-making is sped up as compared to other analytical methods that need many sample preparations and lots of time for analysis (Wang et al. 2021a; b).

3.2.2 Limitations of ELISA in Detecting Antibiotics in Milk

The fundamental restriction of ELISA is the fact that it can react with molecules that are not the target molecule. The way the antibodies are used in the assay could be binding to molecules that have a structure similar to the target antibiotic, which in

Table 1 Overview of electrochemical immunosensors for detecting antibiotic residues in milk

Electrode material	Target antibiotic	Linear range	Detection limit	Reference
Ce-MOF@AgAuNPs nanocomposite with anti-MON antibodies	Monensin (MON)	0.05–250 ng/mL	0.008 ng/mL	Zhou et al. (2024)
rGO-PEI-Ag nanocomposite-modified electrode	Tilmicosin (TMC)	0.01–100 ng/mL	0.0013 ng/mL	Chen et al. (2024)
Graphene Quantum Dots (GQD) with CFX-BSA antibodies	Cephalexin (CFX)	–	0.53 fM	Kolhe et al. (2023)
AgNPs-rGO nanocomposite on gold electrode	Virginiamycin M1 (VIR M1)	0.25–100 ng/mL	0.18 ng/mL	Qian et al. (2023)
Anti-idiotypic nanobody-functionalized SPCE	Aflatoxin M1 (AFM1)	0.25–5.0 ng/mL	0.09 ng/mL	Tang et al. (2022)
SWNTs-COOH/CS on glassy carbon electrode (GCE)	CEX	1–800 ng/mL	45.7 ng/mL (S/N = 3)	Yu et al. (2020)
Gold electrode (with AuNSs)	Kanamycin (KAN), Tobramycin (TOB)	KAN: 1 to 400 nM; TOB: 1 to 10,000 nM	KAN: 0.12 nM; TOB: 0.49 nM	Li et al. (2021)

turn might cause a false-positive result (Grubelnik et al. 2001).This is really a big problem in milk, where there are proteins, peptides, or metabolites which can cause the antibody–antigen binding to be disrupted. By way of illustration, antibiotics or other drugs can yield certain metabolites that have analogous chemical structures, therefore causing misidentification of antibiotic residues (Adrian et al. 2009).

Stability of enzymes is also one of the difficulties involving ELISA. The enzyme antibodies that are used in such testing, like horseradish peroxidase (HRP) and alkaline phosphatase (AP) are very sensitive to temperature and pH. Storing them under suboptimal conditions may lead to the process of enzyme breakdown as well as loss of efficiency and accuracy of the assay (Pollap and Kochana 2019). In milk, endogenous enzymes and proteases may break down the enzyme markers, which may affect the precision of the detection further.

Milk is an intricate combination containing fats, proteins, and sugars, all of which can interfere with ELISA's functioning. Proteins like casein and the fats in milk can become trapped on the microplate's surface, thereby blocking active sites and preventing the antibody–antigen interaction from happening properly (Waritani et al. 2017). This matrix interference often necessitates sample pre-treatment, such as dilution or fat removal, which can extend the testing time (Wang et al. 2013). Moreover, some milk proteins can bind to the antibiotics and make them not available for the immunoassay to detect, which will cause the assay to become less sensitive (Fernández et al. 2011).

3.2.3 Interference Factors Impacting Selectivity of ELISA

A non-specific binding is a common problem in ELISAs that takes place more frequently when samples of milk are being examined. Protein and fat contents in milk may stick to the immunoassay plates' surface, which in turn may cause the specificity to be less between the antibodies and antigens (Wang et al. 2020). One way to control or deal with this issue is the use of blocking agents such as BSA to coat the plate and stop non-specific protein binding. Still, even if these blocking agents enhance specificity, they do not totally fix the matter, and non-specific interactions may still be observed (Contreras-Naranjo and Aguilar 2019).

Enzyme activity in ELISA is essentially very sensitive to variable factors like pH and temperature. Just a small change in pH can change the affinity of the antibody and the antigen making the assay less sensitive to immunological reactions. As an example, a minor pH shift can impair the enzyme and make it to lose the ability to catalyze the reaction that leads to the production of signals (Muhammad et al. 2016). For example, high or low temperatures in the study or storage period might introduce variability by way of faster enzyme degradation or interruption of the bonding of antibody–antigen reaction kinetics (Fernández et al. 2011).

Milk matrix is made up of enzymes, proteins, and different components that may be chemically similar to antibiotics. The analyses of these compounds could be misinterpreted and give rise to false-positives because of a cross-reactivity in the immunoassay (Raykova et al. 2021). For example, some milk proteins or food additives may resemble the molecular structure of the specific antibiotics and thus be wrongly recognized by antibodies in the ELISA (Wang et al. 2020). This makes it necessary to carefully choose antibodies that are highly specific for the target antibiotics to minimize the chance of cross-reactivity with non-target molecules (Feizollahi et al. 2021).

The speed of the antigen–antibody interaction, which is known as the binding kinetics, is also a very important component of the specificity and performance in the ELISAs (Muhammad et al. 2016). The best thing about fast interact is that it makes sure the immunoassay gives the results quite quickly, which is significant for high-throughput testing. On the other hand, a slow and weak interaction between the antibody and the antibiotic can make the test less sensitive and increase probability of incorrect negative results. The following explains examples of ELISA immunosensors used for detecting antibiotics in milk.

The first example presents an innovative approach to enhancing ELISA sensitivity for detecting tylosin, an antibiotic commonly used in veterinary medicine but often found as a residue in food products like milk and honey (Fig. 1a). By integrating a hybridization chain reaction (HCR) with the ELISA system combined with biotin–streptavidin (Bio-SA), researchers achieved a dual signal amplification effect that markedly improves detection sensitivity. The Bio-SA-HCR-enhanced ELISA attained a LOD of 0.39 ng/mL for tylosin, substantially lower than traditional ELISA methods. This approach was validated on real samples of skimmed milk and honey spiked with tylosin, demonstrating its accuracy and suitability for practical applications. Additionally, the method exhibited strong selectivity for tylosin, with minimal cross-reactivity with other antibiotics, such as erythromycin, spiramycin, and seven other interferences, making it a reliable candidate for detecting tylosin even in complex mixtures. When compared to the conventional indirect competitive ELISA, the HCR-ELISA system showed excellent agreement, supporting its potential as a practical and efficient alternative for food safety monitoring of antibiotic residues (Shan et al. 2024).

The research titled "A Bioluminescence Method Based on NLuc-TetR for the Detection of Nine Tetracyclines in Milk" presents a bioluminescent assay for the detection of tetracycline residues in milk, which are considered as potential health hazards (Fig. 1b). The idea of this method revolves around the use of a fusion protein, NLuc-TetR, which consists of the N-terminal of a tetracycline repressor (TetR) and the C-terminal of a nano-luciferase (NLuc). In the situation where tetracyclines attach to TetR, the repressor's ability to hold back NLuc action diminishes, subsequently the bioluminescent response is soon visible. The model is an indirect ELISA (pseudo-ELISA) format, and the study includes the coating competitor BSA-conjugated minocycline for a 96-well plate. It is pointed out in the most important part that the half-maximal inhibitory concentration (IC50) for the quartet of tested tetracyclines is in the range of 0.39 to 1.44 ng/mL which are their key values. The method of the LODs between 2 and 11 pg/mL will give positive results, indicating that the method can detect very low concentration levels of tetracycline in milk. Intra-assay and inter-assay recovery rates for spiked milk samples are, moreover, between 63.7% and 92.2%. Molecular docking research also explains the formation of the complex between NLuc-TetR and tetracyclines and emphasizes the affinity of TetR for tetracyclines, which is the one thing that

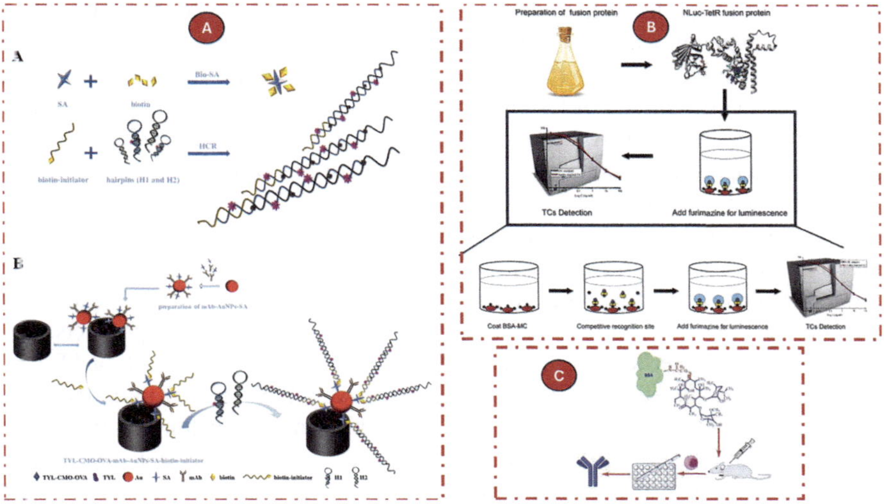

Fig. 1 (**a**) Representation of the use of HCR with biotin-SA signal amplification for tylosin detection. (a) The biotin-SA system is integrated with HCR to enable dual signal amplification. (b) Method for detecting tylosin using HCR combined with biotin-SA signal amplification. (**b**) Illustration of a novel bioluminescence pseudo-ELISA based on NLuc-TetR for the detection of nine tetracyclines in milk; and (**c**) representing a competitive indirect enzyme-linked immunosorbent assay for the determination of erythromycin in milk. (Shan et al. 2024, under the terms of the Creative Commons Attribution-NonCommercial-NoDerivatives License; Wang et al. 2015, License number 5899031484213; Fan et al. 2024, License number 5899040273379)

makes the assay work so well. The NLuc-TetR approach is a big-time innovation in tetracycline milk detection, scaling up the high sensitivity and specificity. This way, the new method is going to make food safer and assure that people will buy only dairy products that are in compliance with the standards set by the regulators, so that finally, consumer safety will really be improved in dairy products (Fan et al. 2024).

Another study presents a reliable method for finding drug residues of erythromycin in milk using a special competitive indirect enzyme-linked immunosorbent assay (ciELISA) (Fig. 1c). Animal medicament erythromycin is widely used, and hence it has the highest possibility of getting food products. For this reason, foodstuff is believed to be contaminated with antibiotics from the treatment of the source animals. The investigation included synthesizing a novel hapten, the carboxy phenyl derivative of erythromycin (ERO-CMO), which was conjugated with the BSA to make immunogens. Finally, hapten structure was proved by high-performance liquid chromatography-tandem mass spectrometry (HPLC-MS/MS). Mice were injected with the hapten to produce monoclonal antibodies (Mabs), resulting in four hybridoma cell lines. The cells developed two specific Mabs, 6C1 and 5B2, which showed the IC50 value of 14.40 and 0.94 μg/L, respectively. These Mabs demonstrated cross-reactivity with 14-membered rings macrolide, but they didn't show interaction with oleandomycin thus ensuring the specificity of the produced Mab. The ciELISA that was developed with 5B2 Mab indicated that there were recovery rates of 76.9%

to 85.7% after tenfold dilution of the milk samples. The LOD for the ciELISA is 0.3 μg/L, which is much lower than the MRLs (40 μg/L) established by the European Union to erythromycin in milk, thus, it is a fast and reliable diagnostic kit for controlling the magnitude of testing. Overall, this study has successfully created a competitive indirect ELISA for the detection of erythromycin in milk, characterizing high sensitivity and specificity, and thereby offering a promising alternative for controlling antibiotic residues in dairy (Wang et al. 2015).

The last example aims to prepare the ELISA-sensitivity immunodiagnostics tests based on prussian blue nanoparticles (PBNPs) for detecting NOR in milk. NOR, a fluoroquinolone antibiotic, poses serious health hazards like antibiotic resistance and allergic reactions through residues in milk and other animal-derived products. The nano-ELISA and the fluorescence-quenching ELISA, which are based on PBNPs, were developed via the study. The nano-ELISA gave a LOD of 0.77 μg/L and IC50 of 18.28 μg/L for NOR detection, while the fluorescence-quenching ELISA had a detection limit of 0.06 μg/L and IC50 of 4.21 μg/L, thus, it was four times more sensitive. Even in real milk samples, the recovery tests were as accurate as the commercial ELISA kits. Based on the result, it was found that both methods are still robust even if they are stored for up to one year at 4 °C. The fluorescence-quenching ELISA, on the other hand, with its higher sensitivity, makes it a great method for the detection of NOR due to signal amplification. The methods are both flexible and scalable compared to the traditional methods and provide more accurate and efficient ways of controlling antibiotic residues in milk, thereby improving food safety (Hu et al. 2025). In the following (Table 2), a few studies around ELISA immunosensors for antibiotic analysis in milk are presented.

3.3 Surface Plasmon Resonance Immunosensors for Antibiotic Detection in Milk

Plasmons are the collective oscillations of free electrons in metals, induced by electromagnetic radiation interacting with fixed positive atomic cores. This interaction leads to charge displacement, causing polarization in the material and generating forward and backward electromagnetic waves known as "bulk plasmons" (Jackson 2007). When the metal is reduced to a thin film, the oscillations become surface waves called SPPs, which propagate at the interface between the metal and a dielectric material. This interface imposes limits on the frequency at which the electrons can oscillate, a phenomenon known as SPR (Yanase et al. 2014).

The surface of the SPR materials, typically a thin metal film (often gold), are functionalized with specific ligands or receptors that can selectively bind to the target analyte. When the analyte interacts with these ligands, it binds to the surface, causing a localized change in the properties of the material interface.

Also, the binding of the analyte alters the refractive index near the metal surface. SPR is highly sensitive to these changes because the surface plasmons' collective oscillations of free electrons at the metal-dielectric interface are influenced by the local refractive index. When an analyte binds to the sensor surface, the refractive

Table 2 Overview of ELISA immunosensors for detecting antibiotic residues in milk

Type of ELISA	Antibiotic target	Sensitivity	Detection method	Reference
Nano-ELISA, Fluorescence Quenching	NOR	LOD:0.06 μg L^{-1}, IC50: 4.21 μg L^{-1}	PBNPs-based ELISA using fluorescence quenching	Hu et al. (2024)
MOF-ELISA	Sulfamethazine (SM2)	LOD: 0.05 ng/mL, IC50: 1.26 ng/mL (NH2-UiO-66)	MOFs-based ELISA with adsorption from milk	Wang et al. (2022)
Multiplexed ELISA	Quinolones, tetracycline, Lincomycin, streptomycin	LOD: 2.42–4.88 μg kg^{-1}, LOQ: 5.36–10.97 μg kg^{-1}	Visual microarray ELISA	Du et al. (2019)
ciELISA, Colloidal Gold Immunochromatographic Assay (GICA)	GEN	LOD (ELISA): 0.54–0.57 μg kg^{-1}, LOD (GICA): 10 μg kg^{-1}	GEN detection in chicken muscle and milk	Wu et al. (2022)
ciELISA	Tylosin	LOD: 0.07 ng/mL, IC50: 1.39 ng/mL, linear range: 0.17–11.0 ng/mL	Tylosin detection in milk and water samples	Yao et al. (2018)
ciELISA	Erythromycin	IC50: 0.94–14.40 μg L^{-1}	ciELISA using against erythromycin	Wang et al. (2015)

index increases, which shifts the angle or wavelength at which SPR occurs (Hiep et al. 2024). As the analyte binds and changes the refractive index at the surface, it causes a measurable shift in the resonance angle (in angle-resolved SPR) or resonance wavelength (in wavelength-resolved SPR). This shift is proportional to the amount of analyte bound to the surface, allowing for real-time, label-free detection.

As the size of metallic particles decreases to the nanoscale, the mobility of electrons is constrained, affecting the material's physical, chemical, and optical properties. When the dimensions of the nanoparticles are much smaller than the wavelength of the incident light, the spatial confinement alters the material's behavior, resulting in a uniform displacement of electrons within the nanoparticles. When exposed to light, the free electrons on the particle surface shift, creating an asymmetric electron cloud distribution. This displacement generates a restoring force between the displaced electrons and positive nuclei, leading to oscillations, which are referred to as localized surface plasmon resonance (LSPR). When the frequency of incident light matches the LSPR frequency, the nanoparticles absorb a significant amount of light, greatly enhancing the electrical field around the particle surface.

If using LSPR, the binding of the analyte causes a local distortion in the electron cloud of the nanoparticle. This distortion changes the plasmonic response, leading to a shift in the LSPR peak, which can be detected optically. LSPR is particularly sensitive to nanoscale changes in the local environment, making it highly effective for detecting small analytes. LSPR is highly sensitive to nanoscale changes and is often used for detecting small molecules or single-molecule interactions (Omair and Talukder 2019).

There are other different methods to use the SPR phenomenon as sensor to analyze materials. Each of these SPR-based methods can be optimized depending on the analyte type, required sensitivity, and specificity, making SPR a versatile tool for various applications in biosensing, environmental monitoring, and diagnostics.

Imaging SPR (iSPR) technique extends conventional SPR by combining it with imaging, allowing for the detection of multiple analytes simultaneously. Different regions of the sensor surface are functionalized with different receptors, enabling the simultaneous monitoring of multiple interactions in a single experiment (Raz et al. 2008).

Electrochemical SPR (EC-SPR) is another type of sensor which combines SPR with electrochemical methods and allows for the simultaneous measurement of optical and electrochemical properties of the analyte. This dual-mode sensing increases the range of detectable analytes and provides additional information about the binding events (Blidar et al. 2019).

Another type of SPR sensor is functionalized SPR Sensors. In this method, the sensor surface is often functionalized with specific receptors, such as antibodies, aptamers, or molecular imprints, to selectively capture specific analytes. This enhances the specificity of the detection process by ensuring that only the target analyte binds to the surface (Yonzon et al. 2004). The combination of the SPR technique and the immunoassay, namely, the SPR immunosensor, has produced unparalleled advantages in the analysis of complex samples, especially in the fields of food analysis (Xia et al. 2017). In 1995, for the first time, Sternesjö, Mellgren, and

Bjorck used SPR immunosensor for detecting an antibiotic in milk (Sternesjo et al. 1995). In this study, SMZ, a sulfamethazine in milk, was analyzed. The detection mechanism involved an antibody-based interaction: a specific anti-SMZ antibody was immobilized on the SPR sensor surface, including carboxymethyl dextran-modified gold film. When milk samples containing SMZ were introduced, the antibiotic molecules bound to these antibodies, causing a measurable change in the refractive index at the sensor surface. This shift was proportional to the concentration of SMZ, allowing for quantitative analysis in real time. The method offered rapid detection without requiring labels or complex sample preparation and the LOD of the method for SMZ in milk was less than 1 ppb. This work marked a significant advancement in food safety testing by enabling a rapid alternative to traditional methods like chromatography. The research showed how SPR biosensors could streamline the detection process, making it easier to monitor antibiotic contamination in dairy products.

SPR-based immunosensors offered a rapid, sensitive, and reliable screening tool for routine monitoring of antibiotic residues in foodstuffs, particularly in dairy products. In 2007, Moeller et al. developed a method that utilizes SPR for the indirect detection of tetracycline residues in milk samples. The assay is based on the interaction between tetracyclines and the Tet repressor protein (TetR), a key player in the resistance mechanism of gram-negative bacteria. In the presence of tetracyclines, these antibiotics bind to TetR, inducing a conformational change that reduces its affinity for the tet operator (tetO), a biotinylated double-stranded DNA sequence immobilized on a streptavidin biosensor chip. During the assay, raw milk samples spiked with tetracyclines are introduced to the sensor. The binding of tetracyclines to TetR results in a measurable change in the SPR signal, which is proportional to the concentration of the antibiotic present. This change allows for the quantification of tetracycline residues down to MRL established by the European Union. The method demonstrates high specificity, as it effectively detects seven commonly used tetracyclines in veterinary medicine without interference from nine antibiotics belonging to five different classes (Moeller et al. 2007).

Fernandez et al. presented an innovative and cost-effective SPR sensor designed for detecting fluoroquinolone (FQ) antibiotics in milk. The detection mechanism relies on the binding of FQ haptenized proteins to specific antibodies on a gold chip activated with a mixed self-assembled monolayer (m-SAM). When fluoroquinolones are present, they compete with the haptenized proteins, leading to a concentration-dependent change in the SPR signal due to variations in the refractive index. This results in a robust LOD of 0.98 μg/L for enrofloxacin in buffer, demonstrating the sensor's sensitivity. One of the key advantages of this biosensor is its reproducibility across multiple channels and different days, ensuring reliable results over time. Additionally, the straightforward sample preparation, which includes fat removal through centrifugation and dilution with water, allows for easy analysis of milk samples (Fernández et al. 2011).

In a competitive assay format used in the SPR biosensors, target antibiotics compete for binding sites on specific antibodies with conjugated analytes. As the concentration of the target antibiotics increases in the sample, the binding of the conjugated analytes decreases, resulting in a measurable change in the SPR signal. This approach allows for the simultaneous detection of multiple antibiotic residues, enhancing the assay's efficiency and throughput. For example, in 2009, Raz et al. developed a microarray biosensor utilizing an imaging surface plasmon resonance (iSPR) platform for the quantitative and simultaneous detection of various antibiotic residues in milk. The biosensor is designed to detect model compounds from four major antibiotic families: aminoglycosides (including Neomycin, GEN, KAN, and Streptomycin), sulfonamides (SMZ), phenicols (Chloramphenicol), and fluoroquinolones (Enrofloxacin), all using a single sensor chip. The detection mechanism employs multiplexing seven immunoassays in a competitive format, allowing for the measurement of target compounds at ppb levels in buffer solutions as well as in diluted milk samples. Notably, the assays for Neomycin, KAN, Streptomycin, Enrofloxacin, and SMZ exhibited sufficient sensitivity to comply with MRLs set by the European Union. The performance of the microarray biosensor was found to be comparable to that of conventional four-channel SPR-based biosensors in terms of sensitivity and robustness. The advantages of the iSPR platform, combined with the efficient immobilization of ligands in a microarray format, enhance the biosensors' sensitivity and specificity. This enables rapid, simultaneous, and label-free detection of multiple antibiotic residues without the need for complex sample preparation. Consequently, this method holds significant promise for automating and streamlining high-throughput food analysis, making it a relevant alternative to existing analytical techniques in monitoring antibiotic residues in food products.

In 2017, Pan et al. developed a label-free SPR immunosensor for detection of harmful enrofloxacin in animal-derived foods, including milk (Pan et al. 2017a). The enrofloxacin-ovalbumin conjugate was immobilized onto the chip (CM500) wrapped with high-density carboxymethylated dextran matrix on the surface, which is covered by a thin gold film (Fig. 2). After loading anti-enrofloxacin antibody, it absorbs on the enrofloxacin under certain conditions and the SPR immunosensor chip gets ready for enrofloxacin analysis. The detection mechanism is such that when the sample containing enrofloxacin enters, the anti-enrofloxacin antibody is separated from enrofloxacin–ovalbumin and causes a decrease in the energy of the reflected light, which creates a specific signal. This technique provided an accurate, sensitive, stable, automated, and high throughput strategy for the detection of enrofloxacin residues in milk samples. The other characteristic of this immunosensor is its reusability at least for 100 times, analyzing by loading anti-enrofloxacin antibody in the chip.

In 2017, Tomassetti et al. developed a direct-flow SPR immunosensor for ampicillin to perform direct, simple, and fast measurements, and the method was compared with a conventional amperometry immunosensor (Tomassetti et al. 2017). In Fig. 3, the SPR flow device is depicted with a primary component, a prism, mounted on a rotating platform. The device operates under automated computer control, with sample solutions introduced into the cell through a peristaltic pump. At the heart of the sensor is a 50 nm thick gold layer, affixed to a glass substrate using an oil that matches the prism's refractive index. The SPR phenomenon is observed when polarized light reflects off the surface and is detected as a function of the resonance

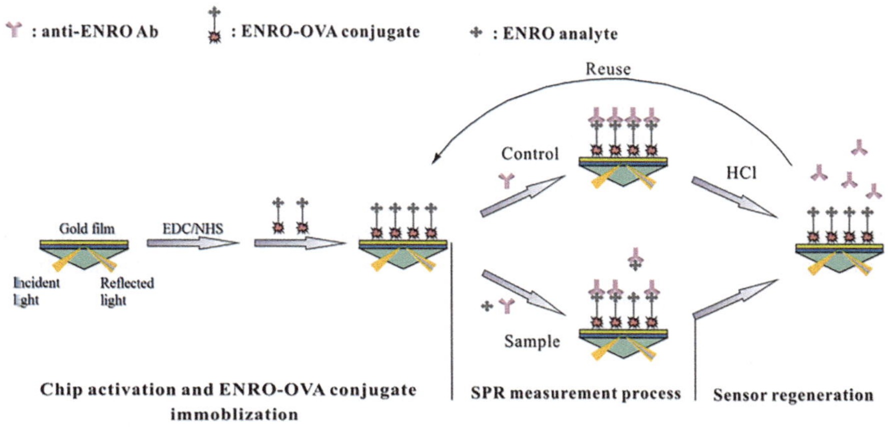

Fig. 2 Schematic diagram of the SPR immunosensor for enrofloxacin detection. Anti -ENRO: anti-enrofloxacin antibody, ENRO-OVA: enrofloxacin–ovalbumin, ENRO: enrofloxacin. (Pan et al. 2017a, under the terms and conditions of the Creative Commons Attribution (CC BY) license)

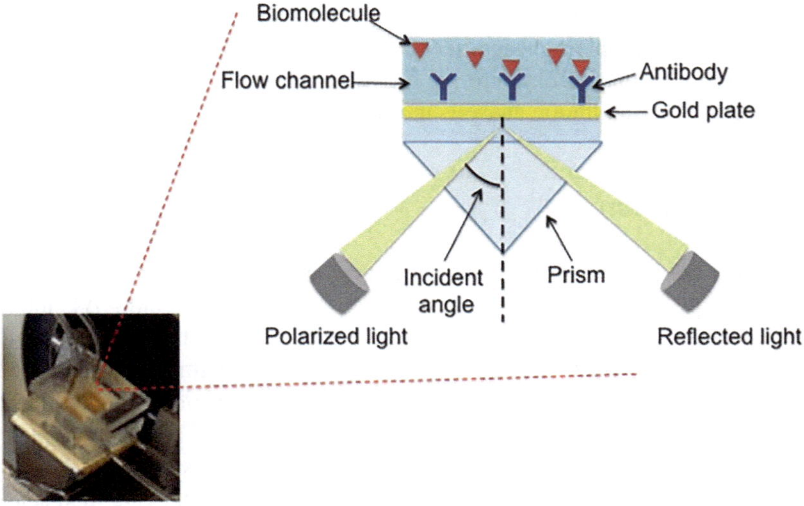

Fig. 3 Measurement device used for SPR operating in flow mode. (Tomassetti et al. 2017, under the terms and conditions of the Creative Commons Attribution (CC BY) license)

angle. When an antibody binds with an analyte, the SPR angle shifts in response to the concentration of the solution. Comparisons between methods indicate that the SPR immunosensor is particularly selective for ampicillin, delivering results with greater ease and speed than those achieved by the traditional competitive immunosensor method.

In recent years, the use of SPR-based immunosensors for antibiotic detection has declined due to several factors, including competition from newer, more practical sensing technologies. Specifically, electrochemical biosensors are gaining popularity because they offer higher sensitivity, faster response times, and better portability for point-of-care applications. These systems, which include aptamer-based and molecularly imprinted sensors, are better suited for the detection of antibiotic residues and can be produced more cost-effectively than SPR setups (Kurç and Türkmen 2022).

Another key reason for this shift is the complexity and cost associated with SPR instrumentation. While SPR offers real-time, label-free detection and kinetic data, it still requires sophisticated optical components, limiting its use outside of specialized laboratories. On the other hand, electrochemical platforms and other optical techniques, such as fluorescence or Raman-based sensors, have become more practical, versatile, and accessible alternatives, especially for monitoring environmental and food matrices (Frigoli et al. 2024).

Moreover, researchers and regulators are increasingly focusing on rapid and inexpensive techniques for tackling antimicrobial resistance (AMR). Electrochemical sensors align with these needs by offering compact and field-deployable solutions for antibiotic monitoring, which is essential in areas like food safety and environmental analysis. Consequently, while SPR remains valuable in research, particularly for high-precision molecular interaction studies, it is being overshadowed by simpler, cheaper technologies better suited for routine antibiotic detection tasks (Chiu 2022). Here, in Table 3, a few studies around SPR-based immunosensors for antibiotic analysis in milk are introduced.

4 Conclusion

This chapter covers one of those critical issues: antibiotic residues present in milk and their implications concerning food safety and public health. As antibiotic resistance becomes an increasing concern, the accurate detection of these residues is essential. This chapter has summarized technology on immunosensors with a specific focus on three main types used for the detection of antibiotics in milk: electrochemical immunosensors, ELISA, and SPR. These principles of immunosensor technology form the backbone of the discussion, from how these sensors rely on antigen–antibody interaction to such sensors showing high sensitivity and specificity. This chapter will discuss each type of immunosensor and its pros and cons regarding ensuring efficacy in the detection of antibiotic residues from complex milk matrices. This chapter further provides information about the challenges faced in this detection process. Additionally, advancements in nanomaterials and signal amplification strategies are discussed, illustrating their potential to enhance sensor performance. In the future, enhancing the detection of antibiotic residues in milk by prioritizing the development and integration of advanced immunosensor technologies is important. This involves investing in research to refine existing immunosensor methods, such as ELISA, SPR, and electrochemical sensors, to improve their sensitivity and specificity for complex milk matrices. Collaboration between academic researchers, industry practitioners,

Table 3 Antibiotics analyzed by SPR based immunosensors, mechanism of the detection, and their analytical information

Analyte	Signal mechanism	LOD (ng/mL)	RSD (%)	Analysis time (min)	Reusability (cycles)	Year	References
Ampicillin	Direct binding SPR	0.1	5.4	20	30	2016	Tomassetti et al. (2016)
Enrofloxacin	Optical Immunosensor	1.2	4.6	6	100	2017	Pan et al. (2017b)
Sulfamethoxazole	Optical Immunosensor	4.3	0.5	30	8	2022	Kumar et al. (2022)
Ampicillin, tetracycline	Fiber-optic SPR sensor	0.01	5.2	15	10	2024	Zhao et al. (2024)
β-Lactam antibiotics	Competitive immunoassay	2.4	–	30	15	2015	Merola et al. (2015)
Sulfonamides	Optical Immunosensor	59.6	3.2–5.2	5	300	2017	Pan et al. (2017a, b)
Fluoroquinolones	Enhanced SPR signal using antibody nanogold probes	0.07	–	33	–	2012	Fernández et al. (2012)

and regulatory agencies will be vital in establishing standardized protocols for routine quality control. Furthermore, incorporating emerging technologies, such as nanomaterials and artificial intelligence, could revolutionize detection processes, enabling real-time monitoring and faster response times to contamination issues. By focusing on these advancements, we can not only meet regulatory requirements but also enhance consumer safety and confidence in dairy products.

References

Abera BD, Ortiz-Gómez I, Shkodra B et al (2021) Laser-induced graphene electrodes modified with a molecularly imprinted polymer for detection of tetracycline in milk and meat. Sensors 22:269

Adrian J, Font H, Diserens J-M et al (2009) Generation of broad specificity antibodies for sulfonamide antibiotics and development of an Enzyme-Linked Immunosorbent Assay (ELISA) for the analysis of milk samples. J Agric Food Chem 57:385–394

Aydin M, Aydin EB, Sezgintürk MK (2021) Advances in immunosensor technology. Adv Clin Chem 102:1–62

Blidar A, Feier B, Tertis M et al (2019) Electrochemical surface plasmon resonance (EC-SPR) aptasensor for ampicillin detection. Anal Bioanal Chem 411:1053–1065

Chaudhary N, Yadav AK, Verma D et al (2024) An electrochemical immunosensor based on a nanostructured lanthanum oxide-substituted reduced graphene oxide interface for ultralow ciprofloxacin detection in milk samples. Mater Adv 5:1597–1613

Chen Y, Zhang J, Liu R, et al (2024) A highly sensitive electrochemical immunosensor based on rGO-PEI-ag nanocomposites for the detection of tilmicosin. Food Chem 140009

Chiu N-F (2022) The current status and future promise of SPR biosensors. Biosensors (Basel) 12

Chomthong K, Kunpatee K, Pimpitak U et al (2024) Label-free simultaneous detection of quinolone antibiotic residues using an origami paper–based electrochemical immunosensor. Sens Actuators B Chem 410:135667

Contreras-Naranjo JE, Aguilar O (2019) Suppressing non-specific binding of proteins onto electrode surfaces in the development of electrochemical immunosensors. Biosensors (Basel) 9:15

de Faria LV, Lisboa TP, Campos N da S et al (2021) Electrochemical methods for the determination of antibiotic residues in milk: a critical review. Anal Chim Acta 1173:338569

Du B, Wen F, Guo X et al (2019) Evaluation of an ELISA-based visualization microarray chip technique for the detection of veterinary antibiotics in milk. Food Control 106:106713

Fan Y, Peng J, Liu J, Wang J (2024) A bioluminescence method based on NLuc-TetR for the detection of nine Tetracyclines in milk. Microchem J 199:109931

Feizollahi A, Rafati AA, Assari P, Joghani RA (2021) Development of an electrochemical sensor for the determination of antibiotic sulfamethazine in cow milk using graphene oxide decorated with Cu–Ag core–shell nanoparticles. Anal Methods 13:910–917

Fernández F, Pinacho DG, Sánchez-Baeza F, Marco MP (2011) Portable surface Plasmon resonance Immunosensor for the detection of fluoroquinolone antibiotic residues in milk. J Agric Food Chem 59:5036–5043

Fernández F, Sánchez-Baeza F, Marco MP (2012) Nanogold probe enhanced surface Plasmon resonance immunosensor for improved detection of antibiotic residues. Biosens Bioelectron 34: 151–158

Frigoli M, Krupa MP, Hooyberghs G et al (2024) Electrochemical sensors for antibiotic detection: a focused review with a brief overview of commercial technologies. Sensors 24:5576

Gaudin V (2021) Contribution of nanomaterials to the development of electrochemical aptasensors for the detection of antimicrobial residues in food products. Chem 9:69

Grubelnik A, Padeste C, Tiefenauer L (2001) Highly sensitive enzyme immunoassays for the detection of β-lactam antibiotics. Food Agric Immunol 13:161–169

Harshavardhan S, Rajadas SE, Vijayakumar KK et al (2019) Electrochemical Immunosensors: working principle, types, scope, applications, and future prospects. Bioelectrochem Interf Eng:343–369

Hiep LT, Thanh KT, Bora T (2024) Surface Plasmon Resonance (SPR) biosensors for antibiotic residue detection. In: Nanoscale matter and principles for sensing and labeling applications. Singapore, Springer Nature Singapore, pp 447–467

Himanshu JK, Lakshmi G, Singh AK, Solanki PR (2024) Reduced graphene oxide-gadolinium oxide-functionalized paper based immunosensor for electrochemical detection of gentamicin. Biosens Bioelectron X 17:100442

Hu G, Yu Q, Zhang Y et al (2024) Development of two novel ELISAs based on Prussian blue nanoparticles for ultrasensitive detection of norfloxacin in milk. Food Chem 141206

Hu G, Yu Q, Zhang Y et al (2025) Development of two novel ELISAs based on Prussian blue nanoparticles for ultrasensitive detection of norfloxacin in milk. Food Chem 463:141206

Li F, Wu Y, Chen D, et al (2021) Sensitive dual-labeled electrochemical aptasensor for simultaneous detection of multi-antibiotics in milk. Int J Hydrogen Energy 46:23301–23309

Jackson JD (2007) Classical electrodynamics, 3rd edn. Wiley, New York

Jayalakshmi K, Paramasivam M, Sasikala M et al (2017) Review on antibiotic residues in animal products and its impact on environments and human health. J Entomol Zool Stud 5:1446–1451

Kim DC, Kang DJ (2008) Molecular recognition and specific interactions for biosensing applications. Sensors 8:6605–6641

Kolhe P, Roberts A, Gandhi S (2023) Fabrication of an ultrasensitive electrochemical immunosensor coupled with biofunctionalized zero-dimensional graphene quantum dots for rapid detection of cephalexin. Food Chem 398:133846

Kumar V, Raghuwanshi SK, Kumar S (2022) Recent advances in carbon nanomaterials based SPR sensor for biomolecules and gas detection—a review. IEEE Sensors J 22:15661–15672

Kurç Ö, Türkmen D (2022) Molecularly imprinted polymers based surface plasmon resonance sensor for sulfamethoxazole detection. Photonic Sens 12. https://doi.org/10.1007/s13320-022-0658-5

Majdinasab M, Mishra RK, Tang X, Marty JL (2020) Detection of antibiotics in food: new achievements in the development of biosensors. TrAC Trends Anal Chem 127:115883

Manasa G, Mascarenhas RJ, Malode SJ, Shetti NP (2022) Graphene-based electrochemical immunosensors for early detection of oncomarker carcinoembryonic antigen. Biosens Bioelectron X 11:100189

Merola G, Martini E, Tomassetti M, Campanella L (2015) Simple and suitable immunosensor for β-lactam antibiotics analysis in real matrixes: milk, serum, urine. J Pharm Biomed Anal 106: 186–196

Moeller N, Mueller-Seitz E, Scholz O, Hillen W, Bergwerff AA, Petz M (2007) A new strategy for the analysis of tetracycline residues in foodstuffs by a surface plasmon resonance biosensor. Euro Food Res Techno 224(3):285–292

Mollarasouli F, Kurbanoglu S, Ozkan SA (2019) The role of electrochemical immunosensors in clinical analysis. Biosensors (Basel) 9:86

Muhammad A, Yusof NA, Hajian R, Abdullah J (2016) Construction of an electrochemical sensor based on carbon nanotubes/gold nanoparticles for trace determination of amoxicillin in bovine milk. Sensors 16

Novakovic Z, Khalife M, Costache V et al (2024) Rapid detection and identification of vancomycin-sensitive bacteria using an electrochemical Apta-sensor. ACS Omega 9:2841–2849

Omair Z, Talukder MA (2019) Sensitivity analysis of gold nanorod biosensors for single molecule detection. Plasmonics 14:1611–1619

Pan M, Li S, Wang J et al (2017a) Development and validation of a reproducible and label-free surface plasmon resonance immunosensor for enrofloxacin detection in animal-derived foods. Sensors 17:1984

Pan M, Wang X, Wang J et al (2017b) Stable and sensitive detection of sulfonamide residues in animal-derived foods using a reproducible surface Plasmon resonance Immunosensor. Food Anal Methods 10:2027–2035

Pollap A, Kochana J (2019) Electrochemical immunosensors for antibiotic detection. Biosensors (Basel) 9:61

Qian W, Zhou J, Chen Y, et al (2023) Label-free electrochemical immunosensor based on staphylococcal protein a and AgNPs-rGO-Nf for sensitive detection of virginiamycin M1. Bioelectrochemistry 153:108489

Raykova MR, Corrigan DK, Holdsworth M et al (2021) Emerging electrochemical sensors for real-time detection of tetracyclines in milk. Biosensors (Basel) 11

Raz SR, Bremer MGEG, Giesbers M, Norde W (2008) Development of a biosensor microarray towards food screening, using imaging surface plasmon resonance. Biosens Bioelectron 24: 552–557

Samanidou V, Nisyriou S (2008) Multi-residue methods for confirmatory determination of antibiotics in milk. J Sep Sci 31:2068–2090

Sanli S (2024) Single-drop electrochemical immunosensor with 3D-printed magnetic attachment for onsite smartphone detection of amoxicillin in raw milk. Food Chem 437:137823

Seth S, Rathinasabapathi P (2022) A short review on detection of antibiotics in milk using nanomaterial-based biosensor. Food Anal Methods 15:2181–2192

Shan S, Lv X, Xiao X et al (2024) Integrated double signal amplification systems with ELISA assay for sensitive detection of tylosin in food. Food Agric Immunol 35:2292971

Sheng W, Li Y, Xu X et al (2011) Enzyme-linked immunosorbent assay and colloidal gold-based immunochromatographic assay for several (fluoro) quinolones in milk. Microchim Acta 173: 307–316

Singh B, Bhat A, Dutta L et al (2023) Electrochemical biosensors for the detection of antibiotics in milk: recent trends and future perspectives. Biosensors (Basel) 13:867

Sternesjo A, Mellgren C, Bjorck L (1995) Determination of sulfamethazine residues in milk by a surface plasmon resonance-based biosensor assay. Anal Biochem 226:175–181

Talib NAA, Salam F, Sulaiman Y (2018) Development of highly sensitive immunosensor for clenbuterol detection by using poly (3, 4-ethylenedioxythiophene)/graphene oxide modified screen-printed carbon electrode. Sensors 18:4324

Tang X, Catanante G, Huang X, et al (2022) Screen-printed electrochemical immunosensor based on a novel nanobody for analyzing aflatoxin M1 in milk. Food Chem 383:132598

Tomassetti M, Conta G, Campanella L et al (2016) A flow SPR immunosensor based on a sandwich direct method. Biosensors (Basel):6

Tomassetti M, Merola G, Martini E et al (2017) Comparison between a direct-flow SPR immunosensor for ampicillin and a competitive conventional amperometric device: analytical features and possible applications to real samples. Sensors 17:819

Wang W, Liu L, Xu L et al (2013) Detection of β-lactamase residues in milk by sandwich ELISA. Int J Environ Res Public Health 10:2688–2698

Wang Z, Mi T, Beier RC et al (2015) Hapten synthesis, monoclonal antibody production and development of a competitive indirect enzyme-linked immunosorbent assay for erythromycin in milk. Food Chem 171:98–107

Wang L, Zhang L, Wang Y et al (2020) Construction of an electrochemical receptor sensor based on graphene/thionine for the sensitive determination of β-lactam antibiotics content in milk. Int J Mol Sci 21:3306

Wang X, Li J, Jian D et al (2021a) Paper-based antibiotic sensor (PAS) relying on colorimetric indirect competitive enzyme-linked immunosorbent assay for quantitative tetracycline and chloramphenicol detection. Sens Actuators B Chem 329:129173

Wang Q, Xue Q, Chen T et al (2021b) Recent advances in electrochemical sensors for antibiotics and their applications. Chin Chem Lett 32:609–619

Wang S, Wang Z, Zhang L et al (2022) Adsorption and convenient ELISA detection of sulfamethazine in milk based on MOFs pretreatment. Food Chem 374:131712

Wang L, Wang H, Huang S et al (2023) Electrochemical sensor for detecting streptomycin in milk based on label-free aptamer chain and magnetic adsorption. Food Chem 403:134399

Waritani T, Chang J, McKinney B, Terato K (2017) An ELISA protocol to improve the accuracy and reliability of serological antibody assays. MethodsX 4:153–165

Wu S-M, Yan C-X, Fan X-H et al (2022) Development of enzyme-linked immunosorbent assay and colloidal gold-based immunochromatographic assay for the rapid detection of gentamicin in chicken muscle and milk. Chin J Anal Chem 50:100142

Xia Y, Su R, Huang R et al (2017) Design of elution strategy for simultaneous detection of chloramphenicol and gentamicin in complex samples using surface plasmon resonance. Biosens Bioelectron 92:266–272

Yanase Y, Hiragun T, Ishii K et al (2014) Surface plasmon resonance for cell-based clinical diagnosis. Sensors 14:4948–4959

Yao C-Y, Yang J-Y, Xu Z-L et al (2018) Indirect competitive enzyme-linked immunosorbent assay for detection of Tylosin in milk and water samples. Chin J Anal Chem 46:1275–1281

Yonzon CR, Jeoung E, Zou S et al (2004) A comparative analysis of localized and propagating surface Plasmon resonance sensors: the binding of Concanavalin a to a monosaccharide functionalized self-assembled monolayer. J Am Chem Soc 126:12669–12676

Yu W, Sang Y, Wang T, et al (2020) Electrochemical immunosensor based on carboxylated single-walled carbon nanotube-chitosan functional layer for the detection of cephalexin. Food Sci Nutr 8:1001–1011

Zhang M, Zhang B, Li T et al (2022) Electrochemical detection of aminoglycoside antibiotics residuals in milk based on magnetic molecularly imprinted particles and metal ions. Food Chem 389:133120

Zhao Z, Yin H, Xiao J et al (2024) Efficient sequential detection of two antibiotics using a fiber-optic surface Plasmon resonance sensor. Sensors 24

Zhou Y, Zhou J, Liu Y, et al (2024) A Novel Label-Free Electrochemical Immunosensor Based on Ce-MOF@ AgAuNPs for Highly Sensitive Detection of Monensin. Food Control 110927

Role of Immunosensors in the Detection of Food-Derived Allergens

Deepali Chaudhary, Shagun Gupta, Rupak Nagraik, and Ankur Kaushal

Abstract

Food allergies are increasing and pose a serious risk to sensitive individuals. This allergy affects mostly underaged children and a small percentage of adults. There is no effective immune therapeutic approach for detecting the presence and protecting susceptible individuals. Conventional approaches such as PCR, ELISA, and chromatography have certain drawbacks, such as their high cost and time-consuming nature and the need for well-skilled individuals. By combining biosensors and nanoparticles, these methods have become more affordable, sensitive, selective, and time-efficient and can replace traditional methods. Food allergies are generally immune responses triggered by contact with specific foods. It is crucial to detect food allergens effectively and practically by monitoring food allergens via a point-of-care biosensor. An immunosensor is an antibody–antigen interaction-based method in which a specific allergen can be detected. In the immunosensor, the antibody–antigen reaction helps to find different allergens in foods such as soybean, peanuts, cow milk, cereal, eggs, and crustaceans. In this chapter, we discuss the different types of allergen-specific antibodies, such as monoclonal antibodies, polyclonal antibodies, and camelid-derived nanobodies, as well as allergens present in various foods. Owing to the efficiency, user-friendliness, and simplicity of immunosensors, they can be used in various foods to detect allergens.

D. Chaudhary · S. Gupta · A. Kaushal (✉)
Department of Bio-Sciences and Technology, Maharishi Markandeshwar (Deemed to be University), Mullana, Ambala, India

R. Nagraik
Department of Biotechnology, Graphic Era (Deemed to be University), Dehradun, Uttarakhand, India

K. Kuča et al. (eds.), *Ensuring Wellness: Immunosensors Monitoring Food Quality and Pathogenicity Risks*, https://doi.org/10.1007/978-3-031-97477-9_10

Keywords

Immunosensor · Food allergens · Antigen–antibody interaction · Food allergies · Nanoparticles

1 Introduction

Five to ten percent of people in wealthy countries suffer from food allergies, which is a growing global public health concern (Pepper et al. 2020). Food allergies are a major health problem since they have become more common in recent years, impacting both consumers and the food industry. According to the law, 14 foods have been classified as allergenic and linked to allergy reactions (Patel et al. 2021; Sicherer and Sampson 2018). The allergen profile and composition of food can be influenced by a number of factors, such as processing, postharvest treatments, climate, cultivar, and growth circumstances (Kurze et al. 2018; Pasquariello et al. 2012). Food processing, in particular, frequently results in a decrease in allergenicity (Bousfiha Bousfiha and Lotfi 2013; Pi et al. 2022), while there have also been reports of no change or even an increase in the quality of foods such as peanuts (Schmitt et al. 2010; Maleki et al. 2000) and fish (Jiang and Rao 2021). Therefore, meals with reduced sensitizing potential and less allergenicity can be obtained by combining various components and processing techniques (Foo and Mueller 2021). A subpopulation of patients with allergies can consume these foods. To reduce the allergenicity of foods, researchers are combining novel and traditional methods (Cheng et al. 2022; Pali-Schöll et al. 2019). For quantification, the method employed must be checked in terms of its effectiveness for processing food, and allergenic food assessment in specific foods is critical (Thomas et al. 2007). The authenticity of food can be checked via various techniques, which rely on the method chosen (Pali-Schöll et al. 2019).

The food allergen Labelling and Consumer Protection Act of 2024 requires food manufacturers to mention any food allergens that are significant on the labels (FDA 2004). In recent years, immunoassays have been the preferred method for detecting and identifying food allergies. The methods of enzyme-linked immunosorbent assay (ELISA), rocket immuno-electrophoresis, and specific immuno-blotting (Andjelković et al. 2017) have been used to determine the proteins in food products, including cow milk (Ren et al. 2014), egg white (Dhanapala et al. 2017), hazelnut (Grishina et al. 2017), peanut (Lexmaulová et al. 2013) soybean (Dhanapala et al. 2017), and wheat (Ji et al. 2017).

Consuming food allergens, coming into contact with them on the skin, or being exposed at work can cause anything from mild urticaria to potentially fatal anaphylactic shock (Abrams and Sicherer 2016; van Ree 2021). Following sensitization, the body releases cells, and the immune process assesses whether food allergies occur and the extent of the damage they inflict. Organs and tissues subsequently counteract these intracellular releases (Morais et al. 2020). When sensitive people consume food containing allergenic proteins, their immune systems develop a pathological reaction known as a food allergy. On the basis of the role of immunoglobulin E (IgE), the mechanism can be divided into three categories: IgE-mediated, non-IgE-mediated (cell-mediated), or a mix of the two (Alessandri et al. 2020). IgE-mediated type I hypersensitivity, ΆΆ immune complex hypersensitivity, ΆΆΆ cytotoxicity hypersensitivity, and ΆV T-cell-mediated hypersensitivity are the four types of food allergies (Baldo and Pham 2021).

The fact that lateral flow assays usually produce only qualitative or semi quantitative results is one of their disadvantages. The detection of target sequences due to the potential for cross-contamination during analysis and significant matrix effects are the main drawbacks of using the PCR method for allergen analysis. Additionally, the use of general sample extraction techniques may result in false-positive results. Owing to their high sensitivity, sensitivity, and reproducibility as well as their practical analytical performance, immuno analytic procedures such as ELISA are the most often used and offer both qualitative and quantitative analyses (Khedri et al. 2018; Wammes et al. 2015). An analytical method that combines electrochemical sensors with antibody antigens is called an electrochemical immunosensor (Aydin et al. 2021). This technique uses immunoglobulins and target molecules for specific binding. This chapter discusses current developments in food allergen detection techniques using immunosensors. It presents what we currently know about food allergens, the difficulties in applying them, contemporary detection techniques, and comprehension.

2 Food Allergies

Food allergens are significant food safety solicitude in developed nations, and avoiding them from one's diet is necessary rather than waiting for effective treatment to stop an allergic reaction. Allergy-induced immune responses differ from those resulting from poisoning, pharmaceutical reactions, and dietary intolerance. The "big eight"—soybeans, milk, eggs, groundnuts, shellfish, tree nuts, cereals, and fish—are the most significant allergic foods (Krisnawati et al. 2022). In general, proteins (allergens) found in nine different food groups, such as eggs, milk, tree nuts, and peanuts, are responsible for approximately 90% of all severe food allergies. Table 1 summarizes the conventional antibodies generated for immunoassay development, and the food allergens recognized are a matter of concern. Figure 1 presents various allergic food materials.

2.1 Milk

According to Nehra et al. (2019), a milk allergy is a common food allergy that can persist for a lifetime. A major public health concern that requires attention is milk allergy, particularly in children. Food allergies, such as milk allergies, are immunological responses to food ingredients, including proteins that result in the production of IgE (Krisnawati et al. 2022). According to Ashley et al. (2018), the human body

Table 1 Immunoassays against food allergens

S. N.	Food	Allergens	Antibody	Detection limit and range	References
1	Egg	Ovomucoid	mAb-based sELISA	0.041 ng/mL and 0.1–6.25 ng/mL	Li et al. (2020)
		Ovalbumin	mAb-based sELISA	0.51 ng/mL and 1.95–500 ng/mL	Peng et al. (2012)
		Ovotransferrin	mAb-based biosensors	23.55 ng/mL and 50–3800 ng/mL	Wang et al. (2018)
		Lysozyme	mAb-based iELISAb	0.264 μg/mL and 0.38–4.8 μg/mL	Vidal et al. (2005); Qiu and Wang (2017)
2	Milk	α-Lactalbumin	mAb-based iELISAb	0.1 ng/mL and 10–500 ng/mL	Jeanson et al. (1999)
		β-Lactoglobulin	mAb-based cELISAc	114 ng/mL and 78–10,000 ng/mL	Orcajo et al. (2019); Hu et al. (2021)
		Casein	mAb-based sELISAa; pAb-based sELISAa	0.04 μg/mL and 0.1–10 μg/mL	Zhou et al. (2013)
3	Peanuts	Ara h 1	mAb-based sELISAa	10 ng/mL	Holden et al. (2005); Hu et al. (2023)
		Ara h 2	mAb-based cELISAc	0.5 ng/mL	Schmitt et al. (2004)
		Ara h 3	mAb-based sELISAa	382–12,676 μg/g	Pandey et al. (2019); Hu et al. (2023)
4	Shrimp	Tropomyosin, TMP	mAb-based sELISAa; pAb-based sELISAa	0.64 ng/mL and 0.08–512 ng/mL	Zhang et al. (2014); Jiao et al. (2024)
		Arginine kinase, AK	mAb-based biosensors	0.11 ng/mL and 1–1000 ng/mL	Wang et al. (2019)
		Sarcoplasmic calcium binding protein, SCP	mAb-based biosensors	0.5 μg/mL	Huang et al. (2024)
5	Tree nut	Lupine protein (Lup an 1)	pAb-based sELISAa	1 μg/g	Holden et al. (2005); Hu et al. (2021)
6	Soyabean	β-Conglycinin, 7S	mAb-based sELISAa;	1.63 ng/mL and 3–100 ng/mL	Hei et al. (2012)

(continued)

Table 1 (continued)

S. N.	Food	Allergens	Antibody	Detection limit and range	References
			pAb-based sELISAa		
		Glycinin, 11S	mAb-based sELISAa; pAb-based sELISAa	1.63 ng/mL and 3–200 ng/mL	Chen et al. (2014)
		Gly m Bd 28 K	pAb-based iELISAb	80–2000 ng/mL	Liu et al. (2013)
		Gly m Bd 30 K	pAb-based sELISAa	0.47 ng/mL	Morishita et al. (2008)
7	Fish	Parvalbumin protein	pAb-based cELISAc	0.59 µg/mL and 0.59–150 µg/mL	Jiang and Rao (2021)
8	Wheat	Gliadin	mAb-based sELISAa	4 ng/mL and 4–40 ng/mL	Blázquez-García et al. (2022); Holden et al. (2005)
		Glutenin	mAb-based cELISAc	25 ng/mL	Tranquet et al. (2015)

Fig. 1 Demonstration of allergenic foods for humans

produces cytokines, histamine, and other chemicals that can result in severe reactions, including anaphylactic shock and other allergic symptoms, such as inflammation.

2.2 Peanut

Allergens to peanuts (*Arachis hypogaea*) have a covert detrimental effect on the health and well-being of millions of people worldwide (Palladino and Breiteneder 2018). According to Li et al. (2020), peanuts are a healthy food type since they contain all the essential components, including protein, carbohydrates, fats, vitamins, and minerals. Among the "top eight food allergies," beans are also a general food that can cause severe food allergies. The incidence of this disease is between 0.5% and 2% worldwide and is continuously increasing (Hilu et al. 2019). The number of pediatric peanut allergy cases that are documented continues to rise. In the United States, the prevalence of peanut allergies has risen by 21% since 2010 and is predicted to continue rising (Cannon 2018). Plant-based food allergies can lead to life-threatening illnesses and are very complex disorders with minor symptoms in the skin, gastrointestinal tract, and mouth. The different types of allergenic compounds have been blamed for the heterogeneity in their clinical characteristics (Meng et al. 2019).

However, approximately 90% of food allergy cases are caused by foods derived from plants (Breiteneder and Radauer 2004; Radauer et al. 2008; Burney et al. 2014). With symptoms ranging from oral irritation to life-threatening allergic responses, cashew nut allergies are becoming more common (Archila et al. 2016; Bastiaan-Net et al. 2020). Nut and peanut allergies are typically life-threatening and lifelong. The majority of adult food allergies are caused by a limited number of allergens (Sicherer and Sampson 2010). According to reports, 4.2% of children with IgE-mediated food allergies have a walnut allergy (Crespo et al. 1995). Twenty percent of tree nut allergy participants have a cashew nut allergy, making it the second most common type of allergy (Sicherer et al. 2001).

2.3 Shellfish

The allergenic protein tropomyosin, found in the muscles of shellfish, is more stable to thermal denaturation than nonallergic substances and is more resistant to peptic acidic digestion, increasing the risk of sensitization. Among all shellfish, i.e., crab, lobster, shrimp, and bivalves, 80% of allergic reactions are linked to shrimp. From 2014 to 2018, the amount of seafood consumed per person (freshwater and marine fish and shellfish) rose from approximately 19.9 kg to 20.5 kg (Torre et al. 2020). By improving food production while considering risk assessment, hazard inspection of good manufacturing practices, and key control points, the detection of traces of tropomyosin can benefit both consumers and producers. In a study conducted by the authors, the devised immunosensor was able to determine traces of tropomyosin

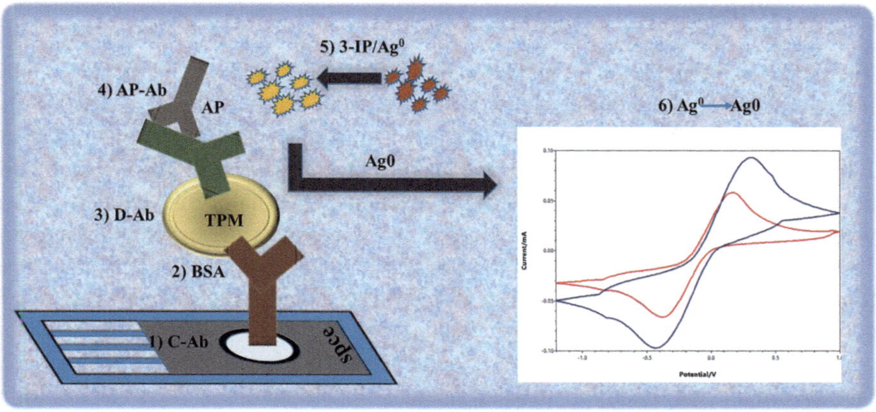

Fig. 2 Schematic illustration of the optimized construction of the immunoassay configuration, immunosensor sensing phase, enzymatic reaction, immunosensor sensing phase and electrochemical detection. (1) C-Ab immobilization, (2) BSA for blocking, (3) incubation of TPM + D-Ab, (4) addition of AP-Ab, (5) enzymatic reaction addition and voltammetric analysis

with accuracy and precision, and the limit of detection was 0.47 ng/mL. It is highly specific and selective, and the developed immunosensor has a stability of 30 days (Torre et al. 2022) (Fig. 2).

Owing to the similarity of allergen proteins from different invertebrates, it is crucial to diagnose shellfish allergies, especially for shrimp (Nugraha et al. 2019). Approximately 10% of children suffer from some form of food allergy, and the prevalence of food allergies has been steadily increasing in recent years. It has been proven that sensitized individual activated B cells produce specific IgE antibodies in response to ingesting or inhaling shellfish allergens. The clinical signs that have been observed during allergy and degranulation are caused by IgE antibodies conjugated with immune effector cells such as basophils and mast cells (Bastiaan-Net et al. 2020). There are several pathways that are responsible for causing shrimp allergens, such as the inhalation of shrimp particles; the organs affected by this allergy are the skin (52–90%), 42% of the respiratory tract, 35% of the digestive tract, and 10% of the cardiovascular system (Burney et al. 2014). The enzymes arginine kinase (AK) and tropomyosin (TMP) are the two most important muscle protein allergens that cause shellfish allergies. Furthermore, the biomarker IgE antibody is important in the allergens tropomyosin and shellfish and can detect 60–80% of patients with shellfish allergies (Jarupalee et al. 2018). Because fish and shellfish allergies are so common, a suitable theranostic approach that combines acceptable management and prevention strategies to remove redundant dietary confinement must be developed. Clinical examination, open or closed oral meal challenge, the skin prick test (SPT), and serum-specific IgE (sIgE) detection are examples of traditional clinical diagnostic techniques (Tong et al. 2018).

2.4 Wheat

The most prevalent ingredient in the foods we eat every day is wheat. Carbohydrates, lipids, cellulose, minerals, proteins, vitamins, and water are only a few of the essential nutritional components found in wheat (Pahr et al. 2014). According to Juhász et al. (2018), wheat is one of the main foods that causes allergies, resulting in both local symptoms and severe systemic reactions. Over half of the children with a wheat allergy suffer from anaphylaxis after ingesting wheat. According to a study of 1000 FA patients, wheat is one of the primary sources of food allergens that cause food-induced anaphylaxis (Frigerio et al. 2019). As a major staple cereal grain that is consumed worldwide, wheat is a great source of nutrients for people. Some wheat constituents, especially proteins, are actually linked to a few significant health issues that have impacted quality of life and patients' health and have the potential to be fatal for a tiny percentage of the population.

2.5 Eggs

Up to 9% of children worldwide often suffer from hypersensitivity to Gallus (chicken) eggs (Rona et al. 2007; Wei-Liang Tan et al. 2017). Records show that it is one of the most prevalent food allergies in children (Savage et al. 2007). According to a 2016 report from the Beating Egg Allergy Trial (BEAT), the primary cause of food allergies linked to IgE in Australian children is egg allergy. The Health Nut study also looks at this. According to a cohort study, the prevalence of egg allergy in one-year-old infants is 9.5%, which is higher than that of other major FAs, such as milk (1.5%) and peanuts (3.1%) (Dang et al. 2019; Peters et al. 2017). Although 60–75% of teenagers spontaneously resolves and tolerates egg allergies, the disease is more prevalent in early infancy because of symptoms such as urticaria, diarrhea, vomiting, and abdominal pain (Palmer et al. 2013; de Maat-Bleeker et al. 1985). Although egg allergy is thought to be a childhood illness, some allergic children continue to have egg allergies as adults do, increasing the risk of latent deadly reactions (Jarupalee et al. 2018).

Approximately 2% of people worldwide suffer from shellfish allergies, which often last a lifetime (Dubiela et al. 2018). Owing to the similarity of allergen proteins from different invertebrates, it is crucial to diagnose shellfish allergies, especially for shrimp (Nugraha et al. 2019). Approximately 10% of children have some form of FA, and the incidence of FAs has been steadily increasing in recent years. Exposure to shellfish allergens through ingestion or inhalation has been shown to trigger the production of certain IgE antibodies from activated B cells of the immune system in sensitized individuals. To cause degranulation and clinical manifestations of allergy symptoms, these IgE antibodies further conjugate with immune effector cells such as mast cells and basophils (Bastiaan-Net et al. 2020). Shrimp allergens can cause reactions through a variety of pathways, such as the inhalation of shrimp particles. The skin (52–90%), respiratory tract (42%), digestive tract (35%), and cardiovascular system (10% anaphylaxis) are among the organs affected by shrimp allergy symptoms (Burney et al. 2014). The two muscle protein allergens that are most significant for shellfish allergies are tropomyosin (TMP) and the enzyme arginine kinase (AK). Additionally, a significant allergen in shellfish, tropomyosin, contains a particular IgE antibody that has been shown to be a reliable biomarker for severe clinical reactions to shellfish and to explain the reactions of approximately 60–80% of patients with shellfish allergies (Jarupalee et al. 2018). The development of suitable theranostic protocols is needed to provide acceptable preventative and management techniques to eliminate needless dietary constraints, given the high incidence of allergies to fish and shellfish. Clinical examination, open or closed oral meal challenge, the skin prick test (SPT), and serum-specific IgE (sIgE) detection are examples of traditional clinical diagnostic techniques (Tong et al. 2018).

3 The Requirement to Detect Food Allergies

The Codex Committee on Food Labelling has been considering allergens in foods since 1993 (reference: CX/FL 93/5). The Food and Agriculture Organization (FAO) Technical Consultation on Food Allergies, held in Rome in 1995, was asked to "provide guidance on the development of science-based criteria to determine which foods or food products should be placed on a list of those foods or food products whose presence should always be declared in the list of ingredients on a food label, because of their allergenic properties" (Scherf and Poms 2016). The Codex Alimentarius recognizes that the following foods and ingredients are known to induce hypersensitivity and must always be disclosed. Gluten-containing cereals include wheat, rye, barley, oats, spelt, and their hybridized strains, as well as crustacea, eggs, fish, peanuts, soybeans, milk, and milk products (Sugawara et al. 2017; Dong et al. 2017; Zheng et al. 2016; Alves et al. 2016; Alves et al. 2017; Sugawara et al. 2018; Ruiz-Valdepeñas Montiel et al. 2016).

Peanuts are a major source of allergies in adults, with 12 allergens identified to date (Krishna et al. 2018). Peanut allergy is one of the most common allergic reactions and the leading cause of anaphylaxis death globally (Sobhan et al. 2018a). According to the FAO/WHO CODEX (Scherf and Poms 2016), only gliadin is currently regulated in gluten-free products. As a result, very sensitive methods must be developed to notify consumers of specific allergens in foods.

4 Role of the Immunosensor

Food allergen biosensors are vital for protecting people with food allergies and guaranteeing food safety; therefore, testing biosensors with food models is necessary for both development and validation. Foods are complex matrices that consist of carbohydrates, proteins, trace elements such as vitamins and polyphenols, and lipids (Fig. 3). These matrices impact the selectivity, sensitivity, and response time. This is a common method for allergy monitoring in food matrices to check for allergenic

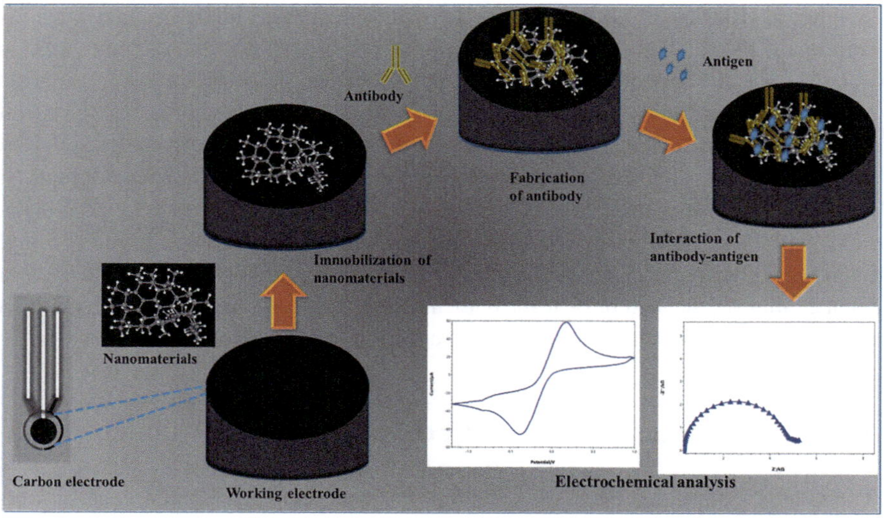

Fig. 3 Schematic representations of the development of an electrochemical immunosensor

ingredients. This technique validates the dependability and resilience of sensors in food matrices.

The validation and development of food allergen biosensors depend on testing real food samples, as they are vital for individuals to detect food allergies and ensure their safety via early diagnosis (Dias et al. 2024). One typical tactic in the development of systems for allergy monitoring is the intentional insertion of allergenic substances into food matrices to simulate real-life situations. This strategy allows the performance of biosensors to be assessed in the presence of food matrix components, verifying their robustness and reliability.

Biosensor applications, which focus on soybean allergenic proteins, are studied in this chapter. Singh and other authors have studied many antibodies that bind to surface plasmon resonance to detect soybean allergens accurately and rapidly (Singh et al. 2014). In another study, the iSPR technique was used with microfluidic platforms made with monoclonal anti-soybean antibodies (mAb 3G12) that were evaluated in food matrices for the quantification of soybean. The limit of detection (LOD) and limit of quantification (LQD) were 5 and 14.8 mg kg^{-1}, respectively, for soybean in biscuits. It has shown good sensitivity and rapid detection of allergens (Rebe Raz et al. 2010). Furthermore, on silica nanoparticles, anti-soybean antibodies were mounted, and soybean proteins were targeted to generate a long-wavelength fluoro-immunoassay (Godoy-Navajas et al. 2011). An LOD and linear dynamic range of 0.05 mg L^{-1} and 0.1–10 mg L^{-1}, respectively, were observed in liquid food samples of soybean proteins such as soybean milk, yogurt, or fruit juices/pulp. The results reported ranged from 2700 to 113,000 mg kg-1 or 24,000 to 123,000 mg L-1. Research on the creation of electrochemical bio platforms for the detection/quantification of soybean in food matrices on the basis of the identification of two allergenic proteins, β-conglycinin and glycinin, was published by Blazquez-García et al. in 2022. The suitability of the bioplatforms was evaluated in cookies and dough with varying proportions of soybean flour, yielding an LOD of 0.0005 mg kg-1 soybean protein in both matrices.

5 Classical Analytical Methods Generally, Used for Allergen Detection in Foods

Food allergies have been found via a variety of techniques. They consist of immunological assays and analyses based on genetic and mass spectrometry (Thomas et al. 2007; Alves et al. 2015). DNA-based methods, biosensors technology, mass spectrometry-based methods, ELISA, LFIA, Nb-based Immunoassays for food allergen surveillance.

5.1 DNA-Based Methods

Specific allergens in foods can be detected by targeting DNA that only codes for an allergic protein or indicates the presence of an allergenic source. After that, suitable polynucleotide probes were developed, and DNA was quantified via polymerase chain reaction (Madrid et al. 2021; Prado et al. 2016). This molecular-based technique is highly specific and sensitive. This DNA-based method also imparts accuracy to prepared food products. The PCR results are dependent on the heat treatment and processing methods used (Platteau et al. 2011; Mano et al. 2017). This eliminates the substance concentrations and health risks indicated on the label (Iniesto et al. 2013; Hildebrandt and Garber 2010).

5.2 Biosensing Technology

A biosensor uses signals that are proportional to the amount of analyte in the reaction to detect biological or chemical processes (Hammond et al. 2016). A biological receptor-antibody, enzyme, nucleic acid, cell, or aptamer that interacts with (2) a particular analyte and (3) a transducer that transforms the biorecognition event into a measurable signal (Campuzano et al. 2020; Alves et al. 2015). Biosensors are classified into three types on the basis of the transducer they use: optical, electrochemical, and piezoelectric. According to Freitas et al. (2021), a biosensor can detect an allergen in a sample when the biological receptor is a particular antibody and the analyte is an allergen that the antibody recognizes. Lactoglobulin in dairy products is one of several individual allergens that biosensors have shown to have high sensitivity and specificity in detecting in food samples, such as peanut Ara h 1 and Ara h 6 in commercially processed foods (Jiang et al. 2021; Sobhan et al. 2018b), lactoglobulin in dairy products (Sun et al. 2020), hazelnut Cor a 14 (Costa et al. 2021) and Sin a 1 in mustard seeds (Gamella et al. 2020). Biosensors are rapid, sensitive, specific, and simple. Biosensors can help in the detection of an allergy in the case of processed foods for specific antibodies. The only drawback of this technology is that it can analyze one allergen per test.

5.3 Mass Spectrometry-Based Methods

To detect proteins in food extracts, a number of mass spectrometry-based methods and strategies can be used (Yakhlef et al. 2021; González-Buitrago et al. 2007; Stanly et al. 2021). Additionally, allergenic protein traces can be found and measured in processed foods (Kirsch et al. 2009; Perner et al. 2019). Bottom-up proteomics is a popular and commonly used method. The design and assessment of "peptide mass fingerprinting" is its foundation; this is achieved by splitting down proteins with a specific protease, often trypsin, and then sorting and identifying the fragments via the proper mass spectrometer. After that, software is used to search pertinent protein-sequence databases via mass patterns. Several allergens and proteins found in the extract sample can be identified. Mass spectrometry is commonly used for the identification of denatured or fragmented proteins. To separate extracts prior to proteolytic digestion, SDS-PAGE is utilized (Yakhlef et al. 2021). This process enables ideal extraction conditions, including the presence of reducing and denaturing agents and extreme pH. The harsh conditions make it easier to extract denatured and insoluble proteins (Röder et al. 2021). Therefore, extracts used for mass spectrometry analysis might have higher protein content than those used for other studies, such as those involving immunochemical techniques. This method has the drawback that the protein sequences searched are not present in the database. Only proteins whose same sequence is contained in the database can be identified by mass spectrometry.

5.4 ELISA

The most commonly used immunoassay for identifying allergens in food is ELISA, which is based on the detection of allergens by a specific antibody, usually IgG (Engvall and Perlmann 1972). Whether a sample contains antigens or particular antibodies can be determined with this test. It is therefore used to identify allergens in food and antibodies in serum. ELISAs have shown excellent potential for sensitivity, specificity, and standardization (Baumert 2014; Immer and Lacorn 2015). Allergen-specific monoclonal and/or polyclonal antibodies are necessary for this experiment. The two methods for performing ELISA described by Madrid et al. (2021) are "sandwich ELISA" and "direct ELISA." By using direct ELISA, it is also possible to detect the presence of IgE antibodies in the serum of allergic patients (Kadooka et al. 2000). A particular IgE functions as the principal antibody in this instance, identifying the immobilized allergen, and then a secondary labeled antibody is used for identification. The sandwich ELISA is performed by adsorbing the antigen-specific antibody into the wells (Schmidt et al. 2012). This IgG antibody will bind to the antigen in the sample under test, which may be a food extract (Lacorn et al. 2022; Koppelman et al. 2021). A primary antibody, typically IgG, that is specific to the allergen being studied is then injected and binds to the antigen.

5.5 LFIA

A lateral flow immuno chromatographic test, sometimes referred to as a lateral flow immunoassay (LFIA), is used to identify how much a chemical is present in a food sample. A band next to a control is formed via a strip containing IgG antibodies that are specific to the target molecule. Upon loading the strip properly, the liquid sample, possibly a food extract, passively flows down it. Affinity chromatography states that if the allergen under investigation exists in the sample, it adheres to the specific antibody. A reporter molecule will release a signal when the target allergen is present in the sample, indicating the development of the antigen–antibody complex. The strip may consist of a single lane (singleplex) or numerous lanes. Although LFIA is faster than ELISA and has been used successfully to detect allergens in food samples, it has similar benefits and limitations to the ELISA technique (Anfossi et al. 2019; Civera et al. 2022).

5.6 Nb-Based Immunoassays for Food Allergen Surveillance

A food allergy is an abnormal immunological reaction that occurs repeatedly after consuming or being exposed to a specific food item. Currently, the basic foundations of established food Allergen detection approaches involve gene- and protein-level determination (Ivens et al. 2016; Tsakali et al. 2019). The majority of gene-level determination techniques, such as real-time fluorescence quantitative PCR and PCR, are unable to reliably determine if allergen proteins are present or missing, and they appear to be useless for dietary allergens with high protein or low DNA content. Furthermore, processing may yield questionable results (Campuzano et al. 2017; Eischeid et al. 2021). Chromatography and immunoassays are the principal technologies used to detect allergen proteins (Flicker et al. 2020). For in situ use, antibody-based immunoassays, particularly mAb-based detection approaches, are the most successful (Sena-Torralba et al. 2020). A number of conventional antibody-based identification methods have been created for food allergy analysis; however, these techniques have several limitations, such as high cost and labor, batch differences during production or manipulation, and significant cross-reactivity with different food allergen proteins.

This chapter discusses the prevalence of allergies caused by the six primary food ingredients—peanuts, milk, shellfish, wheat, nuts, and eggs. Additionally, the immunological response and how to treat it during allergic reactions were investigated. We believe that this chapter can help people better understand food allergies. There are also strategies to avoid food-induced allergies, which include screening children for allergens early on to avoid allergies later in life.

6 Conclusion

Normally, harmless food proteins can cause a food allergy, which is a pathological and potentially lethal immunological response. The need to develop analytical methods to enable accurate detection of trace food allergens has increased because of the incidence of food allergies. As a substitute antibody for food allergy identification, nanobodies have gained increasing interest. The production of particular nanobodies against dietary allergens, such as milk proteins, peanuts, almond, and legumes, has been documented in earlier research. For sensitive detection, corresponding immunoassays, such as photothermal, fluorescence, and electrochemical methods, have been developed. Identifying allergens in food, either in their original form or after processing, is a difficult problem that can be solved in a number of ways (Ciardiello et al. 2013). Allergens can be detected indirectly via genetic techniques. Proteins that belong to known allergenic families can be directly detected via biochemical techniques, such as mass spectrometry-based techniques. Nonetheless, these techniques provide no confirmation that the IgE of allergic patients can identify and bind to these molecules. Immunological approaches, on the other hand, can be used to identify molecules that show the epitopes that specific IgEs recognize and require eliciting IgE-mediated allergic responses. ELISA, a common immunological technique that has been extensively employed for allergen identification, is beneficial because it is useful for identifying a single allergen per test. The multiplex microarray-based allergen method is a revolutionary technique for detecting various allergens that IgE can detect in a single test. These advantages seem to have not been fully utilized. There are few examples in the literature of multiplex allergen systems being used to identify IgE-binding proteins in food extracts, but this technique has been utilized to assess allergens in processed foods. It does, however, show enormous potential for this application. Indeed, multiple characteristics suggest that it could be a highly useful tool for determining how well processing procedures eliminate allergens.

References

Abrams EM, Sicherer SH (2016) Diagnosis and management of food allergy. CMAJ 188(15):1087–1093. https://doi.org/10.1503/cmaj.160124

Alessandri C, Ferrara R, Bernardi ML et al (2020) Molecular approach to a patient's tailored diagnosis of the oral allergy syndrome. Clin Transl Allergy 10(1):22. https://doi.org/10.1186/s13601-020-00329-8

Alves RC, Pimentel FB, Nouws HPA et al (2015) Detection of the peanut allergen Ara h 6 in foodstuffs using a voltammetric biosensing approach. Anal Bioanal Chem 407(23):7157–7163. https://doi.org/10.1007/s00216-015-8879-8

Alves RC, Barroso MF, González-García MB et al (2016) New trends in food allergens detection: toward biosensing strategies. Crit Rev Food Sci Nutr 56(14):2304–2319. https://doi.org/10.1080/10408398.2013.831026

Alves RC, Pimentel FB, Nouws HPA (2017) Improving the extraction of Ara h 6 (a peanut allergen) from a chocolate-based matrix for immunosensing detection: influence of time, temperature and additives. Food Chem 218:242–248. https://doi.org/10.1016/j.foodchem.2016.09.085

Andjelković U, Gavrović-Jankulović M, Martinović T et al (2017) Omics methods as a tool for investigation of food allergies. TrAC Trends Anal Chem 96:107–115. https://doi.org/10.1016/j.trac.2017.07.011

Anfossi L, Di Nardo F, Russo A et al (2019) Silver and gold nanoparticles as multi-chromatic lateral flow assay probes for the detection of food allergens. Anal Bioanal Chem 411(9):1905–1913. https://doi.org/10.1007/s00216-018-1451-6

Ann Allergy (1985) Mar;54(3):245–8

Archila LD, Chow I, McGinty JW et al (2016) Ana o 1 and Ana o 2 cashew allergens share cross-reactive CD_4(+) T cell epitopes with other tree nuts. Clin Exp Allergy 46(6):871–883. https://doi.org/10.1111/cea.12746

Ashley J, D'Aurelio R, Piekarska M et al (2018) Development of a β-Lactoglobulin sensor based on SPR for Milk allergens detection. Biosensors 8(2):32. https://doi.org/10.3390/bios8020032

Aydin M, Aydin EB, Sezgintürk MK (2021) Advances in immunosensor technology. pp 1–62

Baldo BA, Pham NH (2021) Classification and descriptions of allergic reactions to drugs. In: Drug Allergy. Springer International Publishing, Cham, pp 17–57

Bastiaan-Net S, Pina-Pérez MC, Dekkers BJW et al (2020) Identification and in silico bioinformatics analysis of PR_{10} proteins in cashew nut. Protein Sci 29(7):1581–1595. https://doi.org/10.1002/pro.3856

Baumert JL (2014) Detecting and measuring allergens in food. In: Risk management for food allergy. Elsevier, pp 215–226

Blázquez-García M, Arévalo B, Serafín V et al (2022) Ultrasensitive detection of soy traces by immunosensing of glycinin and β-conglycinin at disposable electrochemical platforms. Talanta 241:123226. https://doi.org/10.1016/j.talanta.2022.123226

Bousfiha Bousfiha A, Lotfi A (2013) Effect of heat and enzymatic treatments on human IgE and rabbit IgG sensitivity to white bean allergens. Iran J Allergy Asthma Immunol 12:304–311

Breiteneder H, Radauer C (2004) A classification of plant food allergens☆. J Allergy Clin Immunol 113(5):821–830. https://doi.org/10.1016/j.jaci.2004.01.779

Burney PGJ, Potts J, Kummeling I et al (2014) The prevalence and distribution of food sensitization in European adults. Allergy 69(3):365–371. https://doi.org/10.1111/all.12341

Campuzano S, Yáez-Sedeño P, Pingarrón J (2017) Electrochemical affinity biosensors in food safety. Chem 5(1):8. https://doi.org/10.3390/chemosensors5010008

Campuzano S, Ruiz-Valdepeñas Montiel V, Serafín V et al (2020) Cutting-edge advances in electrochemical affinity biosensing at different molecular level of emerging food allergens and adulterants. Biosensors 10(2):10. https://doi.org/10.3390/bios10020010

Cannon HE (2018) The economic impact of peanut allergies

Chen X, Yang X (2014) Does food environment influence food choices? A geographical analysis through "tweets". Appl Geogr 51:82–89. https://doi.org/10.1016/j.apgeog.2014.04.003

Cheng J, Wang H, Sun D (2022) An overview of tropomyosin as an important seafood allergen: structure, cross-reactivity, epitopes, allergenicity, and processing modifications. Compr Rev Food Sci Food Saf 21(1):127–147. https://doi.org/10.1111/1541-4337.12889

Ciardiello MA, Tamburrini M, Liso M et al (2013) Food allergen profiling: a big challenge. Food Res Int 54(1):1033–1041. https://doi.org/10.1016/j.foodres.2013.03.013

Civera A, Galan-Malo P, Segura-Gil I et al (2022) Development of sandwich ELISA and lateral flow immunoassay to detect almond in processed food. Food Chem 371:131338. https://doi.org/10.1016/j.foodchem.2021.131338

Costa R, Costa J, Sagastizábal I et al (2021) Electrochemical and optical biosensing platforms for the immunorecognition of hazelnut Cor a 14 allergen. Food Chem 361:130122. https://doi.org/10.1016/j.foodchem.2021.130122

Crespo JF, Pascual C, Burks AW et al (1995) Frequency of food allergy in a pediatric population from Spain. Pediatr Allergy Immunol 6(1):39–43. https://doi.org/10.1111/j.1399-3038.1995.tb00256.x

Dang TD, Peters RL, Koplin JJ et al (2019) Egg allergen specific IgE diversity predicts resolution of egg allergy in the population cohort HealthNuts. Allergy 74(2):318–326. https://doi.org/10.1111/all.13572

de Maat-Bleeker F, van Dijk AG, Berrens L (1985) Allergy to egg yolk possibly induced by sensitization to bird serum antigens. Ann Allergy 54(3):245–248

Dhanapala P, Withanage-Dona D, Tang M et al (2017) Hypoallergenic variant of the major egg white allergen gal d 1 produced by disruption of cysteine bridges. Nutrients 9(2):171. https://doi.org/10.3390/nu9020171

Dias C, Costa J, Mafra I et al (2024) Electrochemical immunosensor for point-of-care detection of soybean Gly m TI allergen in foods. Talanta 268:125284. https://doi.org/10.1016/j.talanta.2023.125284

Dong X-X, Yang J-Y, Luo L et al (2017) Portable amperometric immunosensor for histamine detection using Prussian blue-chitosan-gold nanoparticle nanocomposite films. Biosens Bioelectron 98:305–309. https://doi.org/10.1016/j.bios.2017.07.014

Dubiela P, Kabasser S, Smargiasso N et al (2018) Jug r 6 is the allergenic vicilin present in walnut responsible for IgE cross-reactivities to other tree nuts and seeds. Sci Rep 8(1):11366. https://doi.org/10.1038/s41598-018-29656-4

Eischeid AC, Stadig SR, Rallabhandi P (2021) Comparison of real-time PCR and ELISA for the detection of crustacean shellfish allergens. Food Addit Contam Part A 38(4):563–572. https://doi.org/10.1080/19440049.2021.1874061

Engvall E, Perlmann P (1972) Enzyme-Linked Immunosorbent Assay. Elisa J Immunol 109(1):129–135. https://doi.org/10.4049/jimmunol.109.1.129

FDA (2004) US Food and Drug Administration

Flicker S, Zettl I, Tillib SV (2020) Nanobodies—useful tools for allergy treatment? Front Immunol 11. https://doi.org/10.3389/fimmu.2020.576255

Foo ACY, Mueller GA (2021) Abundance and stability as common properties of allergens. Front Allergy 2. https://doi.org/10.3389/falgy.2021.769728

Freitas M, Neves MMPS, Nouws HPA et al (2021) Electrochemical Immunosensor for the simultaneous determination of two Main Peanut allergenic proteins (Ara h 1 and Ara h 6) in food matrices. Foods 10(8):1718. https://doi.org/10.3390/foods10081718

Frigerio J, Pellesi R, Mezzasalma V et al (2019) Development of a DNA barcoding-like approach to detect mustard allergens in wheat flours. Genes (Basel) 10(3):234. https://doi.org/10.3390/genes10030234

Gamella M, Bueno-Díaz C, Ruiz-Valdepeñas Montiel V et al (2020) First electrochemical immunosensor for the rapid detection of mustard seeds in plant food extracts. Talanta 219:121247. https://doi.org/10.1016/j.talanta.2020.121247

Godoy-Navajas J, Aguilar Caballos MP, Gómez-Hens A (2011) Heterogeneous immunoassay for soy protein determination using nile blue-doped silica nanoparticles as labels and front-surface long-wavelength fluorimetry. Anal Chim Acta 701(2):194–199. https://doi.org/10.1016/j.aca.2011.06.005

González-Buitrago JM, Ferreira L, Isidoro-García M et al (2007) Proteomic approaches for identifying new allergens and diagnosing allergic diseases. Clin Chim Acta 385(1–2):21–27. https://doi.org/10.1016/j.cca.2007.07.010

Grishina G, Bardina L, Grishin A (2017) 2D-electrophoresis and immunoblotting in food. Allergy:59–69

Hammond JL, Formisano N, Estrela P et al (2016) Electrochemical biosensors and nanobiosensors. Essays Biochem 60(1):69–80. https://doi.org/10.1042/EBC20150008

Hei W, Li Z, Ma X, He P (2012) Determination of beta-conglycinin in soybean and soybean products using a sandwich enzyme-linked immunosorbent assay. Anal Chim Acta 734:62–68 https://doi.org/10.1016/j.aca.2012.05.009

Hildebrandt S, Garber EAE (2010) Effects of processing on detection and quantification of the parvalbumin gene in Atlantic salmon (Salmo salar). Food Chem 119(1):75–80. https://doi.org/10.1016/j.foodchem.2009.05.074

Hilu KW, Friend SA, Vallanadu V et al (2019) Molecular evolution of genes encoding allergen proteins in the peanuts genus Arachis: structural and functional implications. PLoS One 14(11): e0222440. https://doi.org/10.1371/journal.pone.0222440

Holden L, Christiane K, Fæste A et al (2005) Quantitative Sandwich ELISA for the determination of lupine (Lupinus spp.) in foods. J Agric Food Chem 53(15):5866–5871. https://doi.org/10.1021/jf050631i

Hu Y, Zhang C, Yang F et al (2021) Selection of specific Nanobodies against lupine allergen Lup an 1 for immunoassay development. Foods 10(10):2428. https://doi.org/10.3390/foods10102428

Hu Y, Lin J, Peng L et al (2023) Nanobody-based electrochemical immunoassay for sensitive detection of Peanut allergen Ara h 1. J Agric Food Chem 71(19):7535–7545. https://doi.org/10.1021/acs.jafc.3c00921

Huang Y, Li R, Zhu W et al (2024) Development of a fluorescent multiplexed lateral flow immunoassay for the simultaneous detection of crustacean allergen tropomyosin sarcoplasmic calcium binding protein and egg allergen ovalbumin in different matrices and commercial foods. Food Chem 440:138275. https://doi.org/10.1016/j.foodchem.2023.138275

Immer U, Lacorn M (2015) Enzyme-linked immunosorbent assays (ELISAs) for detecting allergens in food. In: Handbook of food allergen detection and control. Elsevier, pp 199–217

Iniesto E, Jiménez A, Prieto N et al (2013) Real time PCR to detect hazelnut allergen coding sequences in processed foods. Food Chem 138(2–3):1976–1981. https://doi.org/10.1016/j.foodchem.2012.11.036

Ivens KO, Baumert JL, Taylor SL (2016) Commercial Milk Enzyme-Linked Immunosorbent Assay (ELISA) kit Reactivities to purified Milk proteins and Milk-derived ingredients. J Food Sci 81(7). https://doi.org/10.1111/1750-3841.13357

Jarupalee T, Chatchatee P, Komolpis K et al (2018) Detecting allergens from black Tiger shrimp Penaeus monodon that can bind and cross-link IgE by ELISA, Western blot, and a humanized rat basophilic leukemia reporter cell line RS-ATL8. Allergy Asthma Immunol Res 10(1):62. https://doi.org/10.4168/aair.2018.10.1.62

Jeanson S, Dupont D, Grattard N et al (1999) Characterization of the heat treatment undergone by milk using two inhibition ELISAs for quantification of native and heat denatured α-lactalbumin. J Agric Food Chem 47(6):2249–2254. https://doi.org/10.1021/jf9809232

Ji J, Zhu P, Pi F et al (2017) Development of a liquid chromatography-tandem mass spectrometry method for simultaneous detection of the main milk allergens. Food Control 74:79–88. https://doi.org/10.1016/j.foodcont.2016.11.030

Jiao S, Xie X, He Z et al (2024) Lateral flow immunochromatographic assay for competitive detection of crustacean allergen tropomyosin using phage-displayed shark single-domain antibody. J Agric Food Chem 72(3):1811–1821

Jiang X, Rao Q (2021) Effect of processing on fish protein antigenicity and Allergenicity. Foods 10(5):969. https://doi.org/10.3390/foods10050969

Jiang H, Guo Q, Zhang C et al (2021) Microfluidic origami nano-aptasensor for peanut allergen Ara h1 detection. Food Chem 365:130511. https://doi.org/10.1016/j.foodchem.2021.130511

Juhász A, Belova T, Florides CG et al (2018) Genome mapping of seed-borne allergens and immunoresponsive proteins in wheat. Sci Adv 4(8). https://doi.org/10.1126/sciadv.aar8602

Kadooka Y, Idota T, Gunji H et al (2000) A method for measuring specific IgE in sera by direct ELISA without interference by IgG competition or IgG autoantibodies to IgE. Int Arch Allergy Immunol 122(4):264–269. https://doi.org/10.1159/000024408

Khedri M, Ramezani M, Rafatpanah H et al (2018) Detection of food-born allergens with aptamer-based biosensors. TrAC Trends Anal Chem 103:126–136. https://doi.org/10.1016/j.trac.2018.04.001

Kirsch S, Fourdrilis S, Dobson R et al (2009) Quantitative methods for food allergens: a review. Anal Bioanal Chem 395(1):57–67. https://doi.org/10.1007/s00216-009-2869-7

Koppelman SJ, Lardizabal AL, Niemann L et al (2021) Development of a Sandwich enzyme-linked immunosorbent assay for detection and quantification of clam residues in food products. Biomed Res Int 2021:1–9. https://doi.org/10.1155/2021/6685575

Krishna VD, Wu K, Su D et al (2018) Nanotechnology: review of concepts and potential application of sensing platforms in food safety. Food Microbiol 75:47–54. https://doi.org/10.1016/j.fm.2018.01.025

Krisnawati DI, Alimansur M, Atmojo DS et al (2022) Food allergies: Immunosensors and management. Appl Sci 12(5):2393. https://doi.org/10.3390/app12052393

Kurze E, Lo Scalzo R, Campanelli G et al (2018) Effect of tomato variety, cultivation, climate and processing on sola l 4, an allergen from Solanum lycopersicum. PLoS One 13(6):e0197971. https://doi.org/10.1371/journal.pone.0197971

Lacorn M, Dubois T, Gößwein C et al (2022) Validation of the RIDASCREEN® Peanut for determination of Peanut protein in cookies, Milk chocolate, ice cream, trail mix, puffed Rice cereals, and granola Bar: AOAC performance tested method SM 112102. J AOAC Int 105(3): 784–801. https://doi.org/10.1093/jaoacint/qsab168

Lexmaulová H, Gabrovská D, Rysová J et al (2013) ELISA kit for Peanut protein determination: collaborative study. J AOAC Int 96(5):1041–1047. https://doi.org/10.5740/jaoacint.12-474

Li H, Liang X, Zhou B et al (2020) A proteomic analysis of peanut seed at different stages of underground development to understand the changes of seed proteins. PLoS One 15(12): e0243132. https://doi.org/10.1371/journal.pone.0243132

Liu R, Pieniak Z, Verbeke W (2013) Consumers' attitudes and behaviour towards safe food in China: A review. Food Control 33(1):93–104. https://doi.org/10.1016/j.foodcont.2013.01.051

Madrid R, García-García A, Cabrera P et al (2021) Survey of commercial food products for detection of walnut (Juglans regia) by two ELISA methods and real time PCR. Foods 10(2): 440. https://doi.org/10.3390/foods10020440

Maleki SJ, Chung S-Y, Champagne ET et al (2000) The effects of roasting on the allergenic properties of peanut proteins. J Allergy Clin Immunol 106(4):763–768. https://doi.org/10.1067/mai.2000.109620

Mano J, Nishitsuji Y, Kikuchi Y et al (2017) Quantification of DNA fragmentation in processed foods using real-time PCR. Food Chem 226:149–155. https://doi.org/10.1016/j.foodchem.2017.01.064

Meng S, Li J, Chang S et al (2019) Quantitative and kinetic analyses of peanut allergens as affected by food processing. Food Chem X 1:100004. https://doi.org/10.1016/j.fochx.2019.100004

Morais S, Tortajada-Genaro LA, Maquieira Á et al (2020) Biosensors for food allergy detection according to specific IgE levels in serum. TrAC Trends Anal Chem 127:115904. https://doi.org/10.1016/j.trac.2020.115904

Morishita N, Kamiya K, Matsumoto T et al (2008) Reliable enzyme-linked immunosorbent assay for the determination of soybean proteins in processed foods. J Agric Food Chem 56(16):6818–6824

Nehra M, Lettieri M, Dilbaghi N et al (2019) Nano-biosensing platforms for detection of cow's milk allergens: an overview. Sensors 20(1):32. https://doi.org/10.3390/s20010032

Nugraha R, Kamath SD, Johnston E et al (2019) Conservation analysis of B-cell allergen epitopes to predict clinical cross-reactivity between shellfish and inhalant invertebrate allergens. Front Immunol 10. https://doi.org/10.3389/fimmu.2019.02676

Orcajo J, Lavilla M, Martínez-de-Marañón I (2019) Specific and sensitive ELISA for measurement of IgE-binding variations of milk allergen β-lactoglobulin in processed foods. Anal Chim Acta 1052:163–169. https://doi.org/10.1016/j.aca.2018.11.048

Pahr S, Selb R, Weber M et al (2014) Biochemical, biophysical and IgE-epitope characterization of the wheat food allergen, tri a 37. PLoS One 9(11):e111483. https://doi.org/10.1371/journal.pone.0111483

Pali-Schöll I, Meinlschmidt P, Larenas-Linnemann D et al (2019) Edible insects: cross-recognition of IgE from crustacean- and house dust mite allergic patients, and reduction of allergenicity by food processing. World Allergy Organ J 12(1):100006. https://doi.org/10.1016/j.waojou.2018.10.001

Palladino C, Breiteneder H (2018) Peanut allergens. Mol Immunol 100:58–70. https://doi.org/10.1016/j.molimm.2018.04.005

Palmer DJ, Metcalfe J, Makrides M et al (2013) Early regular egg exposure in infants with eczema: a randomized controlled trial. J Allergy Clin Immunol 132(2):387–392.e1. https://doi.org/10.1016/j.jaci.2013.05.002

Pandey AK, Varshney RK, Sudini H, Pandey MK (2019). An improved enzyme-linked immunosorbent assay (ELISA) based protocol using seeds for detection of five major peanut allergens Ara h 1, Ara h 2, Ara h 3, Ara h 6, and Ara h 8. Front Nutr 6:68

Pasquariello MS, Palazzo P, Tuppo L et al (2012) Analysis of the potential allergenicity of traditional apple cultivars by multiplex biochip-based immunoassay. Food Chem 135(1): 219–227. https://doi.org/10.1016/j.foodchem.2012.04.075

Patel N, Adelman DC, Anagnostou K et al (2021) Using data from food challenges to inform management of consumers with food allergy: a systematic review with individual participant data meta-analysis. J Allergy Clin Immunol 147(6):2249–2262.e7. https://doi.org/10.1016/j.jaci.2021.01.025

Pepper AN, Assa'ad A, Blaiss M et al (2020) Consensus report from the Food Allergy Research & Education (FARE) 2019 Oral Immunotherapy for Food Allergy Summit. J Allergy Clin Immunol 146(2):244–249. https://doi.org/10.1016/j.jaci.2020.05.027

Perner SP, Heupel L, Zimmermann L et al (2019) Investigation of reduced ELISA recovery of almond and hazelnut traces from roasted nut samples by SDS-PAGE and mass spectrometry. J AOAC Int 102(5):1271–1279. https://doi.org/10.1093/jaoac/102.5.1271

Peng D, Ye S, Wang Y et al (2012) Development and validation of an indirect competitive enzyme-linked immunosorbent assay for the screening of tylosin and tilmicosin in muscle, liver, milk, honey and eggs. J Agric Food Chem 60(1):44–51

Peters RL, Koplin JJ, Gurrin LC et al (2017) The prevalence of food allergy and other allergic diseases in early childhood in a population-based study: HealthNuts age 4-year follow-up. J Allergy Clin Immunol 140(1):145–153.e8. https://doi.org/10.1016/j.jaci.2017.02.019

Pi X, Yang Y, Sun Y et al (2022) Recent advances in alleviating food allergenicity through fermentation. Crit Rev Food Sci Nutr 62(26):7255–7268. https://doi.org/10.1080/10408398.2021.1913093

Platteau C, De Loose M, De Meulenaer B et al (2011) Quantitative detection of hazelnut (Corylus avellana) in cookies: ELISA versus real-time PCR. J Agric Food Chem 59(21):11395–11402. https://doi.org/10.1021/jf202167b

Prado M, Ortea I, Vial S et al (2016) Advanced DNA- and protein-based methods for the detection and investigation of food allergens. Crit Rev Food Sci Nutr 56(15):2511–2542. https://doi.org/10.1080/10408398.2013.873767

Qiu S, Wang J (2017) The prediction of food additives in the fruit juice based on electronic nose with chemometrics. Food Chem 230:208–214. https://doi.org/10.1016/j.foodchem.2017.03.011

Radauer C, Bublin M, Wagner S et al (2008) Allergens are distributed into few protein families and possess a restricted number of biochemical functions. J Allergy Clin Immunol 121(4):847–852.e7. https://doi.org/10.1016/j.jaci.2008.01.025

Rebe Raz S, Liu H, Norde W et al (2010) Food allergens profiling with an imaging surface Plasmon resonance-based biosensor. Anal Chem 82(20):8485–8491. https://doi.org/10.1021/ac101819g

Ren QR, Zhang H, Guo HY et al (2014) Detection of cow milk adulteration in yak milk by ELISA. J Dairy Sci 97(10):6000–6006. https://doi.org/10.3168/jds.2014-8127

Röder M, Wiacek C, Lankamp F et al (2021) Improved sensitivity of allergen detection by Immunoaffinity LC-MS/MS using ovalbumin as a case study. Foods 10(12):2932. https://doi.org/10.3390/foods10122932

Rona RJ, Keil T, Summers C et al (2007) The prevalence of food allergy: a meta-analysis. J Allergy Clin Immunol 120(3):638–646. https://doi.org/10.1016/j.jaci.2007.05.026

Ruiz-Valdepeñas Montiel V, Campuzano S, Torrente-Rodríguez RM et al (2016) Electrochemical magnetic beads-based immunosensing platform for the determination of α-lactalbumin in milk. Food Chem 213:595–601. https://doi.org/10.1016/j.foodchem.2016.07.004

Savage JH, Matsui EC, Skripak JM et al (2007) The natural history of egg allergy. J Allergy Clin Immunol 120(6):1413–1417. https://doi.org/10.1016/j.jaci.2007.09.040

Scherf KA, Poms RE (2016) Recent developments in analytical methods for tracing gluten. J Cereal Sci 67:112–122. https://doi.org/10.1016/j.jcs.2015.08.006

Schmidt SD, Mazzella MJ, Nixon RA et al (2012) Aβ measurement by enzyme-linked immunosorbent assay. Methods Mol Biol. 849:507–527

Schmitt DA, Nesbit JB, Hurlburt BK et al (2010) Processing can Alter the properties of Peanut extract preparations. J Agric Food Chem 58(2):1138–1143. https://doi.org/10.1021/jf902694j

Schmitt DA, Cheng H, Maleki SJ, Burks AW (2004) Competitive Inhibition ELISA for Quantification of Ara h 1 and Ara h 2 the major allergens of peanuts. J AOAC Int 87(6):1492–1497. https://doi.org/10.1093/jaoac/87.6.1492

Sena-Torralba A, Pallás-Tamarit Y, Morais S et al (2020) Recent advances and challenges in food-borne allergen detection. TrAC Trends Anal Chem 132:116050. https://doi.org/10.1016/j.trac.2020.116050

Sicherer SH, Sampson HA (2010) Food allergy. J Allergy Clin Immunol 125(2):S116–S125. https://doi.org/10.1016/j.jaci.2009.08.028

Sicherer SH, Sampson HA (2018) Food allergy: a review and update on epidemiology, pathogenesis, diagnosis, prevention, and management. J Allergy Clin Immunol 141(1):41–58. https://doi.org/10.1016/j.jaci.2017.11.003

Sicherer SH, Furlong TJ, Muñoz-Furlong A et al (2001) A voluntary registry for peanut and tree nut allergy: characteristics of the first 5149 registrants☆. J Allergy Clin Immunol 108(1):128–132. https://doi.org/10.1067/mai.2001.115755

Singh S, Chaudhary S, Agarwal A et al (2014) Chapter 15 – antibodies: monoclonal and polyclonal. In: Verma AS, Singh ABT-AB (eds). Academic Press, San Diego, pp 265–287

Sobhan A, Oh J-H, Park M-K et al (2018a) Single walled carbon nanotube based biosensor for detection of peanut allergy-inducing protein ara h1. Korean J Chem Eng 35(1):172–178. https://doi.org/10.1007/s11814-017-0259-y

Sobhan A, Oh J-H, Park M-K et al (2018b) Detection of Peanut Allergen Ara h 6 in commercially processed foods using a single-walled carbon nanotube–based biosensor. J AOAC Int 101(5):1558–1565. https://doi.org/10.5740/jaoacint.18-0041

Stanly C, Kim H, Antonucci G et al (2021) Crosstalk between the immune system and plant-derived Nanovesicles: a study of allergen transporting. Front Bioeng Biotechnol 9. https://doi.org/10.3389/fbioe.2021.760730

Sugawara K, Kuramitz H, Shinohara H (2017) Fabrication of micromagnetic beads with molecular recognition/electron-transfer peptides for the sensing of ovalbumin. Anal Chim Acta 958:30–37. https://doi.org/10.1016/j.aca.2016.12.028

Sugawara K, Kadoya T, Kuramitz H (2018) Magnetic beads modified with an electron-transfer carbohydrate-mimetic peptide for sensing of a galactose-dependent protein. Anal Chim Acta 1001:158–167. https://doi.org/10.1016/j.aca.2017.11.047

Sun X, Li C, Zhu Q et al (2020) A label-free photoelectrochemical immunosensor for detection of the milk allergen β-lactoglobulin based on Ag2S -sensitized spindle-shaped BiVO4/BiOBr heterojunction by an in situ growth method. Anal Chim Acta 1140:122–131. https://doi.org/10.1016/j.aca.2020.10.021

Thomas K, Herouet-Guicheney C, Ladics G et al (2007) Evaluating the effect of food processing on the potential human allergenicity of novel proteins: international workshop report. Food Chem Toxicol 45(7):1116–1122. https://doi.org/10.1016/j.fct.2006.12.016

Tong WS, Yuen AW, Wai CY et al (2018) Diagnosis of fish and shellfish allergies. J Asthma Allergy 11:247–260. https://doi.org/10.2147/JAA.S142476

Torre R, Costa-Rama E, Nouws HPA et al (2020) Screen-printed electrode-based sensors for food spoilage control: bacteria and biogenic amines detection. Biosensors 10(10):139. https://doi.org/10.3390/bios10100139

Torre R, Freitas M, Costa-Rama E et al (2022) Food allergen control: tropomyosin analysis through electrochemical immunosensing. Food Chem 396:133659. https://doi.org/10.1016/j.foodchem.2022.133659

Tranquet O, Lupi R, Echasserieau-Laporte V et al (2015) Characterization of antibodies and development of an indirect competitive immunoassay for detection of deamidated gluten. J Agric Food Chem 63(22):5403–5409

Tsakali E, Chatzilazarou A, Houhoula D et al (2019) A rapid HPLC method for the determination of lactoferrin in milk of various species. J Dairy Res 86(2):238–241. https://doi.org/10.1017/S0022029919000189

van Ree R (2021) Grand challenges in food allergy. Front Allergy. https://doi.org/10.3389/falgy.2021.668479

Vidal ML, Gautron J, Nys Y (2005) Development of an ELISA for quantifying lysozyme in hen egg white. J Agric Food Chem 53(7):2379–2385

Wammes AEM, van Delft FL, van Hest JCM et al (2015) Surface plasmon resonance (SPR) sensors for detecting allergens in food. In: Handbook of food allergen detection and control. Elsevier, pp 229–244

Wang W, Zhu X, Teng S et al (2018) Development and validation of a surface Plasmon resonance biosensor for specific detection of porcine serum albumin in food. J AOAC Int 101(6):1868–1872. https://doi.org/10.5740/jaoacint.17-0356

Wang Y, Rao Z, Zhou J et al (2019) A chiral assembly of gold nanoparticle trimer-based biosensors for ultrasensitive detection of the major allergen tropomyosin in shellfish. Biosens Bioelectron 132:84–89. https://doi.org/10.1016/j.bios.2019.02.038

Wei-Liang Tan J, Valerio C, Barnes EH et al (2017) A randomized trial of egg introduction from 4 months of age in infants at risk for egg allergy. J Allergy Clin Immunol 139(5):1621–1628.e8. https://doi.org/10.1016/j.jaci.2016.08.035

Yakhlef M, Giangrieco I, Ciardiello MA (2021) Potential allergenicity of Medicago sativa investigated by a combined IgE -binding inhibition, proteomics and in silico approach. J Sci Food Agric 101(3):1182–1192. https://doi.org/10.1002/jsfa.10730

Zhang H, Lu Y, Ushio H, Shiomi K (2014) Development of sandwich ELISA for detection and quantification of invertebrate major allergen tropomyosin by a monoclonal antibody. Food Chem 150:151–157. https://doi.org/10.1016/j.foodchem.2013.10.154

Zheng M, Wu JH, Louie JCY et al (2016) Typical food portion sizes consumed by Australian adults: results from the 2011–12 Australian National Nutrition and Physical Activity Survey. Sci Rep 6(1):19596

Zhou Y, Song F, Li YS et al (2013) Double-antibody based immunoassay for the detection of β-casein in bovine milk samples. Food Chem 141(1):167–173. https://doi.org/10.1016/j.foodchem.2013.03.010

Role of Immunosensors in the Detection of Pesticides in Different Food Matrixes

Harsh Kumar, Shivani Guleria, Neetika Kimta, Rajni Dhalaria, Daljeet Singh Dhanjal, and Kamil Kuča

Abstract

The use of pesticides to protect crop yields has been a long-standing practice, spanning several decades. Numerous types of pesticides have been synthesized, with many designed to target multiple organisms or processes. However, the persistence of pesticide residues in both raw and processed food products remains a significant concern. To address this, highly sensitive detection methods have been developed to identify pesticide components at very low levels with high accuracy. Point-of-care testing is increasingly critical for rapid detection, and biosensors play a crucial role in this regard. Furthermore, the development of immunosensors has proven invaluable for pesticide detection. In this chapter, we

H. Kumar
Centre of Advanced Technologies, Faculty of Science, University of Hradec Kralove, Hradec Kralove, Czech Republic

S. Guleria
Department of Biotechnology, TIFAC-Centre of Relevance and Excellence in Agro and Industrial Biotechnology (CORE), Thapar Institute of Engineering and Technology, Patiala, India

N. Kimta
School of Biological and Environmental Sciences, Shoolini University of Biotechnology and Management Sciences, Solan, India

R. Dhalaria
Department of Biosciences, Minerva PG College of Arts, Science and Commerce, Indora, Kangra, Himachal Pradesh, India

D. S. Dhanjal
School of Bioengineering and Biosciences, Lovely Professional University, Phagwara, Punjab, India

K. Kuča (✉)
Biomedical Research Centre, University Hospital in Hradec Kralove, Hradec Kralove, Czech Republic
e-mail: kamil.kuca@uhk.cz

K. Kuča et al. (eds.), *Ensuring Wellness: Immunosensors Monitoring Food Quality and Pathogenicity Risks*, https://doi.org/10.1007/978-3-031-97477-9_11

intend to provide a comprehensive overview of different aspects associated with pesticides and its occurrence in food products, along with their potential health effects as well as available techniques for detection.

Keywords

Detection methods · Limit of detection (LOD) · Pesticides · Maximum residue limits (MRLs) · Health effects

1 Introduction

Pesticides are nowadays extensively used in agricultural practices to curb and regulate the pests and weed infestations that substantially reduce the crop losses and increase the yield of agricultural products (Fang et al. 2009; Meng et al. 2013; Odukkathil and Vasudevan 2013). But, overuse of these pesticides has led to severe environmental implications as they have crossed permissible limit in food products. The presence of pesticide residues in food items has been determined to cause various health complications (Sivasankaran et al. 2007; Ricking and Schwarzbauer 2012; Riah et al. 2014; Singh et al. 2016; Yang et al. 2019). These chemicals are also classified as neurotoxins as they inhibit the activity of acetylcholinesterase (John and Shaike 2015; Kaur and Goyal 2019), and can cause damage to essential organs like the liver, red blood cells, and alters the endocrine, immune, and reproductive systems (Rawlings et al. 1998; Suwalsky et al. 2005; Farag et al. 2010; Singh et al. 2018; Yang et al. 2018). Some pesticides are even categorized as carcinogens and mutagens (De Flora et al. 1993). Although the pesticides' residue levels in food are within tolerance limits, their long-term exposure will be chronic and harmful to human health (Yan et al. 2018). Owing to the extensive usage of pesticides in agriculture, the development of efficient detection techniques has become a necessity. High-Performance Liquid Chromatography (HPLC), Mass Spectrometry (MS), and other conventional methods offer high sensitivity, automation, and reproducibility, but are very expensive, time-consuming, and unsuitable for real-time monitoring (Velasco-Garcia and Mottram 2003; Jiang et al. 2008; LeDoux 2011; Sherma 2015; Grimalt and Dehouck 2016).

Biosensors, serving as point-of-care devices, have been determined to have numerous benefits like accuracy, convenience, rapid analysis, mobility, and on-site testing abilities (Yan et al. 2018; Patel et al. 2019). Biosensors have gained significant interest among researchers for detection of pesticide residue by converting the detection of the target analytes to measurable physical signal such as electrical, magnetic, or optical signal (Baruah and Dutta 2009; Verma 2017; Verdian 2018). Furthermore, nanomaterial-based biosensors exhibit the ability to detect the traces of pesticide residues as they provide effective signal amplification (Tan et al. 2012; Kaphle et al. 2018; Verdian 2018). The next section covers and discusses the development of diverse immunosensors for on-site detection of pesticides in different food matrices.

2 An Overview on the Types of Pesticides Used in Food Production

In modern agricultural practices, pesticides are classified based on different characteristics like their chemical makeup (i.e., biopesticide, synthetic, organic, or inorganic), and mode of action based on the target pests (i.e., herbicide, fungicide, or insecticide). The most widely used classifications are based on the type of organism the pesticide is designed to control or combat and the chemical nature of the pesticide (Lazarević-Pašti et al. 2025). Pesticides have become the necessary component for safeguarding crops from numerous threats, such as weeds (herbicides), insects (insecticides), rodents (rodenticides), and fungi (fungicides) (Fig. 1). The chemical constituents are made in diverse forms, depending on their specific purpose and the type of pest that they are intended to control.

3 Current Status of Pesticides' Residues in Food

The occurrence and concentration of pesticide residues present in foods are a real public health concern, especially at the time when pesticides are being extensively used in agricultural practices. These pesticides cannot be extracted or washed off as they remain either on or inside the food, allowing the consumers to get exposed to these harmful chemicals (Costas-Ferreira et al. 2022). Fruits and vegetables by their nature are highly susceptible to pesticide residues, with commons examples of fruits like apples, grapes, strawberries, tomatoes, and leafy greens such as spinach are often found to be positive for pesticides detection (Lazarević-Pašti et al. 2025). These pesticide residues can be found on the outer surface and interior of the fruit. Similarly, products obtained from animals like eggs, meat, and milk may contain pesticide residues owing to chances of their consuming contaminated feed or by directly getting exposed to pesticides (Bajwa and Sandhu 2014). Washing and

Fig. 1 Pesticide classification based on the target organism/site. (Lazarević-Pašti et al. 2025, under the terms and conditions of the Creative Commons Attribution (CC BY) license)

peeling reduces residues on the surface of the food but are less effective at eliminating pesticides that have penetrated inside the food and cannot be removed merely by physical approach (Đorđević and Đurović-Pejčev 2016). Key groups of pesticides that have been commonly accorded in food and food-products are organochlorines, organophosphates, carbamates, and pyrethroids.

For agricultural operations, the utilization of pesticides is highly regulated with explicit guidelines regarding the chemicals use and limit on crops. Developing countries are facing various challenges in implementing these regulations to control the level of pesticide residue in food products though (Sapbamrer et al. 2023). The main reason for non-implementation of these regulations in developing countries is improper infrastructure, lack of resources, and regulatory bodies that can effectively monitor pesticide usage. Owing to this, illiterate farmers can end up using pesticides in excessive amounts, which will eventually result in the presence of these chemicals in food beyond the safe limits (Lazarević-Pašti et al. 2025). Furthermore, poor training and knowledge and mishandling of pesticide by farmers is also a cause of the deteriorating situation, leading to contamination of food (Lazarević-Pašti et al. 2025). As a result, local markets and the economy are severely impacted. For example, in 2023, large quantities of fruit for export were rejected at the border crossing in the period from 2023 onwards due to pesticide residues that were not permitted (Lazarević-Pašti et al. 2025). In another example, in 2024, frozen cherries exported from Serbia were refused entry into the Netherlands because of high levels of an EU-banned pesticide, dimethoate (Lazarević-Pašti et al. 2025). Similarly, entry of peaches and nectarines consignments in 2023, Croatia and Montenegro rejected consignments at the gate because of the presence of chlorpyrifos (another banned pesticide by the EU) (Lazarević-Pašti et al. 2025). Serbian plums also were not received at the border, as they were contaminated with the residues of chlorpyrifos or other unauthorized substances (Lazarević-Pašti et al. 2025).

Additionally, chlorpyrifos is allowed to be strictly used for sugar beets in Serbia, while it is banned for usage on other crops. On the contrary, it is still being used extensively in nations like the USA (Lazarević-Pašti et al. 2025). Thus, proper regulatory enforcement on the pesticide's usage is required to counterbalance the negative implication of these pesticides and safeguard human health as well as the environment.

4 Health Effects of Food Containing Residual Pesticides

Extensive research being conducted on pesticide has led to the identification of neurotoxicity of organophosphate and carbamate insecticides, which function by inhibiting acetylcholinesterase, an enzyme necessary to maintain the normal function of the nervous system. The long-term exposure of these chemicals has been determined to lead to variety of neurological deficits like cognitive impairment and increased risk of neurodegenerative diseases such as Parkinson's and Alzheimer's (Colovic et al. 2013; Vale and Lotti 2015; Xie et al. 2024). Additionally, few insecticides have been determined to induce cancer, especially colorectal cancer, which is raising the suspicion of their carcinogenic potential. On the other hand, other insecticides have been reported to disrupt the endocrine system and affect the reproductive system, leading to infertility and birth defects (Bhardwaj et al. 2018; Abolhassani et al. 2019).

Herbicides have the ability to interfere and mimic as the endogenous hormone in the body and thereby can inhibit growth, reproduction, and metabolic processes (Mnif et al. 2011). Moreover, few herbicides on their long-term exposure have been determined to be associated with complications like sperm malformation and congenital defects (Beyuo et al. 2024). Even though there is limited evidence that long-term, low-level exposure to herbicides increases the risk of development of certain cancers, it is still unclear due to limited evidence. In order to be conclusive about this association, additional research studies are required in this direction (Beyuo et al. 2024).

Fungicides are toxic and acutely poisonous chemicals, and their consumption in large quantities can lead to acute poisoning. It has been found that exposure of fungicides to animals via dermal absorption, inhalation, and ingestion leads to symptoms like, diarrhoea, headache, nausea, vomiting, and in severe cases, neurological impairment and organ failure (Gupta 2018; Kadirov et al. 2023). Organizations like the International Agency for Research on Cancer (IARC) and the US Environmental Protection Agency (EPA) have recognized a few fungicides as probable or potential carcinogenic chemicals. It has been well comprehended that long-term exposure to these fungicides has been responsible for increased incidence of cancer development (Schreinemachers et al. 1999).

Bromethalin and Strychnine are also toxic chemicals and ingestion of these exhibits symptoms like hypersensitivity to light or noise, seizures, muscle tremors, and even death in highly severe cases (Flores and Singh 2023; Gray et al. 2011). Similarly, exposure of zinc phosphide can lead to delayed effect symptoms like ataxia, nervousness, restlessness, seizures, and vomiting. In severe cases, the exposure to these pesticides can cause coma or death due to respiratory failure (Gray et al. 2011).

5 Briefing About the Routine Methods to Detect Pesticide Residues in Different Food

Various methods have been developed for the analysis of pesticide residues in food, all of which have its own advantages and disadvantages. The determination of pesticide residues is challenging because of the complexity of food matrix and low concentration of pesticides (Bashir et al. 2025). One of the most extensively used methods is solvent extraction for the pesticide residues purification from food, whereas their quantification is done via methods like Gas Chromatography-Mass Spectrometry (GC-MS) and Liquid Chromatography-Mass Spectrometry (LC-MS) (Bashir et al. 2025). Yet these procedures are time-consuming, complicated, exhaustive, and necessitate an elaborate processing and preparation of the samples. Considering these difficulties, only a group of researchers have modified the original

Quick, Easy, Cheap, Effective, Rugged, and Safe (QuEChERS) method to achieve more sensitive pesticide determination in various foods. This procedure has also been proven effective for the analysis of pesticides, and it was found to be easy to operate than traditional GC-MS or LC-MS method (Table 1).

6 Biosensor Based Technology to Detect the Pesticide Residues in Food

Biosensors are being considered as useful tools for monitoring pesticide residue in food source due to their sensitivity, portability, and on-site real-time analysis. A laccase enzyme-based sensor cross-linked on a gold electrode has been applied for analyzing carbamate and formetanate hydrochloride in mangoes and grapes with an LOD of $0.02 \pm 2.6 \times 10^{-4}$ mg/kg (Ribeiro et al. 2014). Additionally, an OP hydrolase-derived sensor was developed to detect OP pesticide residues with sensitivity of 3–32 μS/ppb, and LOD values of diazinon, malathion, chlorpyrifos, and profenofos were 40, 20, 30, and 40 ppb, respectively (Mulyasuryani and Prasetyawan 2015). An acetylcholinesterase (AChE)-based biosensor was constructed for the determination of monocrotophos and other OPs with an LOD 0.1 mg/mL applicable for laboratory (Mukherjee et al. 2019). Additionally, its sensitivity was improved by replacing AChE with organophosphate hydrolase and it was also used to detect organophosphate residues in fruits and vegetables in the field (Mukherjee et al. 2021). In addition, aptasensors have been created to detect acetamiprid in cabbage leaves (LOD: 0.7 nM), tomatoes, cucumbers, and cabbage (LOD: 10.03 nmol/L), malathion (LOD: 0.4 pM) in apples, and chlorpyrifos (LOD: 0.036 ng/mL) in apples, and other pesticides (Balkrishna et al. 2024).

Amperometric biosensors, which are usually employed in detecting organophosphorus insecticides, particularly on olive plants, involve the AChE immobilization onto screen-printed electrodes sandwiched in a sol–gel composite. The enzymatic activity is estimated by measuring the liberation of thiocholine produced during the enzymatic hydrolysis of acetylthiocholine chloride mediated by cobalt phthalocyanine (Oujji et al. 2013). In addition, a flow biosensor for milk testing has also been developed using genetically engineered AChE enzymes (B394, B4, and wild-type B131). This biosensor encompasses the screen-printed carbon electrode (SPE) held inside a flow cell, and has been proven effective for detecting three Ops, i.e., chlorpyrifos-oxon (CPO), ethyl paraoxon (EPOx), and malaoxon (MOx) in milk samples (Mishra et al. 2012).

7 Immunosensor Applications in the Detection of Pesticides in Food

Immunosensors have gained significant attention as well as interest of researchers due to their ability to rapidly detect the pesticide residues in food. Immunosensors is a generic term, which is interchangeably used for conventional solid-phase

Table 1 Pesticides reported in different food analyzed by modified QuEChERS method (Bashir et al. 2025, license number 6013010931356)

Food Type	Analytes identified	Few examples	Limit of detection (LOD (ppm))	Instrument
Barley	27 Herbicides	Pyridate; Bifenox; Isoxaben; Metoxuron; Florasulam	0.023; 0.003; 0.003; 0.002; 0.0026	Liquid Chromatography with tandem mass spectrometry (LC-MS/MS)
Radish	107 pesticides	Alachlor; Bromacil; Diazinon; Fenthion; Simazine	0.002; 0.02; 0.004; 0.002; 0.01	Gas chromatography-mass spectrometry (GC-MS)
Barley	16 Herbicides	Trifluralin; Terbuthylazine; Prosulfocarb; Pendimethalin; Diflufenican	0.002; 0.016; 0.0015; 0.0021; 0.0011	Gas Chromatography–time-of-flight mass spectrometry (GC-TOF-MS)
Paddy	203 pesticides	Triazophos; Alachlor; Bifenox; Diazinon; Endosulfan; Indanofan	0.015; 0.005; 0.04; 0.008; 0.04; 0.03	GC-MS
Cabbage	107 pesticides	Aldrin; Bromacil; Endrine; Phosmet; Propanil	0.004; 0.02; 0.004; 0.015; 0.02	GC-MS
Olive	7 pesticides	Simazine; Endosulfan; Endosulfate; Deltamethrin	0.012; 0.005; 0.004; 0.01	GC-MS
Wheat	24 pesticides	Aldrin; Dicofol; Dieldrin; Lindane; Tetradifon	0.0025; 0.003; 0.0025; 0.003; 0.0025	GC-MS
Banana	128 pesticides	Aldicarb; Allethrin; Phosmet; Ethion	0.005; 0.005; 0.005; 0.005	Ultra-high performance liquid chromatography–tandem mass spectrometry (UHPLC-MS/MS)
Honey	116 pesticides	Triforin; Ethion; Allethrin; Carbaryl; Imazalil	0.005; 0.005; 0.005; 0.005; 0.005	UHPLC-MS/MS
Orange juice	74 pesticides		0.003; 0.003;	UHPLC-MS/MS

(continued)

Table 1 (continued)

Food Type	Analytes identified	Few examples	Limit of detection (LOD (ppm))	Instrument
		Acetamiprid; Metalaxyl; Triadimefon; Diuron	0.0076; 0.003	
Milk	14 pesticides	Vinclozolin; 2,4-DDT; Dieldrin; Bifenthrin; Fenbuconazole	0.009; 0.0085; 0.0052; 0.0061; 0.004	Gas chromatography-electron capture detector (GC-ECD)

immunoassay-based sensors, that utilizes antibodies as target-specific detection agents. When coupled with an appropriate physicochemical transduction mechanism, the antibodies convert recognition events into measurable signals, allowing the detection of pesticides of interest. Immunosensors can be electrochemical, optical, or piezoelectric (Balkrishna et al. 2024) (Table 2).

For instance, an uncompetitive immunoassay-based optical immunosensor for triazophos detection in Chinese cabbage, cucumber, and apples was reported. This sensor was modified with two triazophos-specific monoclonal antibodies (mAbs) to achieve the direct assay of triazophos. The developed biosensor exhibited excellent selectivity with a LOD of 0.096 ng/mL and a linear detection range of 0.98–8.29 ng/mL. This immunosensor extends a practical, fast, and accurate method for real-time monitoring of triazophos to meet its MRLs (maximum residue limits) (Guo et al. 2018). A new piezoelectric immunosensor was developed for competitive immunoassay of carbaryl in food commodities. Based on the very large value of the electromechanical coupling factor (kt) for these Bi-base piezoelectric ceramics led to the development of an immunosensor with improved sensitivity with respect of those of standard quartz-based biosensors. In the range of sensitivity, these results are comparable to those of the commercial resonators of low- and high-frequency quartz crystals as well as the ELISA methods (Fernández-Benavides et al. 2019). Furthermore, an SPR immunosensor was developed to detect six pesticides, including: azoxystrobin, boscalid, chlorfenapyr, imazalil, isoxathion, and nitenpyram in tomato samples simultaneously. The pesticide detection is between 8.5–110 ng/mL and 3.5–50 ng/mL, respectively. Results of the analysis obtained by the SPR immunosensor and traditional HPLC or LC-MS/MS methods were in good agreement (Miyake et al. 2020).

An electrochemical immunoassay based on interdigitated array microelectrodes (IDAMs) was established for rapid, accurate, and highly sensitive analysis of chlorpyrifos. Cao et al. (2015) prepared a sensor platform based on gold microelectrode for attachment of anti-chlorpyrifos mAbs by protein A. In this fabricated sensor, chlorpyrifos is bound to the mAbs which causes change in impedance at the interface of IDAMs that are recorded using electrochemical impedance

Table 2 Pesticides analysis in different food matrices using immunosensors (Kumar et al. 2024, license number 6013180030605)

Type of sensor	Type of antibody	Pesticide	Food matrix	LOD	Linear range
Electrochemical	Anti-monocrotophos (monoclonal)	Monocrotophos	Cabbage (spiked)	0.49 ppm	1 ppt–1 ppm
Electrochemical	Anti-parathion (monoclonal)	Parathion	Cucumber, orange, and Chinese cabbage (all spiked)	2.2 pg/mL	0.01 ng/mL–100 ng/ml
Electrochemical	Anti-atrazine (monoclonal)	Atrazine	Plum juice, rice extract (all spiked)	0.08 ng/L	50 ng/L–30 μg/L
Electrochemical	Anti-glyphosate (monoclonal)	Glyphosate	Tea, soy milk (all spiked)	8.6 pg/mL	28.9 pg/mL–200 pg/ml
Electrochemical	Anti-carbaryl (monoclonal)	Carbaryl	Flour of wheat, oats, and corn (all spiked)	0.08 μg/kg	0.8 μg/kg–32.3 μg/kg
Optical	Anti-imidacloprid (monoclonal)	Imidacloprid	Apple juice	9.5 nM	10 nM–400 nM
Gravimetric	Anti-carbaryl (monoclonal)	Carbaryl	Honey (spiked)	0.03 μg/L	0.5 μg/L–100 μg/L

spectroscopy. In optimal conditions, the maximum impedance change was observed with chlorpyrifos concentration ranging from 100 to 105 ng/mL. This biosensor is extensively used for analyzing real samples like Chinese chives, cabbage, and lettuce. Guerrero-Esteban et al. (2021) developed an electrochemiluminescence (ECL)-immunosensor using electrografted carbon nanodots for rapid assessment of glyphosate in soy milk and tea. This sensor encompasses magnetic nanoparticles in a magnetic particle-based immunoassay in which mAbs (anti-glyphosate immunoglobulin G) were immobilized onto the magnetic particles. The utilization of this ECL sensors offers a highly sensitive and effective method for detection of glyphosate in food products.

8 Conclusion

The use of pesticides in agriculture has become inevitable due to the development of resistance by pests, microbes, and weeds that has resulted in significant crop losses. Notably, the persistence of these chemicals in both plant-based and animal-based food products poses serious health risks to humans. To address this challenge, regulatory bodies around the world have established maximum residue limits (MRLs) for various pesticides in food. As a result, many food testing laboratories have been established to monitor pesticide levels and ensure compliance with these MRLs using different testing methods. Among these methods, immunosensors stand out as an ideal choice for point-of-care testing. While many researchers have developed immunosensors for detecting pesticide residues in different foods, their widespread use at the commercial level is still highly recommended to fully realize their true potential.

References

Abolhassani M, Asadikaram G, Paydar P et al (2019) Organochlorine and organophosphorous pesticides may induce colorectal cancer; a case-control study. Ecotoxicol Environ Saf 178:168–177
Bajwa U, Sandhu KS (2014) Effect of handling and processing on pesticide residues in food-a review. J Food Sci Technol 51:201–220
Balkrishna A, Kumari A, Kumar A et al (2024) Biosensors for detection of pesticide residue, mycotoxins and heavy metals in fruits and vegetables: a concise review. Microchem J 205:111292
Baruah S, Dutta J (2009) Nanotechnology applications in pollution sensing and degradation in agriculture: a review. Environ Chem Lett 7:191–204
Bashir K, Shikha S, Rattu G et al (2025) Pesticide residues and their detection techniques in foods using sensors-a review. J Food Sci Technol 62:221–239
Beyuo J, Sackey LN, Yeboah C et al (2024) The implications of pesticide residue in food crops on human health: a critical review. Discov Agric 2(1):123
Bhardwaj JK, Mittal M, Saraf P et al (2018) Pesticides induced oxidative stress and female infertility: a review. Toxin Rev 39:1–13
Cao Y, Sun X, Guo Y et al (2015) An electrochemical immunosensor based on interdigitated array microelectrode for the detection of chlorpyrifos. Bioprocess Biosyst Eng 38:307–313
Colovic MB, Krstic DZ, Lazarevic-Pasti TD et al (2013) Acetylcholinesterase inhibitors: pharmacology and toxicology. Curr Neuropharmacol 11(3):315–335
Costas-Ferreira C, Durán R, Faro LR (2022) Toxic effects of glyphosate on the nervous system: a systematic review. Int J Mol Sci 23(9):4605
De Flora S, Viganò L, D'Agostini F et al (1993) Multiple genotoxicity biomarkers in fish exposed in situ to polluted river water. Mutat Res Toxicol 319:167–177
Đorđević T, Đurović-Pejčev R (2016) Food processing as a means for pesticide residue dissipation. Pestic Fitomed 31(3–4):89–105
Fang H, Yu Y, Chu X et al (2009) Degradation of chlorpyrifos in laboratory soil and its impact on soil microbial functional diversity. J Environ Sci 21:380–386
Farag AT, Radwan AH, Sorour F et al (2010) Chlorpyrifos induced reproductive toxicity in male mice. Reprod Toxicol 29:80–85
Fernández-Benavides DA, Cervera-Chiner L, Jiménez Y et al (2019) A novel bismuth-based lead-free piezoelectric transducer immunosensor for carbaryl quantification. Sens Actuators B: Chem 285:423–430
Flores MM, Singh B (2023) Neuromuscular blocking agents. In: Anesthesia and analgesia in laboratory animals. Academic Press, London, pp 187–201
Gray SL, Lee JA, Hovda LR et al (2011) Potential zinc phosphide rodenticide toxicosis in dogs: 362 cases (2004–2009). J Am Vet Med Assoc 239(5):646–651

Grimalt S, Dehouck P (2016) Review of analytical methods for the determination of pesticide residues in grapes. J Chromatogr A 1433:1–23

Guerrero-Esteban T, Gutiérrez-Sánchez C, Martínez-Periñán E et al (2021) Sensitive glyphosate electrochemiluminescence immunosensor based on electrografted carbon nanodots. Sens Actuators B Chem 330:129389

Guo Y, Liu R, Liu Y et al (2018) A non-competitive surface plasmon resonance immunosensor for rapid detection of triazophos residue in environmental and agricultural samples. Sci Total Environ 613:783–791

Gupta RC (2018) Non-anticoagulant rodenticides. In: Veterinary toxicology. Academic Press, London, pp 613–626

Jiang X, Li D, Xu X et al (2008) Immunosensors for detection of pesticide residues. Biosens Bioelectron 23:1577–1587

John EM, Shaike JM (2015) Chlorpyrifos: pollution and remediation. Environ Chem Lett 13:269–291

Kadirov M, Yusufova I, Azamatova F (2023) Liver biochemistry. Neutralization of toxic substances in the body. The role of the liver in metabolism. Sci Innov 2(6):97–100

Kaphle A, Navya PN, Umapathi A et al (2018) Nanomaterials for agriculture, food and environment: applications, toxicity and regulation. Environ Chem Lett 16:43–58

Kaur R, Goyal D (2019) Toxicity and degradation of the insecticide monocrotophos. Environ Chem Lett 17:1299–1324

Kumar H, Dhalaria R, Guleria S et al (2024) Immunosensors in food, health, environment, and agriculture: a review. Environ Chem Lett 22:2573–2605

Lazarević-Pašti T, Milanković V, Tasić T et al (2025) With or without you?—A critical review on pesticides in food. Food Secur 14(7):1128

LeDoux M (2011) Analytical methods applied to the determination of pesticide residues in foods of animal origin: a review of the past two decades. J Chromatogr A 1218:1021–1036

Meng X, Wei J, Ren X et al (2013) A simple and sensitive fluorescence biosensor for detection of organophosphorus pesticides using H_2O_2- sensitive quantum dots/bi-enzyme. Biosens Bioelectron 47:402–407

Mishra RK, Dominguez RB, Bhand S et al (2012) A novel automated flow-based biosensor for the determination of organophosphate pesticides in milk. Biosens Bioelectron 32:56–61

Miyake S, Hirakawa Y, Yamasaki T et al (2020) Simultaneous detection of six different types of pesticides by an immunosensor based on surface plasmon resonance. Anal Sci 36(3):335–340

Mnif W, Hassine AIH, Bouaziz A et al (2011) Effect of endocrine disruptor pesticides: a review. Int J Environ Res Public Health 8(6):2265–2303

Mukherjee S, Pal S, Pal A et al (2019) UIIS^{Scan} 1.1: a field portable high-throughput platform tool for biomedical and agricultural applications. J Pharm Biomed Anal 174:70–80

Mukherjee S, Pal S, Paria P et al (2021) On-spot biosensing device for organophosphate pesticide residue detection in fruits and vegetables. Curr Res Biotechnol 3:308–316

Mulyasuryani A, Prasetyawan S (2015) Organophosphate hydrolase in conductometric biosensor for the detection of organophosphate pesticides. Anal Chem Insights 10:23–27

Odukkathil G, Vasudevan N (2013) Toxicity and bioremediation of pesticides in agricultural soil. Rev Environ Sci Biotechnol 12:421–444

Oujji NB, Bakas I, Istamboulie G et al (2013) Sol–gel immobilization of acetylcholinesterase for the determination of organophosphate pesticides in olive oil with biosensors. Food Control 30:657–661

Patel H, Rawtani D, Agrawal YK (2019) A newly emerging trend of chitosan-based sensing platform for the organophosphate pesticide detection using acetylcholinesterase: a review. Trends Food Sci Technol 85:78–91

Rawlings NC, Cook SJ, Waldbillig D (1998) Effects of the pesticides carbofuran, chlorpyrifos, dimethoate, lindane, triallate, trifluralin, 2,4-D, and pentachlorophenol on the metabolic endocrine and reproductive endocrine system in ewes. J Toxicol Environ Health: Part A 54:21–36

Riah W, Laval K, Laroche-Ajzenberg E et al (2014) Effects of pesticides on soil enzymes: a review. Environ Chem Lett 12:257–273

Ribeiro FWP, Barroso MF, Morais S et al (2014) Simple laccase-based biosensor for formetanate hydrochloride quantification in fruits. Bioelectrochemistry 95:7–14

Ricking M, Schwarzbauer J (2012) DDT isomers and metabolites in the environment: an overview. Environ Chem Lett 10:317–323

Sapbamrer R, Kitro A, Panumasvivat J et al (2023) Important role of the government in reducing pesticide use and risk sustainably in Thailand: current situation and recommendations. Front Public Health 11:1141142

Schreinemachers DM, Creason JP, Garry VF (1999) Cancer mortality in agricultural regions of Minnesota. Environ Health Perspect 107(3):205–211

Sherma J (2015) Review of advances in the thin layer chromatography of pesticides: 2012−2014. J Environ Sci Health: Part B Pestic Food Contam Agric Wastes 50:301–316

Singh S, Singh N, Kumar V et al (2016) Toxicity, monitoring and biodegradation of the fungicide carbendazim. Environ Chem Lett 14:317–329

Singh S, Kumar V, Chauhan A et al (2018) Toxicity, degradation and analysis of the herbicide atrazine. Environ Chem Lett 16:211–237

Sivasankaran MA, Reddy SS, Govindaradjane S et al (2007) Organochlorine residuals in groundwater of Pondicherry region. J Environ Sci Eng 49:7–12

Suwalsky M, Rodríguez C, Villena F et al (2005) Human erythrocytes are affected by the organochloride insecticide chlordane. Food Chem Toxicol 43:647–654

Tan CW, Tan KH, Ong YT et al (2012) Energy and environmental applications of carbon nanotubes. Environ Chem Lett 10:265–273

Vale A, Lotti M (2015) Organophosphorus and carbamate insecticide poisoning. Handb Clin Neurol 131:149–168

Velasco-Garcia MN, Mottram T (2003) Biosensor technology addressing agricultural problems. Biosyst Eng 84:1–12

Verdian A (2018) Apta-nanosensors for detection and quantitative determination of acetamiprid: a pesticide residue in food and environment. Talanta 176:456–464

Verma ML (2017) Nanobiotechnology advances in enzymatic biosensors for the agri-food industry. Environ Chem Lett 15:555–560

Xie PP, Zong ZQ, Qiao JC et al (2024) Exposure to pesticides and risk of colorectal cancer: a systematic review and meta-analysis. Environ Pollut 345:123530

Yan X, Li H, Su X (2018) Review of optical sensors for pesticides. Trends Anal Chem 103:1–20

Yang Y, Wu N, Wang C (2018) Toxicity of the pyrethroid bifenthrin insecticide. Environ Chem Lett 16:1377–1391

Yang F-W, Li Y-X, Ren F-Z et al (2019) Toxicity, residue, degradation and detection methods of the insecticide triazophos. Environ Chem Lett 17:1769–1785

Application of Immunosensors in the Detection of Plant Pathogens

Sahil Kumar, Shagun Gupta, Rishi Thakur, and Ankur Kaushal

Abstract

Pathogenic plant microorganisms cause significant yield and economic losses in the agricultural sector. Plant diseases are caused by numerous pathogenic microbes, namely, fungi, bacteria, viruses, phytoplasmas, and nematodes. Sustaining maximum productivity requires rapid and early detection of diseases to avoid their spread and minimize associated productivity losses. Although chemical control is used to manage various diseases, it also has adverse effects on the environment and poses a heavy financial burden on farmers. Hence, early detection of diseases is considered the best approach for disease management and prevention. Conventional pathogen detection methods often involve methods based on microscopic visualization or techniques such as enzyme-linked immunosorbent assay (ELISA) and polymerase chain reaction (PCR). These methods are time-consuming and require specialized instruments/lab setups; hence, they are not considered suitable for in situ detection. In recent years, immunosensors have presented promising alternatives for pathogen detection. In this chapter, we present a possible application of immunosensors for plant pathogen detection, with various examples showing their efficiency. Improved and accurate plant pathogen detection by immunosensors can be used in integrated disease management.

Keywords

Plant pathogens · Immunosensor · Disease management

S. Kumar · S. Gupta · A. Kaushal (✉)
Department of Bio-Sciences and Technology, Maharishi Markandeshwar (Deemed to be) University, Mullana, Ambala, India

R. Thakur
Department of Chemistry and Biochemistry, College of Basic Sciences, CSK Himachal Pradesh Agriculture University, Palampur, Himachal Pradesh, India

K. Kuča et al. (eds.), *Ensuring Wellness: Immunosensors Monitoring Food Quality and Pathogenicity Risks*, https://doi.org/10.1007/978-3-031-97477-9_12

1 Introduction

Owing to their adverse impact on agricultural productivity, plant diseases are considered an important risk for global food security (Rohr et al. 2019). There are several plant pathogens, namely, fungi, bacteria, viruses, and nematodes that cause various diseases in plants (Vikal et al. 2023; Kaur et al. 2023). Sustaining productivity is crucial for satisfying the needs of the expending population (Širić et al. 2022). The yield is influenced by multiple biotic (diseases, insects, pests, weeds) and abiotic factors (climatic conditions, soil moisture and fertility, topography, and water quality) (Ngoune Liliane and Shelton Charles 2020; Shakya et al. 2021; Shahare et al. 2023; Vashishth et al. 2023). The Food and Agriculture Organization (FAO) reported 20%–40% global food production losses due to pests and diseases (Ahmad et al. 2023). Reducing agricultural losses by mitigating plant diseases is considered the primary and most efficient approach for disease management (HE et al. 2016). Additionally, effective post-harvest disease management is crucial to minimize losses, maintain quality, and ensure the safety of agricultural produce (Bisht et al. 2021). Disease is an outcome of the host, a favourable environment, and the pathogen (Fig. 1) (Jones and Barbetti 2012). Typically, plants exhibit symptoms after infection, but many diseases produce similar signs, making accurate diagnosis of the disease difficult. The initial and asymptomatic stages often go undetected, resulting in significant economic losses (Buja et al. 2021). Early diagnosis of plant diseases enables timely intervention, reducing crop losses, economic impacts, and reliance on chemicals, thereby improving yield and sustainability (John et al. 2023).

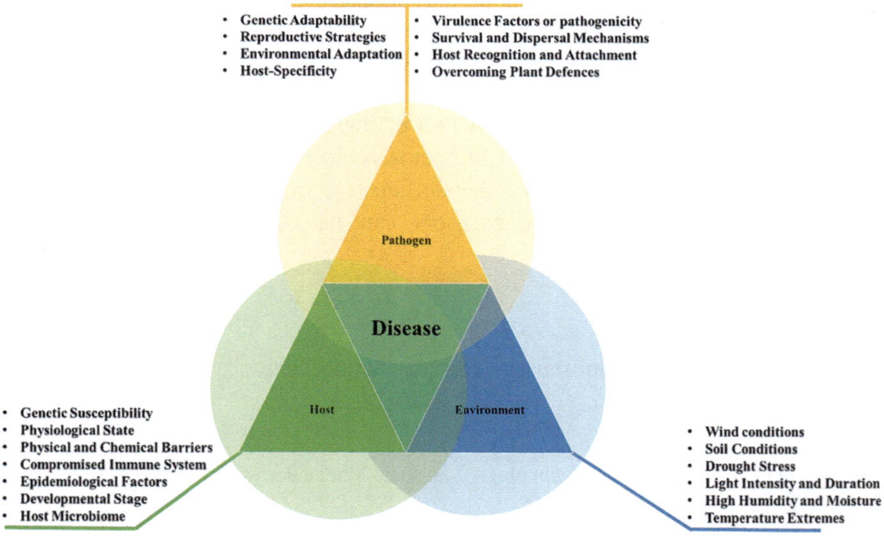

Fig. 1 A schematic representation of the disease triangle illustrates the interactions and relationships among the host, pathogen, and environment, highlighting the factors that contribute to disease development

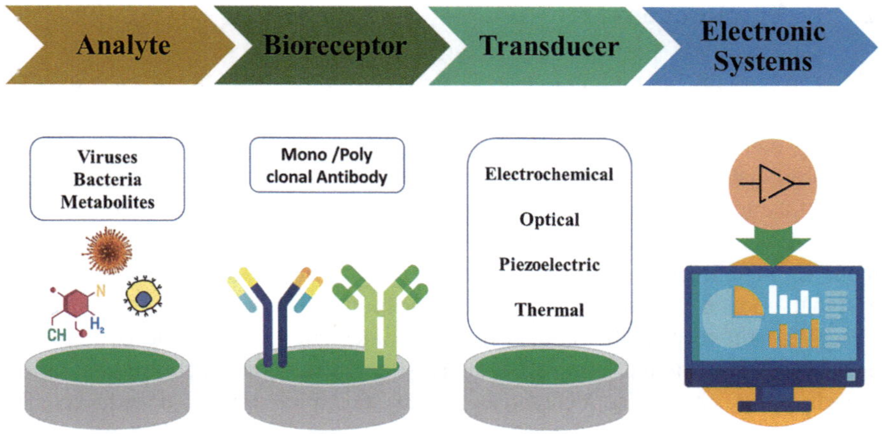

Fig. 2 Illustration showing the basic principle of immunosensor components

Diagnosis is clearly a critical step in precision agriculture for effective and precise disease management (Zou et al. 2020). Many diagnostic techniques worldwide target genetic material and antigens related to pathogens as direct indicators of their presence in plant systems (Khater et al. 2017). Generally, laboratory-based methods such as culture, polymerase chain reaction (PCR), and enzyme-linked immunosorbent assay (ELISA) are considered the gold standard methods for accurate disease diagnosis (Kakkar et al. 2023). However, all these methods have several drawbacks, including being expensive, laborious, time-consuming, and prone to cross-reactivity with limited sensitivity (Martin et al. 2000). Considering these limitations, significant research in recent years has been conducted on biosensors for disease detection, as they have high specificity and sensitivity. Technology that is simple, specific, and reliable and provides rapid diagnosis is always needed (Yager et al. 2006).

A biosensor is an analytical device that produces electrical signals in response to biological interactions, utilizing components such as enzyme, cells, antibodies, and DNA (Naresh and Lee 2021; Manjunathan et al. 2024). The electrical signal conveys information about physiological changes caused by the interaction of a bioreceptor with an analyte. Immunosensors are a type of affinity-based biosensor in which antibody-dependent bioreceptors are used to detect different analytes (Fig. 2) (Prajapati and Shrivastav 2023). Once binding occurs, a transducer converts this interaction into a measurable signal, which can be electrochemical, optical, piezoelectric, or thermal, depending on the type of transducer used (Polat et al. 2022). Immunosensors offer many advantages over conventional methods in terms of disease diagnostics in point-of-care testing (POCT) (Hou et al. 2023). In this context, we summarize recent findings and advancements in the implementation of immunosensors for plant disease detection.

2 Immunosensors

The term 'immune' in immunosensors is derived from immunoglobulins, commonly known as antibodies. It belongs to the immunoglobulin superfamily and consists of 2 heavy and 2 light chains (Lefranc and Lefranc 2020). The immunosensor is a detection system that quantifies antigen–antibody interactions or, in some cases, detects stable immunocomplexes formed during interactions (Karunakaran et al. 2015). Detection via an immunosensor relies on direct and indirect detection methods. In indirect detection techniques, an enzyme linked to an antibody generates a signal upon a binding event. Direct detection is often faster and more cost-effective, as it records the antigen–antibody interaction without the need for any enzyme (Porstmann and Kiessig 1992). This rapid detection is achieved by various transducers, such as electrochemical, optical, and piezoelectric transducers, that produce signals proportional to the concentration of the analyte (Feyziazar et al. 2022). These transducers are currently extensively utilized for the detection of pathogens. To the best of our knowledge, the application of these transducers in the field of plant pathogen detection is discussed in the following paragraph.

2.1 Electrochemical Immunosensors

The fundamental principle of an electrochemical biosensor is to detect changes occurring near the electrode surface, where electron flow takes place between the electrode and the electrolytes in the solution (Leva-Bueno et al. 2020). A biosensor consists of a bioreceptor that identifies the target analyte and a transducer that converts this recognition event into a measurable electrical signal. The bioreceptors can be tissue, cells, enzymes, antibodies, nucleic acids, microorganisms, or organelles (Noori and Abdulameer 2023). Common transducing elements include electrochemical, optical, and piezoelectric or combinations of these. Transducers in electrochemical immunosensors are used to detect interactions between antigens and antibodies where they recognize changes in electrochemical signals (Felix and Angnes 2018). Electrochemical variations in different parameters, such as potential, current, ion concentration, conductance, capacitance or impedance, exist (Kumar et al. 2023). Electrochemical transducers can be classified into four classes, amperometric, conductometric, impedimetric, and potentiometric, on the basis of the type of electrochemical signal used (Kumar et al. 2024). Nanotechnology is an emerging breakthrough with vast potential across various fields, including agriculture, electronics, healthcare, and the food industry. Applications include nanosensors, nano-packaging, and nano-fertilizers (Chakroborty et al. 2023). Nanoparticles (NPs) are also employed as signal amplifiers, enhancing detection by making signal variations more identifiable, and numerous reports highlight their utilization in plant pathogen detection (Cardoso et al. 2022). For example, nanoparticle-based biosensors have been designed that amplify analytical signals and significantly increase the range of detection. Wide applications of nanoparticles, specifically in immunosensors, have been reported over the last few decades (Rauf et al. 2022; Pramanik et al. 2023).

Some of the electrochemical immunosensors mentioned below have been shown to be effective in plant disease diagnosis.

Graphene oxide (GO) is among the most common carbon nanoparticles utilized for biosensor development because of its high conductivity, chemical stability, strong mechanical strength, and biocompatibility (Mishra et al. 2021; Eivazzadeh-Keihan et al. 2022; Rauf et al. 2022). A graphene oxide-based electrochemical immunosensor was used to detect groundnut bud necrosis orthotospovirus. The sensor shows excellent sensitivity and a low limit of detection of 5.7 ng/ml and is able to detect targeted viruses in multiple hosts (Chaudhary et al. 2021). On the other hand, a novel label-free electrochemical immunosensor based on a bamboo-like multiwall carbon nanotube-ionic liquid nanocomposite deposited on a glassy carbon electrode was developed. Many possibilities exist regarding the application of immunosensors in the field of plant disease diagnostics (Haji-Hashemi et al. 2018). An immunosensor was constructed to target Candidatus Phytoplasma Aurantifoli (CPA), the main factor involved in witches' broom disease. Polyclonal anti-CPA was fabricated on a gold electrode using gold nanoparticles. The developed sensor shows good sensitivity and seems appropriate for CPA detection in a real sample matrix (Ebrahimi et al. 2019). Similarly in an another approch, a gold nanoparticle-based immunometric immunosensor for the detection of *Pseudomonas syringae* pv. *lachrymans* (*Psl*) was presented. This label-free electrochemical sensor detects 337 CFU/ml without interfering with the negative probe or other bacteria (Cebula et al. 2019). Among all types of pathogens, viruses are often the most silent, as they typically remain unnoticed and undetected (Barba et al. 2015). The symptoms caused by viruses are usually very common, making them difficult to identify specifically (Barbedo 2016). Immunosensors targeting viruses have become a major focus in the field of plant pathogen detection. A mercaptoundecanoic acid (MUA)- and 3-mercaptopropionic acid (MPA)-modified gold electrode-based immunosensor with good sensitivity was proposed to detect *citrus tristeza virus* by targeting the coat protein (Haji-Hashemi et al. 2017). Recently, biosensors have gained momentum, utilizing interactions for high specificity and versatility, whereas immunosensors, which rely on antigen–antibody interactions, achieve exceptional specificity owing to their unique chemical composition (Barbedo 2016). Early detection and prediction of disease is always preferred under integrated pest management of any crop; hence, an immuno-impedance biosensor was developed to monitor ascospores for accurate forecasting of Sclerotinia stem rot in Canola. This immunosensor offers early detection of ascospores in plants, aiding in effective disease management. Alternatively, immunosensors utilizing a glassy carbon electrode have been developed for the detection of Prunus necrotic ringspot virus (PNRSV). The sensor clearly differentiates between healthy and infected samples via the use of an anti-PNRSV IgG polyclonal antibody (Jarocka et al. 2013). Viruses are challenging pathogens to detect because of the complexity of their infection mechanism and the often-indistinguishable nature of their symptoms. Impedimetric immunosensors are highly useful electrochemical immunosensors, and one was specifically developed to detect plum pox virus in plant extracts from plum and tobacco. The sensor was based on 1,6-hexanedithiol, gold nanoparticles, and anti-PPV IgG polyclonal antibody-modified gold electrodes. The proposed electrochemical immunosensor showed a good detection limit of 10 pg/mL. The sensor was also found to be very specific when tested, and no interference from the components of the plant extract was observed (Jarocka et al. 2011). Timely diagnosis of plant viruses is crucial for effective control. An immunosensor was developed for the detection of capsicum chlorosis virus (CaCV) in bell pepper via nano-Au/CMWCNT electrodes. Compared with direct antigen-coated enzyme-linked immunosorbent assay (DAC-ELISA), the sensor was 800–1000 times more sensitive, offering a rapid and economic biosensor for the detection of the CaCV virus in bell pepper leaves (Sharma et al. 2017). Hence, an electrochemical immunosensor was developed that targets the coat protein of papaya ringspot virus (PRSV). The developed sensor presented rapid and cost-effective detection capabilities (Valekunja et al. 2016). An electrochemical immunosensor was developed for the detection of bacteria from the Pseudomonas genus. The sensor shows high specificity when tested against non-target bacteria and has an LOD of 0.119 CFU (Kowalski et al. 2022). An electrochemical immunosensor based on gold nanoparticles and reduced graphene oxide sheets was developed for the detection of potato aucuba mosaic virus (PAMV). The authors claimed to have achieved limits of detection (LODs) of 0.96×10^{-12} M and 0.065×10^{-12} M, respectively (Verma et al. 2024). A novel label-free electrochemical immunosensor was developed for the detection of important pathogenic viruses (chicken chlorotic dwarf virus) in chickpea. Cyclic voltammetry and electrochemical impedance spectroscopy were utilized to discover the changes in the fabricated electrode. In this sensor, a glassy carbon electrode was modified with an ionic liquid (IL), reduced graphene oxide (rGO), and Prussian blue (PB) to form a AuNPs/PB-rGO-IL/GCE electrode, which has excellent electrochemical activity. The presented electrochemical immunosensor has a detection limit of 0.01 μg/mL with good reproducibility, stability, and durability (Habibi et al. 2022). A competitive immunosensor using carbon-based screen-printed electrodes (SPECs) with multiwalled carbon nanotubes (CNTs) was developed for the detection of *Botrytis cinerea*, which causes grey mould in apples. The sensor used enzyme-labelled antibodies fabricated on a multiwalled carbon nanotube-modified screen-printed electrode. The sensor achieved a detection limit of 0.02 g/mL with a processing time of 30 min, surpassing that of ELISA. It enables early detection before symptoms appear and helps in the timely management of diseases and crop sustainability (Fernández-Baldo et al. 2009). Floriculture is a thriving sector of agriculture, leading to much research in the field of flower production and management (De and Singh 2016). As disease diagnosis is an important step in the management of any crop, a sensitive label-free electrochemical immunosensor was developed for the simultaneous detection of the following *Phalaenopsis* viruses: Odontoglossum ringspot virus (ORSV) and Cymbidium mosaic virus (CymMV). A nano/micro hybrid gold electrode prepared via lithography and self-assembled monolayers facilitated detection via differential pulse voltammetry. The biosensor achieved a detection limit of 0.5 ng/mL, effectively identifying viruses in leaf sap diluted 100-fold and offering rapid diagnostic ability (Lin et al. 2022). Owing to their striking characteristics, carbon-based nanoparticles are widely utilized to make

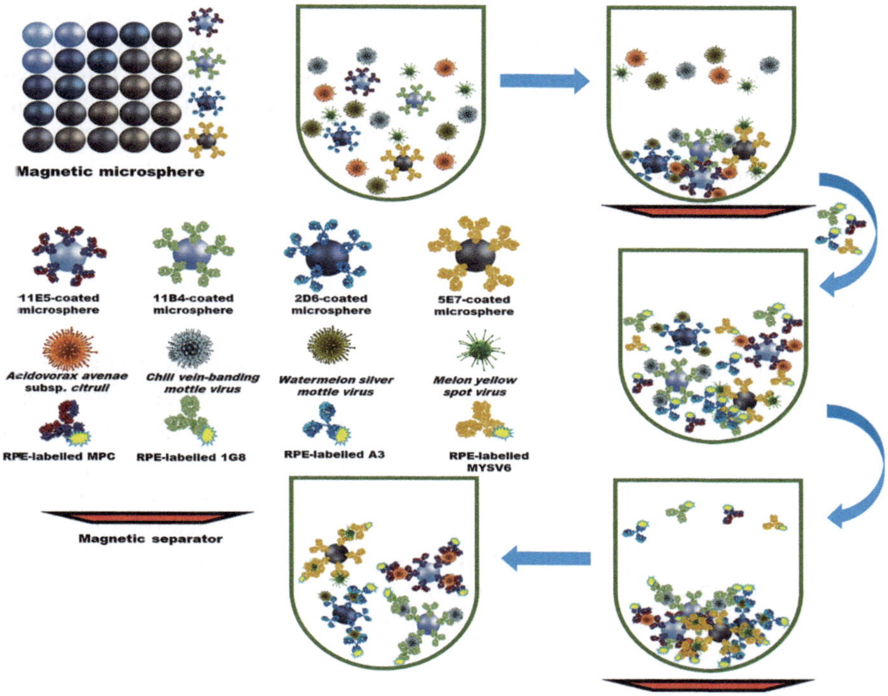

Fig. 3 Schematic illustration showing multiplexing in plant viral disease by using magnetic microspheres

diagnostic sensors and disease treatment therapies (Yeganeh et al. 2022). Polypyrrole- and functionalized multiwall carbon nanotube (fMWCNT)-modified screen-printed carbon electrode-based immunosensors were presented for onsite detection of bacterial blight disease in rice (Razali et al. 2022). Similarly, a highly sensitive detection system was developed to detect *Pantoea srewartii* with the aim of controlling associated agricultural economic losses. Compared to the conventional enzyme-linked immunosorbent assay (ELISA), the developed immunosensor has a 20-fold lower LOD, with a value of 7.8×10^3 cfu/mL (Zhao et al. 2014).

A voltammetric immunosensor based on magnetic beads and a sandwich immunoassay with enzyme labelling was designed for rapid and early detection of pathogens (*Phakopsora pachyrhizi*) causing soybean rust. The sensor targets fungal mycelium, enabling swift disease detection, has high specificity, and achieves a detection limit of 90 ng/mL, offering a field-ready disposable diagnostic tool for rust (Mendes et al. 2012). The authors also presented another electrochemical immunosensor for the detection of soybean rust with a detection limit of 385 ng/mL (Mendes et al. 2009). Magnetic microsphere immunoassays have already been reported as a novel means of multiplexing in plant pathogen detection. Fluorescently coded microspheres are conjugated with specific antibodies to target multiple analytes (Fig. 3) (Charlermroj et al. 2013). A microfluidic immunosensor, coupled with an electrochemical technique, was presented for rapid and sensitive quantification of grey mould-causing pathogens. Grey mould caused by *Botrytis* cinerea affects many crops, including fruits such as apples, grapes, and pears. This sensor carries micromagnetic beads and screen-printed carbon electrodes labelled with anti-*B. cinera* antibody. Using a competitive immunoassay, this sensor achieved a detection limit of 0.008 μg/mL within 40 min, enabling early detection of asymptomatic pathogens of grey mould (Fernández-Baldo et al. 2010).

2.2 Optical Immunosensors

Optical sensors include colorimetric biosensors, fluorescence-based assays, and surface plasmon resonance-based biosensors. Colorimetric sensors allow the immediate detection of pathogens in a sample by colour change. Moreover, optical immunosensors rely on target molecules or antibody interactions. A surface plasmon resonance (SPR) immunosensor was developed for the detection of black Sigatoka disease in banana plantations (Luna-Moreno et al. 2019). Similarly, a label-free chemiresistive immunosensor, utilizing polypyrrole nanoribbons, was developed for detecting cucumber mosaic virus (Chartuprayoon et al. 2013). The utilization of SPR immunosensors has been shown frequently in the field of plant pathogen diagnostics. Since fungal pathogens are major causes of plant diseases, an SPR-based immunosensor was developed for the rapid detection of soybean rust caused by *Phakopsora pachyrhizi*, which uses anti-*P. pachyrhizi* antibodies immobilized on a gold substrate via a self-assembled monolayer. The sensor demonstrated a linear response range of 3.5–28.0 μg/mL with a detection limit of 800 ng mL^{-1}, indicating an amazing specificity for soybean rust. *Tilletia indica* is a quarantined pathogen that is challenging to differentiate from other bunt fungi because of its similar teliospores. An innovative SPR platform was developed on the basis of a nanogold lateral flow immune-dipstick assay. This portable immunosensor has wide applicability in seed certification and plant quarantine regulations through rapid on-site testing (Singh et al. 2010). A surface plasmon resonance (SPR) immunosensor using a monoclonal antibody (phyt/G1470 mAb) was developed, which showed high specificity and minimal cross-reactivity. This sensor was developed to target late blight of potatoes caused by *Phytophthora infestans* and demands early detection for effective control. The immunosensor had a detection limit of 2.2×10^6 sporangia/mL and a rapid 75-mi analysis time; thus, it outperforms existing methods (Skottrup et al. 2007). This is novel work showing the implementation of surface plasmon resonance (SPR) immunosensors for the detection of *Tilletia indica* teliospores, the causal agent of Karnal bunt disease in wheat. With the use of a rabbit polyclonal anti-teliospore antibody, the label-free SPR system detected as few as 2.5 teliospores (625 pg) with high sensitivity. Cross-reactivity analysis revealed some overlap, which was justified by its polyclonal nature and antigenic homology to the pathogen (Singh 2012). This gold-coated chip-based immunosensor targets HF1 (the cell wall protein of *P. flijiensis*) with an anti-HF1 antibody. The sensor shows an amazing sensitivity of 0.0021 units of reflectance per ng mL^{-1} and achieves a detection limit of

11.7 μg mL^{-1}. In another attempt, a diatom-based photoluminescence immunosensor was developed to detect Karnal bunt disease in wheat. Diatoms consisting of a nanoporous three-dimensional biogenic silica material were used in this immunosensor because of their unique photoluminescence under UV light (Mishra et al. 2020). Postharvest agricultural production losses have always been a major concern, and a nanostructured immunosensor was developed from early and in situ detection of *Xanthomonas arboricola* in walnuts. This microfluidic immunosensor is very promising, as it has a limit of detection of 1.5×10^2 CFU mL^{-1}. The integration of microfluidic techniques and optics has emerged as optofluidics, which has high applicability in diagnostics (Regiart et al. 2017). Rice is an important staple crop among various countries worldwide, and tungro disease is considered a major constraint for production in South and Southeast Asia. With this consideration, an immunosensor was developed that targets rice tungro spherical virus (RTSV) and rice tungro bacilliform virus (RTBV) with an anti-RTSV/anti-RTBV antibody. These results support the potential application of immunosensors in virus detection (Uda et al. 2017).

2.2.1 Fluorescence Immunosensors

Fluorescence immunosensors were developed to detect *Xanthomonas oryzae* pv. *Oryzae (Xoo)* in rice leaves. Changes in the intensity of fluorescence resulting from the interaction between AntiXoo and *Xoo* cells were recorded (Awaludin et al. 2020). A paper-based immunosensor for the early detection of soybean rust was developed using a nitrocellulose membrane modified with fluorescent nanoparticle-labelled polyclonal antibodies. It offers high sensitivity, with a detection limit of 2.2 ng/mL, outperforming traditional diagnostic techniques such as ELISA and PCR (Miranda et al. 2013). A novel FRET-based biosensor was developed for the detection of Rhizomania. It is caused by Beet necrotic yellow vein virus (BNYVV), is transmitted by *Polymyxa betae*, and is a devastating disease in sugar beet. This sensor is a conjugated nano-based immunosensor utilizing quantum dots (CdTe-QDs) biofunctionalized with specific antibodies against *P. betae*. This FRET technology-derived sensor demonstrated 100% sensitivity, specificity, and stability, marking the first nano biosensor for the detection of plant pathogenic fungi (Safarpour et al. 2012).

2.2.2 Colorimetric Immunosensors

A cost-effective, rapid colorimetric sensor based on magnetic nanozymes was developed for virus detection in sugarcane. This method overcomes the limitations of traditional diagnostic methods for sugarcane yellow leaf curl virus (ScYLV), revolutionizing sugarcane production and enhancing global food security (Thangavelu et al. 2024). A portable biosensor utilizing lateral flow immunoassays and thin-layer chromatography enables field testing with sensitivity on par with the gold standards of diagnostics. Recent advances in optical techniques, including efficient detectors integrated into portable devices, offer accessible and cost-effective solutions for analytical applications, with promising practical use in diagnostics (Pohanka 2020).

2.2.3 Piezoelectric Immunosensors

Piezoelectric immunosensors utilize quartz crystal balance (QCM) technology to detect biomolecular interactions through changes in resonance frequency (Jandas et al. 2021). By immobilizing specific antibodies on the sensor surface, they enable label-free, real-time detection of target analytes with high sensitivity (Ray et al. 2010). These sensors are widely applicable, but our focus is on their potential for detecting only plant pathogens. *Tilletia indica*, a fungal pathogen causing Karnal bunt disease in wheat crop. A label-free immunosensor using an anti-teliosporic monoclonal antibody immobilized on 1-mercaptoundecanoic acid (11-MUA) was developed to target pathogens with high accuracy. The sensor showed limits of detection of 4.4 pg mL^{-1} and 12.5 pg mL^{-1} when analysed by a quartz crystal microbalance (QCM) and cyclic voltammetry (CV), respectively (Sharma et al. 2022). The QCM-based immunosensor was developed for the detection of plant growth-promoting rhizobacteria (PGPRs), which are rich in lipopolysaccharides and induce immune responses, particularly in Cd-resistant strains at Cd-affected sites. A piezoelectric immunosensor was developed using anti-Z9 or anti-S2 antibodies, offering high sensitivity and specificity. Compared with ELISA, the immunosensor was faster and more sensitive, enabling future PGPR detection automation (Agrawal 2012).

3 Conclusion and Future Aspects

Immunosensors are emerging as powerful tools for the rapid, sensitive, and specific detection of plant pathogens, offering several advantages over conventional diagnostic techniques. Traditional methods such as PCR, ELISA, and culture are time-consuming, expensive, and often lack the sensitivity and specificity required for early disease detection. On the other hand, immunosensors provide real-time, on-site disease diagnosis with improved efficiency. They work by detecting antigen–antibody interactions, which are highly specific, ensuring that the sensor can target the nearly exact pathogen of interest with minimal cross-reactivity. The development of various types of immunosensors, including electrochemical, optical, fluorescence, colorimetric, and piezoelectric-based sensors, has broadened their applicability in plant pathogen detection. Electrochemical sensors, particularly those enhanced with zero- or multidimensional nanoparticles, have demonstrated significant advancements in sensitivity and detection limits. Optical sensors, such as surface plasmon resonance (SPR) and fluorescence-based sensors, can be used for rapid detection, making them suitable for field diagnostics. Fluorescence resonance energy transfer (FRET) technology is another promising avenue, showing great potential for pathogen detection with high accuracy. Colorimetric sensors, which provide visible results, and piezoelectric sensors, which offer real-time monitoring, further increase the potential of immunosensors for plant disease management. These advancements highlight the growing role of immunosensors in precision agriculture, allowing farmers and researchers to quickly identify and manage plant diseases, minimize crop losses, and maximize food security. The future of immunosensors in plant disease detection looks promising, with ongoing research focused on improving their sensitivity, specificity, and cost-effectiveness. Advances in nanotechnology, particularly the use of other nanoparticles and nanocomposites, are expected to enhance the performance of immunosensors, enabling faster and more accurate detection. Moreover, the integration of immunosensors with other advanced techniques and portable devices could further enable field-based diagnostics, providing farmers with accessible tools/devices for real-time disease monitoring. There is a growing need for collaborative research between agricultural research institutes and biosensor researchers to facilitate the rapid and precise development of sensors. Moreover, there is significant potential to expand the use of multiplexed sensors, which can detect multiple pathogens at once, providing a more comprehensive approach to plant disease management. As research continues, the combination of immunosensors with artificial intelligence and machine learning could lead to even smarter diagnostic tools, increasing the speed and accuracy of disease predictions and interventions.

References

Agrawal R (2012) Rapid detection of cadmium-resistant plant growth promotory rhizobacteria: a perspective of ELISA and QCM-based immunosensor. J Microbiol Biotechnol 22(6):849–855. https://doi.org/10.4014/jmb.1108.08055

Ahmad A, Saraswat D, El Gamal A (2023) A survey on using deep learning techniques for plant disease diagnosis and recommendations for development of appropriate tools. Smart Agric Technol 3:100083. https://doi.org/10.1016/j.atech.2022.100083

Awaludin N, Abdullah J, Salam F et al (2020) Fluorescence-based immunoassay for the detection of Xanthomonas oryzae pv. oryzae in rice leaf. Anal Biochem 610:113876. https://doi.org/10.1016/j.ab.2020.113876

Barba M, Ilardi V, Pasquini G (2015) Control of pome and stone fruit virus diseases. Adv Virus Res 91:47–83

Barbedo JGA (2016) A review on the main challenges in automatic plant disease identification based on visible range images. Biosyst Eng 144:52–60. https://doi.org/10.1016/j.biosystemseng.2016.01.017

Bisht B, Bhatnagar P, Gururani P et al (2021) Food irradiation: effect of ionizing and non-ionizing radiations on preservation of fruits and vegetables– a review. Trends Food Sci Technol 114:372–385. https://doi.org/10.1016/j.tifs.2021.06.002

Buja I, Sabella E, Monteduro AG et al (2021) Advances in plant disease detection and monitoring: from traditional assays to in-field diagnostics. Sensors 21(6):2129. https://doi.org/10.3390/s21062129

Cardoso RM, Pereira TS, Facure MHM et al (2022) Current progress in plant pathogen detection enabled by nanomaterials-based (bio)sensors. Sens Actuators Rep 4:100068. https://doi.org/10.1016/j.snr.2021.100068

Cebula Z, Żołędowska S, Dziąbowska K et al (2019) Detection of the plant pathogen Pseudomonas syringae pv. lachrymans on antibody-modified gold electrodes by electrochemical impedance spectroscopy. Sensors 19(24):5411. https://doi.org/10.3390/s19245411

Chakroborty S, Pal K, Nath N et al (2023) Sustainable synthesis of multifunctional nanomaterials from rice wastes: a comprehensive review. Environ Sci Pollut Res 30(42):95039–95053. https://doi.org/10.1007/s11356-023-29235-9

Charlermroj R, Himananto O, Seepiban C et al (2013) Multiplex detection of plant pathogens using a microsphere immunoassay technology. PLoS One 8(4):e62344. https://doi.org/10.1371/journal.pone.0062344

Chartuprayoon N, Rheem Y, Ng JCK et al (2013) Polypyrrole nanoribbon based chemiresistive immunosensors for viral plant pathogen detection. Anal Methods 5(14):3497. https://doi.org/10.1039/c3ay40371h

Chaudhary M, Verma S, Kumar A et al (2021) Graphene oxide based electrochemical immunosensor for rapid detection of groundnut bud necrosis orthotospovirus in agricultural crops. Talanta 235:122717. https://doi.org/10.1016/j.talanta.2021.122717

De LC, Singh DR (2016) Floriculture industries, opportunities and challenges in Indian Hills. Int J Hortic. https://doi.org/10.5376/ijh.2016.06.0013

Ebrahimi M, Norouzi P, Safarnejad MR et al (2019) Fabrication of a label-free electrochemical immunosensor for direct detection of Candidatus Phytoplasma Aurantifolia. J Electroanal Chem 851:113451. https://doi.org/10.1016/j.jelechem.2019.113451

Eivazzadeh-Keihan R, Bahojb Noruzi E, Chidar E et al (2022) Applications of carbon-based conductive nanomaterials in biosensors. Chem Eng J 442:136183. https://doi.org/10.1016/j.cej.2022.136183

Felix FS, Angnes L (2018) Electrochemical immunosensors – a powerful tool for analytical applications. Biosens Bioelectron 102:470–478. https://doi.org/10.1016/j.bios.2017.11.029

Fernández-Baldo MA, Messina GA, Sanz MI et al (2009) Screen-printed immunosensor modified with carbon nanotubes in a continuous-flow system for the Botrytis cinerea determination in apple tissues. Talanta 79(3):681–686. https://doi.org/10.1016/j.talanta.2009.04.059

Fernández-Baldo MA, Messina GA, Sanz MI et al (2010) Microfluidic immunosensor with micromagnetic beads coupled to carbon-based screen-printed electrodes (SPCEs) for determination of Botrytis cinerea in tissue of fruits. J Agric Food Chem 58(21):11201–11206. https://doi.org/10.1021/jf1025604

Feyziazar M, Amini M, Jahanban-Esfahlan A et al (2022) Recent advances on the piezoelectric, electrochemical, and optical biosensors for the detection of protozoan pathogens. TrAC Trends Anal Chem 157:116803. https://doi.org/10.1016/j.trac.2022.116803

Habibi MM, Norouzi P, Hashemian E et al (2022) Development of a sensitive label-free electrochemical immunosensor for detection of chickpea chlorotic dwarf virus. Diam Relat Mater 128:109203. https://doi.org/10.1016/j.diamond.2022.109203

Haji-Hashemi H, Norouzi P, Safarnejad MR et al (2017) Label-free electrochemical immunosensor for direct detection of Citrus tristeza virus using modified gold electrode. Sensors Actuators B Chem 244:211–216. https://doi.org/10.1016/j.snb.2016.12.135

Haji-Hashemi H, Habibi MM, Safarnejad MR et al (2018) Label-free electrochemical immunosensor based on electrodeposited Prussian blue and gold nanoparticles for sensitive detection of citrus bacterial canker disease. Sensors Actuators B Chem 275:61–68. https://doi.org/10.1016/j.snb.2018.07.148

He D, Zhan J, Xie L (2016) Problems, challenges and future of plant disease management: from an ecological point of view. J Integr Agric 15(4):705–715. https://doi.org/10.1016/S2095-3119(15)61300-4

Hou F, Sun S, Abdullah SW et al (2023) The application of nanoparticles in point-of-care testing (POCT) immunoassays. Anal Methods 15(18):2154–2180. https://doi.org/10.1039/D3AY00182B

Jandas PJ, Prabakaran K, Luo J et al (2021) Effective utilization of quartz crystal microbalance as a tool for biosensing applications. Sensors Actuators A Phys 331:113020. https://doi.org/10.1016/j.sna.2021.113020

Jarocka U, Wąsowicz M, Radecka H et al (2011) Impedimetric immunosensor for detection of plum pox virus in plant extracts. Electroanalysis 23(9):2197–2204. https://doi.org/10.1002/elan.201100152

Jarocka U, Radecka H, Malinowski T et al (2013) Detection of prunus necrotic ringspot virus in plant extracts with impedimetric immunosensor based on glassy carbon electrode. Electroanalysis 25(2):433–438. https://doi.org/10.1002/elan.201200470

John MA, Bankole I, Ajayi-Moses O et al (2023) Relevance of advanced plant disease detection techniques in disease and pest management for ensuring food security and their implication: a review. Am J Plant Sci 14(11):1260–1295. https://doi.org/10.4236/ajps.2023.1411086

Jones RAC, Barbetti MJ (2012) Influence of climate change on plant disease infections and epidemics caused by viruses and bacteria. CABI Rev:1–33. https://doi.org/10.1079/PAVSNNR20127022

Kakkar S, Gupta P, Kumar N et al (2023) Progress in fluorescence biosensing and food safety towards point-of-detection (PoD) system. Biosensors 13(2):249. https://doi.org/10.3390/bios13020249

Karunakaran C, Pandiaraj M, Santharaman P (2015) Immunosensors. In: Biosensors and bioelectronics. Elsevier, Amsterdam, pp 205–245

Kaur R, Gupta S, Tripathi V et al (2023) Microbiome based approaches for the degradation of polycyclic aromatic hydrocarbons (PAHs): a current perception. Chemosphere 341:139951. https://doi.org/10.1016/j.chemosphere.2023.139951

Khater M, de la Escosura-Muñiz A, Merkoçi A (2017) Biosensors for plant pathogen detection. Biosens Bioelectron 93:72–86. https://doi.org/10.1016/j.bios.2016.09.091

Kowalski M, Brodowski M, Dziabowska K et al (2022) Electrochemical detection of plant pathogens using boron-doped carbon nanowalls immunosensor. IEEE Sensors J 22(8): 7562–7571. https://doi.org/10.1109/JSEN.2022.3157037

Kumar N, Ansari MNM, Upadhyay S et al (2023) Blending of a 3D cloud-like morphology with a 1D structure in a VO 2/MXene nanocomposite to enhance the charge storage capability. J Mater Chem C 11(48):17022–17033. https://doi.org/10.1039/D3TC02738D

Kumar S, Poria R, Kala D et al (2024) Recent advances in ctDNA detection using electrochemical biosensor for cancer. Discov Oncol 15(1):517. https://doi.org/10.1007/s12672-024-01365-7

Lefranc M-P, Lefranc G (2020) Immunoglobulins or antibodies: IMGT® bridging genes, structures and functions. Biomedicines 8(9):319. https://doi.org/10.3390/biomedicines8090319

Leva-Bueno J, Peyman SA, Millner PA (2020) A review on impedimetric immunosensors for pathogen and biomarker detection. Med Microbiol Immunol 209(3):343–362. https://doi.org/10.1007/s00430-020-00668-0

Lin WP, Wang WJ, Lee CH et al (2022) A two-in-one immunoassay biosensor for the simultaneous detection of Odontoglossum ringspot virus and Cymbidium mosaic virus. Sensors Actuators B Chem 350:130875. https://doi.org/10.1016/j.snb.2021.130875

Luna-Moreno D, Sánchez-Álvarez A, Islas-Flores I et al (2019) Early detection of the fungal banana black Sigatoka pathogen Pseudocercospora fijiensis by an SPR immunosensor method. Sensors 19(3):465. https://doi.org/10.3390/s19030465

Manjunathan J, Revathi M, Sowmya H et al (2024) Recent advancements in nanotechnological approaches for pollution monitoring and environmental sustainability. Clean Techn Environ Policy 26(11):3667–3683. https://doi.org/10.1007/s10098-023-02676-z

Martin RR, James D, Lévesque CA (2000) Impacts of molecular diagnostic technologies on plant disease management. Annu Rev Phytopathol 38(1):207–239. https://doi.org/10.1146/annurev.phyto.38.1.207

Mendes RK, Ferreira DCM, Carvalhal RF et al (2009) Development of an electrochemical immunosensor for Phakopsora pachyrhizi detection in the early diagnosis of soybean rust. J Braz Chem Soc 20(4):795–801. https://doi.org/10.1590/S0103-50532009000400023

Mendes RK, Laschi S, Stach-Machado DR et al (2012) A disposable voltammetric immunosensor based on magnetic beads for early diagnosis of soybean rust. Sensors Actuators B Chem 166–167:135–140. https://doi.org/10.1016/j.snb.2012.02.004

Miranda BS, Linares EM, Thalhammer S et al (2013) Development of a disposable and highly sensitive paper-based immunosensor for early diagnosis of Asian soybean rust. Biosens Bioelectron 45:123–128. https://doi.org/10.1016/j.bios.2013.01.048

Mishra M, Singh SK, Bhardwaj A et al (2020) Development of a diatom-based photoluminescent immunosensor for the early detection of Karnal Bunt disease of wheat crop. ACS Omega 5(14): 8251–8257. https://doi.org/10.1021/acsomega.0c00551
Mishra A, Bhatt R, Bajpai J et al (2021) Nanomaterials based biofuel cells: a review. Int J Hydrog Energy 46(36):19085–19105. https://doi.org/10.1016/j.ijhydene.2021.03.024
Naresh V, Lee N (2021) A review on biosensors and recent development of nanostructured materials-enabled biosensors. Sensors 21(4):1109. https://doi.org/10.3390/s21041109
Ngoune Liliane T, Shelton Charles M (2020) Factors affecting yield of crops. In: Agronomy – climate change and food security. IntechOpen, London
Noori HN, Abdulameer AF (2023) A review of biosensors; definition, classification, properties, and applications. Iraqi J Sci:5665–5690. https://doi.org/10.24996/ijs.2023.64.11.18
Pohanka M (2020) Colorimetric hand-held sensors and biosensors with a small digital camera as signal recorder, a review. Rev Anal Chem 39(1):20–30. https://doi.org/10.1515/revac-2020-0111
Polat EO, Cetin MM, Tabak AF et al (2022) Transducer technologies for biosensors and their wearable applications. Biosensors 12(6):385. https://doi.org/10.3390/bios12060385
Porstmann T, Kiessig ST (1992) Enzyme immunoassay techniques an overview. J Immunol Methods 150(1–2):5–21. https://doi.org/10.1016/0022-1759(92)90061-W
Prajapati V, Shrivastav P (2023) Bioreceptors for antigen–antibody interactions. In: Biosensors nanotechnology. Wiley, Hoboken, pp 371–394
Pramanik B, Sar P, Bharti R et al (2023) Multifactorial role of nanoparticles in alleviating environmental stresses for sustainable crop production and protection. Plant Physiol Biochem 201:107831. https://doi.org/10.1016/j.plaphy.2023.107831
Rauf A, Abu-Izneid T, Khalil AA et al (2022) Nanoparticles in clinical trials of COVID-19: an update. Int J Surg 104:106818. https://doi.org/10.1016/j.ijsu.2022.106818
Ray S, Mehta G, Srivastava S (2010) Label-free detection techniques for protein microarrays: prospects, merits and challenges. Proteomics 10(4):731–748. https://doi.org/10.1002/pmic.200900458
Razali H, Awaludin N, Husin NH et al (2022) Development of an electrochemical immunosensor strip for early detection of rice bacterial leaf blight (Blb) disease and its application on a portable device. Malaysian J Anal Sci 26(6):1191–1204
Regiart M, Rinaldi-Tosi M, Aranda PR et al (2017) Development of a nanostructured immunosensor for early and in situ detection of Xanthomonas arboricola in agricultural food production. Talanta 175:535–541. https://doi.org/10.1016/j.talanta.2017.07.086
Rohr JR, Barrett CB, Civitello DJ et al (2019) Emerging human infectious diseases and the links to global food production. Nat Sustain 2(6):445–456. https://doi.org/10.1038/s41893-019-0293-3
Safarpour H, Safarnejad MR, Tabatabaei M et al (2012) Development of a quantum dots FRET-based biosensor for efficient detection of Polymyxa betae. Can J Plant Pathol 34(4):507–515. https://doi.org/10.1080/07060661.2012.709885
Shahare Y, Singh MP, Singh P et al (2023) A comprehensive analysis of machine learning-based assessment and prediction of soil enzyme activity. Agriculture 13(7):1323. https://doi.org/10.3390/agriculture13071323
Shakya AK, Ramola A, Kandwal A et al (2021) Soil moisture sensor for agricultural applications inspired from state of art study of surfaces scattering models & semi-empirical soil moisture models. J Saudi Soc Agric Sci 20(8):559–572. https://doi.org/10.1016/j.jssas.2021.06.006
Sharma A, Kaushal A, Kulshrestha S (2017) A Nano-Au/C-MWCNT based label free amperometric immunosensor for the detection of capsicum chlorosis virus in bell pepper. Arch Virol 162(7):2047–2052. https://doi.org/10.1007/s00705-017-3293-5
Sharma P, Chauhan R, Pande V et al (2022) Rapid sensing of Tilletia indica – Teliospore in wheat extract by a piezoelectric label free immunosensor. Bioelectrochemistry 147:108175. https://doi.org/10.1016/j.bioelechem.2022.108175

Singh S (2012) Development of surface plasmon resonance (SPR) based immuno-sensing system for detection of fungal teliospores of Karnal Bunt (Tilletia indica), a quarantined disease of wheat. J Biosens Bioelectron 03(04). https://doi.org/10.4172/2155-6210.1000125

Singh S, Singh M, Agrawal VV et al (2010) An attempt to develop surface plasmon resonance based immunosensor for Karnal bunt (Tilletia indica) diagnosis based on the experience of nano-gold based lateral flow immuno-dipstick test. Thin Solid Films 519(3):1156–1159. https://doi.org/10.1016/j.tsf.2010.08.061

Širić I, Eid EM, Taher MA, El-Morsy MHE et al (2022) Combined use of spent mushroom substrate biochar and pgpr improves growth, yield, and biochemical response of cauliflower (Brassica oleracea var. botrytis): a preliminary study on greenhouse cultivation. Horticulturae 8(9):830. https://doi.org/10.3390/horticulturae8090830

Skottrup P, Nicolaisen M, Justesen AF (2007) Rapid determination of Phytophthora infestans sporangia using a surface plasmon resonance immunosensor. J Microbiol Methods 68(3): 507–515. https://doi.org/10.1016/j.mimet.2006.10.011

Thangavelu RM, Janiga PK, Ramasamy N et al (2024) Magnetic nanozyme-enhanced rapid immunoflow-through assay for the femtomolar detection of sugarcane yellow leaf virus. ACS Agric Sci Technol 4(7):759–767. https://doi.org/10.1021/acsagscitech.4c00274

Uda MNA, Hasfalina CM, Samsuzanaa AA et al (2017) Immunosensor development formatting for tungro disease detection using nano-gold antibody particles application. AIP Conf Proc 1808(1): 020057

Valekunja RB, Kamakoti V, Peter A et al (2016) The detection of papaya ringspot virus coat protein using an electrochemical immunosensor. Anal Methods 8(48):8534–8541. https://doi.org/10.1039/C6AY02201D

Vashishth DS, Bachheti A, Bachheti RK et al (2023) Reducing herbicide dependency: impact of Murraya koenigii leaf extract on weed control and growth of wheat (Triticum aestivum) and chickpea (Cicer arietinum). Agriculture 13(9):1678. https://doi.org/10.3390/agriculture13091678

Verma N, Prajapati P, Tiwari BS et al (2024) Exploring shape-dependent nanomatrix immobilization strategy for developing a PAMV viral electrochemical immunosensor. Potato Res 67(1): 237–254. https://doi.org/10.1007/s11540-023-09635-7

Vikal S, Gautam YK, Kumar AA et al (2023) Bioinspired palladium-doped manganese oxide nanocorns: a remarkable antimicrobial agent targeting phyto/animal pathogens. Sci Rep 13(1): 14039. https://doi.org/10.1038/s41598-023-40822-1

Yager P, Edwards T, Fu E et al (2006) Microfluidic diagnostic technologies for global public health. Nature 442(7101):412–418. https://doi.org/10.1038/nature05064

Yeganeh FE, Yeganeh AE, Far BF et al (2022) Synthesis and characterization of tetracycline loaded methionine-coated NiFe2O4 nanoparticles for anticancer and antibacterial applications. Nanomaterials 12(13):2286. https://doi.org/10.3390/nano12132286

Zhao Y, Liu L, Kong D et al (2014) Dual amplified electrochemical immunosensor for highly sensitive detection of Pantoea stewartii sbusp. stewartii. ACS Appl Mater Interfaces 6(23): 21178–21183. https://doi.org/10.1021/am506104r

Zou Y, Mason MG, Botella JR (2020) Evaluation and improvement of isothermal amplification methods for point-of-need plant disease diagnostics. PLoS One 15(6):e0235216. https://doi.org/10.1371/journal.pone.0235216

Detection of Microbial Toxins in Different Food Matrices

Navneet Kumar Upadhyay

Abstract

The demand for various marketed foods has increased recently due to their improved health-promoting properties and nutritional values. Fortification and fermentation processes are the most prominent processes to prepare the marketed food. The consumption of these products is increasing across the world. Generally, fermented foods are safe for human consumption. It is defined as food product through the conversion of the components by some specific enzymatic processes and with controlled microbial growth. Although it is safe, some microbial toxins may contaminate the food matrices due to improper storage, starter cultures, transportation, and processing conditions. There are different microbial toxins may present such as fungal toxins (Beauvericin, enniatins, citrinin, patulin, aflatoxins, ochratoxin A, fumonisins, zearalenone, and trichothecenes), bacterial toxins (enterotoxins, endotoxins, shiga or verotoxins, and emetic toxins), algal toxins (microcystins, nodularins, anatoxins, saxitoxins, ciguatoxins, cyanotoxins, and phycotoxins) and biogenic amines (putrescine, phenyl-ethylamine, histamine, tyramine, and cadaverine). These toxins can cause various foodborne diseases that pose a grave threat to human health. Thus, there is keen interest in detecting microbial toxins in marketed food products to eliminate these toxins from various food matrices. This chapter detailed about different conventional techniques for the detection of microbial toxins in food items. This chapter also emphasizes the role of immunosensors for the detection of microbial toxins in food matrices.

N. K. Upadhyay (✉)
Amity Institute of Pharmacy, Amity University Rajasthan, Jaipur, Rajasthan, India
e-mail: nkupadhyay@jpr.amity.edu

K. Kuča et al. (eds.), *Ensuring Wellness: Immunosensors Monitoring Food Quality and Pathogenicity Risks*, https://doi.org/10.1007/978-3-031-97477-9_13

Keywords

Microbial toxins · Marketed food · Algal toxins · Bacterial toxins · Fungal toxins · Immunosensors · Conventional techniques

1 Introduction

Food quality and safety is an essential issue for agencies and workers of food industries to increased consumer awareness about the related risks to public health (Gupta et al. 2021). Microbial toxins in foods are considered one of the most critical and major problems in food product safety today. Despite the various advanced techniques for food preservation and food safety, major disease outbreaks linked to foodborne microbial toxins showing significant risks to public health. Toxins and pathogen contamination in food have a negative impact on human health and result in significant economic losses. Consequently, highly sensitive, fast, reliable, and cost-effective analytical methods need to be developed to protect the food chain against toxins (Aladhadh 2023).

Certain fungi and molds produce mycotoxins, which are toxic substances capable of causing disease and death in both humans and animals (Atungulu et al. 2018). Bacteria toxins can be classified as either exotoxins or endotoxins. Algal toxins refer to the toxins that algae in fresh water produce. Biogenic amines (BAs) are molecules with biological functions that can accumulate in foods at toxic concentrations. All these natural microbes in food products can produce toxins under favorable relative humidity, pH, temperature, and nutrient availability conditions (Núñez and Rodriguez 2020). Various microbial toxins and their sources are shown in Table 1. The presence of such microbial toxins in food poses considerable health risks to humans and livestock with both acute and chronic effects. The direct or indirect exposure to spoiled food products may cause illnesses ranging from common diseases (e.g., mild headache, vomiting, and diarrhoea) to serious neurological/hepatic diseases (e.g., paralysis, organ failure, and even death) (Hodgson 2012). The concerns regarding toxins have necessitated regulatory actions to govern inter-state and international commerce related to food commodities that are prone to toxin contamination. Consequently, accurate analysis of toxins has become a critical need to ensure product quality, implement regulatory enforcements, and ensure compliance with national and international food standards (Gupta et al. 2021). The prompt and accurate identification of microbial toxins has gained importance in addressing disease outbreaks linked to contaminated food products. Consequently, there is a growing focus on creating affordable and efficient sensors for detecting microbial toxins in food matrices (Gupta et al. 2021; Aladhadh 2023). Also some sensitive, selective, quick, and cost-effective methods are required to detect the toxins (Yadav et al. 2023). So, this chapter aims to define and characterize microbial toxins in different food matrices, and their detection methods specifically emphasize on conventional techniques and immunosensors, as depicted in Fig. 1.

Table 1 Various microbial toxins and their sources

Microbe	Toxin's name	Source	References
Fungal	Beauvericin	*Beauveria bassiana* and *Fusarium* species	Caloni et al. (2020)
	Enniatins	*Fusarium* fungi	Prosperini et al. (2017)
	Citrinin	*Aspergillus, Penicillium,* and *Monascus*	Navale et al. (2021)
	Patulin	*Penicillium expansum, Aspergillus* sp., *Byssochlamys fulva*, and *B. nivea*	Ma et al. (2021)
	Aflatoxins	*Aspergillus flavus* and *Aspergillus parasiticus,*	Mir et al. (2021)
	Ochratoxin A	*Aspergillus ochraceus, A. carbonarius, A. niger, Penicillium verrucosum*	Wang et al. (2022)
	Fumonisins	*Fusarium moniliforme (F. verticillioides), F. proliferatum*	Soriano et al. (2005)
	Zearalenone	*Fusarium graminearum*	Zinedine et al. (2007)
	Lysergic acid	*Claviceps purpurea*	Seneca (2007)
	Trichothecenes	*Myrothecium, Spicellum, Stachybotrys, Cephalosporium, Trichoderma, Trichothecium*	Huang et al. (2019) and Altomare et al. (2021)
	Tentoxin	*Fungus Alternaria alternata*	Lamberth (2010)
Bacterial	Enterotoxins	*Many strains of Salmonella*	Patriarca et al. (2014)
	Cholera toxins	*Vibro cholera*	Ahuja et al. (2023)
	Endotoxins	Gram-negative bacteria, such as *E. coli* and *Salmonella*	Lerouge (2010)
	Shiga	*Shigella dysenteriae* and several *E. coli strains*	Sandvig et al. (2015)
	Verotoxins	*Verotoxigenic E. coli*	Motiwala (2014)
	Emetic toxins (cereulide)	*Bacterium Bacillus cereus*	Agata et al. (2002) and Rouzeau et al. (2020)
	Botulinum toxins	*Clostridium botulium*	Hill et al. (2015) and Ahuja et al. (2023)
Algal	Microcystins	*Cyanobacteria* (blue-green algae) (*Microcystic aeruginosa*)	Dawson (1998), Dörr et al. (2010), and Yadav et al. (2023)
	Nodularins	*Cyanobacterium Nodularia spumigena*	Dörr et al. (2010) and Gupta et al. (2013)
	Anatoxins	*Cyanobacterial genera Anabaena, Planktothrix*, and *Aphanizomenon*	Constable et al. (2017) and Dörr et al. (2010)
	Saxitoxins	*Cyanobacteria, Aphanizomenon, Anabaena, Lyngbya, Cylindrospermopsis*	Dörr et al. (2010) and He et al. (2016)
	Ciguatoxins	*Dinoflagellate, Gambierdiscus toxicus,* and *related dinoflagellates*	Dörr et al. (2010)
	Cyanotoxins	*Cyanobacteria* (blue-green algae)	
	Phycotoxins	Eukaryotic and prokaryotic algal organisms	McCall et al. (2022)

(continued)

Table 1 (continued)

Microbe	Toxin's name	Source	References
	Domoic acid	Diatoms	Ahuja et al. (2023)
	Azaspiracid	*Azadinium poporum*	Okamoto and Fleming (2005)
	Brevetoxin	*Karenia brevis*	Beasley et al. (2023)
	Okadaic acid	*Halichondria melanodocia* and *Halichondria okadai*	Chang et al. (2013)
Bio-genic amines	Putrescine	Synthesized by the enzymatic decarboxylation of certain amino acids in all living microorganisms	del Rio et al. (2024)
	Phenyl-ethylamine		
	Histamine		
	Tyramine		
	Cadaverine		

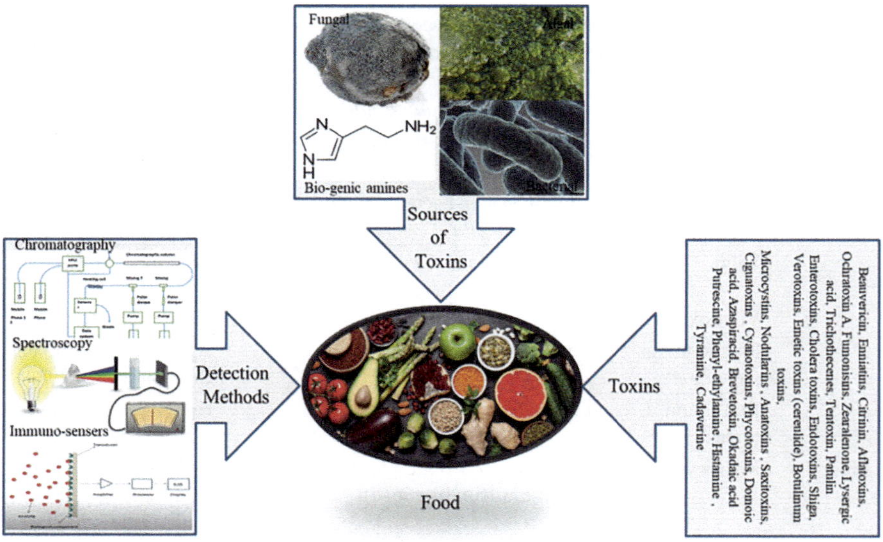

Fig. 1 Schematic representation of the different types of toxins, sources of toxins and their detection methods in food matrices

2 Conventional Methods for Detection of Microbial Toxins in Different Food Items

2.1 Chromatographic Techniques

The content of microbial toxins like biogenic amine in various foods has been extensively researched due to their potential toxicity. Histamine, phenyl ethylamine, tyramine, tryptamine, putrescine, cadaverine, spermine, and spermidine represent the most significant biogenic amines. Biogenic amines are present in various food products, including meat, sausages, milk, chocolate, cheese, fish, and certain beverages. Monitoring the levels of biogenic amines in food and beverages is crucial due to their significance for human health and food safety (Önal 2007). Advanced analytical techniques such as high performance thin layer chromatography (HPTLC), over-pressured layer chromatography (OPLC), supercritical fluid chromatography mass spectrometry (SFC-MS), capillary electrophoresis (CE) (Miklós et al. 2020), high performance liquid chromatography (HPLC), gag chromatography (GC), thin layer chromatography (TLC), inductively coupled plasma mass spectrometry (ICP-MS), ultra performance liquid chromatography tandem mass spectrometry (UPLC-MS/MS), high-performance liquid chromatography fluorescence detection (HPLC-FLD), high performance liquid chromatography-high resolution mass spectrometry (HPLC-HMRS), Hydrophilic interaction liquid chromatography (HILIC), mass spectrometry (MS), multistage mass spectrometry (MS_n) (De Girolamo et al. 2022), LC-MS and GC-MS and various forms of mass spectrometry have become indispensable tools for the detection and quantification of food toxins (Haddad et al. 2019). These powerful analytical tools are widely used for detecting and measuring microbial toxins due to their exceptional sensitivity and ability to distinguish between different substances, even in complex mixtures. With the development of advanced techniques like High-resolution mass spectrometry (HRMS) and tandem mass spectrometry, its performance has greatly improved. These innovations have enhanced the precision and lowered the detection limits, allowing for the identification of harmful compounds in food at extremely low concentrations. MS is a strong analytical method that can identify and quantify a variety of chemicals, including different toxins in food. This approach produces very sensitive and precise data, making it an important tool for food safety and quality management (Tevell Åberg et al. 2013). Shiga toxins, may also be examined using MS (Silva 2018). Monitoring mycotoxin levels has become critical due to the widespread contamination of food items (Rodríguez et al. 2014). MS is a powerful tool routinely also used as hyphenated with chromatographic techniques such as LC-MS, GC-MS for the detection of microbial toxins in different food matrices. LC-MS, GC-MS, and TLC are used to separate and detect various seafood-originated toxins, domoic acid, paralytic shellfish toxins, alkaloids, furocoumarins, cyanogenic glycosides, marine toxins, mycotoxins, and Aflatoxin B1, etc. The detection of multi elementals like heavy metals with high speed ICP-MS also a very specific tool (Jangampalli and Matcha 2018; Ahuja et al. 2023). HPLC is also the most frequently used analytical technique for separation and determination of biogenic amines (Önal 2007). HPLC with either UV or mass spectrometric detection after electrospray ionization (ESI) or thermospray ionization (TSI) is used to identify beauvericin in corn and meat (Uhlig and Ivanova 2004).

Microtoxins, including patulin, are quantified in different apple products using HPLC-UV at a wavelength of 276 nm, while ochratoxin A is measured by HPLC-fluorescence with excitation and emission wavelengths of 333 nm and 420 nm, respectively (Al-Hazmi 2010). Other LC-based approaches like HPLC-FLD, HPLC-HMRS, and UPLC-MS/MS are also used for identifying fungal and microbial toxins in food matrices (Bharti et al. 2024). Citrinin in red rice and dietary supplements made from red yeast rice may be determined quickly and easily using the QuEChERS extraction followed by the UHPLC-MS/MS analytical technique, which lowers the time and expense of the analytical process (Ponz-Perelló et al. 2024).

A study by Haddad et al. (2019) detailed a screening method (HILIC) for the detection of anatoxin-a, cylindrospermopsin, saxitoxin, nodularin, and microcystins in fish tissue. Microcystins, nodularin, anatoxin-a, cylindrospermopsin, and saxitoxin were examined through HILIC–MS/MS, utilizing a zwitterionic hydrophilic interaction LC column for the first time, whereas nodularin and microcystins were assessed using reverse phase liquid chromatography tandem mass spectrometry (RPLC-MS/MS) (Haddad et al. 2019; De Girolamo et al. 2022). Their high sensitivity, specificity, and ability to analyze complex food matrices make them essential in ensuring food safety. These advancements continue to support regulatory compliance and safeguard public health by providing reliable tools for routine monitoring of microbial toxins in different food products.

2.2 Spectroscopic Technique

Every toxin has a unique core structure with distinct patterns, which allows it to absorb light at a certain wavelength and be detected if monitored at that particular wavelength. This concept may be assessed via ultra violet fluorescence (UV-FL) spectroscopy and is used to identify microbial toxins like aflatoxins in different food samples (Ahuja et al. 2023). Aflatoxin B1 and ochratoxin A show maximal absorption (λmax) at 365 and 380 nm (Singh et al. 2021). Both fluorescence and ultra violet visible (UV-Vis) spectroscopy methods are inexpensive and simple to use. Furthermore, toxins in food samples may be detected with great sensitivity using fluorescence spectroscopy. Despite these benefits, many toxins are not showing very selective and specific wavelength for detection, posing a significant constraint in their detection through this technique. The intricate structure of food toxins are further clarified by other spectroscopic techniques, such as nuclear magnetic resonance (NMR) spectroscopy (Ray 2021).

Alternative techniques for identifying mycotoxins in food, such near infrared spectroscopy (NIR), have grown in popularity since they are quick, easy, chemical-free, and non-destructive. NIR spectroscopy, which uses NIR (14,000–4000 cm^{-1}) wavelengths that cause vibration of C–H, O–H, N–H, and C=O bonds, may also help identify toxins in food. Using NIR spectroscopy method, diarrheal shellfish toxins and enniatins were recently found in mussels (Liu et al. 2022) and barley grain (Caramês et al. 2022). Carotenoids and astaxanthin in *Haematococcus pluvialis* can be quantified using fluorescence spectrophotometric analysis (Bharti et al. 2024). Overall, spectroscopic methods contribute to food safety by providing early and reliable detection of microbial toxins in food samples.

3 Role of Immunosensor for Detection of Microbial Toxins in Various Food Matrices

Globally, food poisoning has impacted around 48 million individuals and resulted in 0.1 million hospital admissions per year (Das et al. 2017). The transmission of foodborne illness from farm to fork is primarily attributed to four sources: bacterial pathogens, plant toxins, mycotoxins, and phycotoxins (Malhotra et al. 2014). The rapid quantification of toxins in feed and food products is crucial for the prevention and management of various foodborne diseases, both acute and chronic. Technological advancements in bioelectronics, biosensors, microfluidics, and miniaturization techniques have significantly enhanced the sensitivity and performance of food analyte detection compared to conventional methods (Bhardwaj et al. 2022).

In order to detect okadaic acid, a strong marine neurotoxin present in seafood, Hayat et al. (2011) described a competitive indirect enzymatic immunosensor that employs an electrochemical method and includes superparamagnetic nanobeads as a critical component. Streptavidin-coated magnetic beads were used as a solid support to immobilize the biotinylated okadaic acid. To adjust different experimental settings, first colorimetric measurements were carried out. For electrochemical detection, the study team used differential pulse voltammetry. The produced immunosensors had an inhibitory concentration at 50% (IC_{50}) (mid-point) value of 3.1 μg/L, a detection limit of 0.3 μg/L, and an analysis time of 60 min. By spiking mussel extracts with three distinct okadaic acid concentrations (precisely 52 μg/kg, 25 μg/kg, and 17 μg/kg), the recovery was evaluated. Interestingly, recovery values for spiked samples before and during the extraction process were comparable at these concentrations (Hayat et al. 2011).

Another sensitive electrochemical immunosensor type was developed by Zhang et al. (2016) to detect aflatoxin B1 (AFB1) in maize powder made from chitosan, or single-walled carbon nanotubes. The immunosensor's design makes use of indirect competitive binding between free AFB1 and bovine serum albumin, which is immobilized covalently on nanotubes or chitosan that are put on a glassy carbon electrode, and a fixed amount of anti-AFB1 antibodies. Additionally, the anti-mouse immunoglobulin G, a secondary antibody labeled by the alkaline phosphatase enzyme, was adhered to the glassy carbon electrode surface in order to react with the main antibody. Ultimately, the alkaline phosphatase-assisted catalytic hydrolysis of the α-naphthyl phosphate (substrate) occurs, and the electrochemical signal generated may be effectively detected by the immunosensors. Maize powder naturally contaminated with AFB1 at concentrations of 121.5 pg/mL, 40.5 pg/mL, and

13.5 pg/mL were assessed in order to confirm the viability of the proposed immunosensor for detecting AFB1, considering the testing's accessibility (Zhang et al. 2016).

Using a portable sensor based on the surface plasmon resonance phenomena (Table 2), Moon et al. (2018) created an on-site AFB1 toxin detection device in grains. Cysteine-protein G, bovine serum albumin, and the AFB1 antibody were first used to successively modify the surface plasmon resonance gold chip surface in order to detect the toxin AFB1. The AFB1-bovine serum albumin complex and test sample were first run down the gold chip. Because the AFB1-bovine serum albumin combination binds to the gold chip, the surface plasmon resonance response is increased in the absence of AFB1. However, since the smaller AFB1 molecule binds preferentially on the gold chip's surface rather than the larger AFB1-bovine serum albumin molecule, the surface plasmon resonance response showed slight alterations in the presence of the toxin. These portable competitive immunosensors generated from surface plasmon resonance showed a meager quantification of 16.3 ppb and a detection limit of 2.5 ppb (Moon et al. 2018). Rehmat et al. (2019) used tiny surface plasmon resonance immunosensors to assess the levels of ochratoxin A in coffee. The cross-linked chitosan and carboxymethyl chitosan nano-matrix substrate were used to create these biosensors. Chitosan and its carboxymethyl derivative were used in the spin coating process to create the gold-coated surface plasmon resonance devices. The ochratoxin A-bovine serum albumin complex was immobilized using surface plasmon resonance nano-matrix chips to create a competitive inhibition immunoassay. Ochratoxin A in coffee samples and buffer, biological receptors in the form of antibodies (monoclonal ochratoxin A) were used. Coffee samples were found to have detection limits of 3.8 ng/mL for carboxymethyl chitosan and 5.7 ng/mL for chitosan substrates (Rehmat et al. 2019).

In some instances, the food might contain multiple toxins from the same microorganisms. Multichannel electrodes are essential in these circumstances because they enable scientists or researchers to include certain antibodies for simultaneous detection (Kumar et al. 2024).

4 Conclusion

In conclusion, food quality and safety are essential for public health, especially against microbial toxins. Even though food preservation technologies have improved, microbial toxins still pose health hazards and economic losses. These toxins, produced by fungi, molds, bacteria, algae, and biogenic amines, can cause a wide range of diseases in humans and animals, influenced by environmental factors such as humidity, pH, temperature, and nutrient availability. The development of cost-effective, efficient sensors and the application of both conventional techniques and immunosensors are critical steps toward ensuring safer food systems and minimizing toxin-related outbreaks. The detection of microbial toxins in food matrices is critical for ensuring food safety and protecting public health. Conventional methods such as chromatographic (HPLC, GC, LC-MS, GC-MS, TLC) and

Table 2 Detection of microbial toxins from different food matrices using different types of immunosensors (Kumar et al. 2024) (License Number 6013180376948)

Study country	Type of sensor	Type of antibody	Toxin	Food matrices	Limit of detection	Linear range
China	Electrochemical	Anti-aflatoxin B1 (monoclonal)	Aflatoxin B1	Rice bran oil	2.2 pg/mL	10–8000 pg/mL
Argentina	Electrochemical	Anti-zearalenone (monoclonal)	Zearalenone	Corn (spiked)	4.40 μg/kg	10–1000 μg/kg
China	Optical	Anti-aflatoxin B1 (monoclonal)	Aflatoxin B1	Peanut	0.58 ng/g	0.001–500 ng/mL
China	Optical	Anti-tetrodotoxin (monoclonal)	Tetrodotoxin	Shellfish	0.05 ng/mL	0.5–40 ng/mL
China	Electrochemical	Anti-aflatoxin B1	Aflatoxin B1	Peanut oil (spiked)	0.03 ng/mL	1×10^{-3}–1×10^{3} ng/mL
Spain	Electrochemical	Anti-aflatoxin B1, B2, G1, and G2 (all monoclonal)	Aflatoxin B1, B2, G1, and G2	Pistachio (spiked)	0.01 μg/L	0.01–2 μg/L
China	Electrochemical	Anti-tetrodotoxin (monoclonal)	Tetrodotoxin	Puffer fish, Asteroidea, Nassariidae, and Gobiidae (all spiked)	1.0×10^{-5} μg/mL	1.0×10^{-4} – 1.0×10^{3} μg/mL
	Optical				1.8×10^{-4} μg/mL	1.05×10^{-4}–3.8×10^{-1} μg/mL
Norway	Optical	Anti-*Staphylococcal* enterotoxin B	*Staphylococcal* enterotoxin B	Apple juice (spiked)	0.1 fM	Not specified
France	Electrochemical	Anti-aflatoxin B1 (monoclonal)	Aflatoxin B1	Rice milk (spiked)	50 fg/mL	0.5–2.5 pg/mL
France	Optical	Anti-aflatoxin B1 (monoclonal)	Aflatoxin B1	Wheat (spiked)	0.003 nM	0.01–66.6 nM
China	Electrochemical		Aflatoxin B1	Lotus seed (spiked)	0.01 ng/mL	0.05–100 ng/mL

(continued)

Table 2 (continued)

Study country	Type of sensor	Type of antibody	Toxin	Food matrices	Limit of detection	Linear range
		Anti-aflatoxin B1 (monoclonal)				
Saudi Arabia	Electrochemical	Anti-aflatoxin B1 (monoclonal)	Aflatoxin B1	Peanuts (spiked)	6.9 pg/mL	1 pg–100 ng/mL
United States of America	Optical	Anti-cyclopiazonic acid (monoclonal)	Cyclopiazonic acid	Maize, cheese (all spiked)	17 µg/kg (maize), 6 µg/kg (cheese)	0.1–5 ng/mL
Spain	Electrochemical	Anti-azaspiracids (polyclonal)	Azaspiracids	Mussels	0.6 µg/L	Not specified
China	Electrochemical	Zearalenones (monoclonal)	Zearalenones	Corn flour and corn-based baby food (all spiked)	1.5 pg/mL	0.005–50 ng/mL

spectroscopic techniques (UV-Vis, fluorescence, NIR, NMR) remain fundamental due to their high sensitivity, specificity, and ability to analyze complex samples. Advances like high-resolution and tandem mass spectrometry have further enhanced detection capabilities, enabling the identification of even trace levels of toxins in different food samples. Complementing these, immune-sensor technologies, leveraging biosensors and electrochemical detection, offer rapid, sensitive, and portable alternatives for real-time monitoring of toxins such as okadaic acid and aflatoxin B1. Together, these approaches form a comprehensive toolkit essential for routine monitoring, regulatory compliance, and safeguarding food quality in an increasingly complex global food supply chain.

References

Agata N, Ohta M, Yokoyama K (2002) Production of *Bacillus cereus* emetic toxin (cereulide) in various foods. Int J Food Microbiol 73:23–27

Ahuja V, Singh A, Paul D et al (2023) Recent advances in the detection of food toxins using mass spectrometry. Chem Res Toxicol 36:1834–1863

Aladhadh M (2023) A review of modern methods for the detection of foodborne pathogens. Microorganisms 11:1111

Al-Hazmi NA (2010) Determination of patulin and ochratoxin A using HPLC in apple juice samples in Saudi Arabia. Saudi J Biol Sci 17:353–359

Altomare C, Logrieco AF, Gallo A (2021) Mycotoxins and mycotoxigenic fungi: risk and management. A challenge for future global food safety and security. In: Zaragoza Ó, Casadevall A (eds) Encyclopedia of mycology. Elsevier, Oxford, pp 64–93

Atungulu GG, Mohammadi-Shad Z, Wilson S (2018) Chapter 2 - Mycotoxin Issues in Pet Food. In: Ricke SC, Atungulu GG, Rainwater CE, Park SH, (eds) Food and Feed Safety Systems and Analysis, Academic Press, pp 25–44

Beasley V, Carmichael W, Haschek WM et al (2023) Chapter 5 – phycotoxins. In: Haschek WM, Rousseaux CG, Wallig MA, Bolon B, Heinz-taheny KM, Rudmann DG, Mahler BW (eds) Haschek and Rousseaux' s handbook of toxicologic pathology, 4th edn. Academic Press, Amsterdam, pp 305–391

Bhardwaj H, Rajesh, Sumana G (2022) Recent advances in nanomaterials integrated immunosensors for food toxin detection. J Food Sci Technol 59:12–33

Bharti A, Jain U, Chauhan N (2024) Progressive analytical techniques utilized for the detection of contaminants attributed to food safety and security. Talanta Open 10:100368

Caloni F, Fossati P, Anadón A et al (2020) Beauvericin: the beauty and the beast. Environ Toxicol Pharmacol 75:103349

Caramês ETDS, Piacentini KC et al (2022) Rapid assessment of enniatins in barley grains using near infrared spectroscopy and chemometric tools. Food Res Int 161:111759

Chang NC, Lin ACM, Hsu CC et al (2013) Okadaic acid, a bioactive fatty acid from Halichondria okadai, stimulates lipolysis in rat adipocytes: the pivotal role of perilipin translocation. Evid Based Complement Alternat Med 2013:545739

Das Q, Islam MR, Marcone MF et al (2017) Potential of berry extracts to control foodborne pathogens. Food Control 73:650–662

Dawson RM (1998) The toxicology of microcystins. Toxicon 36:953–962

De Girolamo A, Lippolis V, Pascale M (2022) Overview of recent liquid chromatography mass spectrometry-based methods for natural toxins detection in food products. Toxins 14:328

del Rio B, Fernandez M, Redruello B et al (2024) New insights into the toxicological effects of dietary biogenic amines. Food Chem 435:137558

Dörr FA, Pinto E, Soares RM et al (2010) Microcystins in south American aquatic ecosystems: occurrence, toxicity and toxicological assays. Toxicon, Highlights in Toxinology: Biodiversity in Toxins – Tools for Biological Research and Drug Development 56:1247–1256

Constable PD, Hinchliff KW, Done SH et al (2017) General systemic states. In Veterinary Medicine, 11th edn. Elsevier, USA, pp 43-112

Gupta V, Ratha SK, Sood A et al (2013) New insights into the biodiversity and applications of cyanobacteria (blue-green algae)—prospects and challenges. Algal Res 2:79–97

Gupta R, Raza N, Bhardwaj SK (2021) Advances in nanomaterial-based electrochemical biosensors for the detection of microbial toxins, pathogenic bacteria in food matrices. J Hazard Mater 401: 123379

Haddad SP, Bobbitt JM, Taylor RB et al (2019) Determination of microcystins, nodularin, anatoxin-a, cylindrospermopsin, and saxitoxin in water and fish tissue using isotope dilution liquid chromatography tandem mass spectrometry. J Chromatogr A 1599:66–74

Hayat A, Barthelmebs L, Marty JL (2011) Enzyme-linked immunosensor based on super paramagnetic nanobeads for easy and rapid detection of okadaic acid. Anal Chim Acta 690:248–252

He X, Liu YL, Conklin A et al (2016) Toxic cyanobacteria and drinking water: impacts, detection, and treatment. Harmful Algae, Global Expansion of Harmful Cyanobacterial Blooms: Diversity, Ecology, Causes, and Controls 54:174–193

Hill KK, Xie G, Foley BT et al (2015) Genetic diversity within the botulinum neurotoxin-producing bacteria and their neurotoxins. Toxicon 107(Part A):2–8

Hodgson E (2012) Chapter fourteen – toxins and venoms. In: Hodgson E (ed) Progress in molecular biology and translational science, toxicology and human environments. Academic Press, Amsterdam, pp 373–415

Huang D, Cui L, Dai M et al (2019) Mitochondrion: a new molecular target and potential treatment strategies against trichothecenes. Trends Food Sci Technol 88:33–45

Jangampalli AP, Matcha B (2018) Analysis of aflatoxin B1 in contaminated feed, media, and serum samples of Cyprinus carpio L. by high-performance liquid chromatography. Food Qual Saf 2: 199–204

Kumar H, Dhalaria R, Guleria S (2024) Immunosensors in food, health, environment, and agriculture: a review. Environ Chem Lett 22:2573–2605

Lamberth C (2010) Amino acid chemistry in crop protection. Tetrahedron 66:7239–7256

Lerouge S (2010) 13 – Sterilisation and cleaning of metallic biomaterials. In: Niinomi M (ed) Metals for biomedical devices, Woodhead publishing series in biomaterials. Woodhead Publishing, Elsevier, pp 303–326

Liu Y, Qiao F, Xu L et al (2022) Fast detection of diarrhetic shellfish poisoning toxins in mussels using NIR spectroscopy and improved twin support vector machines. Front Mar Sci 9:907378

Ma TF, Chen YP, Shen Y (2021) Chapter seven – progress in the applications of surface plasmon resonance for food safety. In: Chen Y-P, Ma T-F (eds) Comprehensive analytical chemistry, surface plasmon resonance in bioanalysis. Elsevier, Amsterdam, pp 237–275

Malhotra BD, Srivastava S, Ali A et al (2014) Nanomaterial-based biosensors for food toxin detection. Appl Biochem Biotechnol 174:880–896

McCall JR, Brown AP, Sausman KT (2022) Chapter 9 – microalgae nanotechnology and drug development. In: Hussain CM (ed) Handbook of microbial nanotechnology. Academic Press, Amsterdam, pp 169–190

Miklós G, Angeli C, Ambrus Á et al. (2020) Detection of Aflatoxins in Different Matrices and Food-Chain Positions. Front Microbiol 11:1–21

Mir SA, Dar BN, Shah MA et al (2021) Application of new technologies in decontamination of mycotoxins in cereal grains: challenges, and perspectives. Food Chem Toxicol 148:111976

Moon J, Byun J, Kim H et al (2018) On-site detection of aflatoxin B1 in grains by a palm-sized surface plasmon resonance sensor. Sensors 18:598

Motiwala AS (2014) Verotoxigenic *Escherichia coli*. In: Batt CA, Tortorello ML (eds) Encyclopedia of food microbiology, 2nd edn. Academic Press, Oxford, pp 680–685

Navale V, Vamkudoth KR, Ajmera S (2021) *Aspergillus* derived mycotoxins in food and the environment: prevalence, detection, and toxicity. Toxicol Rep 8:1008–1030

Núñez F, Rodriguez JM (2020) Novel approaches to minimising mycotoxin contamination. Toxins 12:216

Okamoto K, Fleming LE (2005) Algae. In: Wexler P (ed) Encyclopedia of toxicology, 2nd edn. Elsevier, New York, pp 68–76

Önal A (2007) A review: current analytical methods for the determination of biogenic amines in foods. Food Chem 103:1475–1486

Patriarca A, Vaamonde G, Pinto VF (2014) Alternaria. In: Batt CA, Tortorello ML (eds) Encyclopedia of food microbiology, 2nd edn. Academic Press, Oxford, pp 54–60

Ponz-Perelló P, Esteve TFA, Cortés MÁ et al (2024) Development and validation of an analytical method for determination of citrinin in red rice and red yeast rice-based food supplements by ultra-high performance liquid chromatography tandem mass spectrometry. Food Chem 455: 139941

Prosperini A, Berrada H, Ruiz MJ et al (2017) A review of the mycotoxin enniatin B. Front Public Health 5:304

Ray S (2021) Sensory properties of foods and their measurement methods. In: Khan MS, Shafiur Rahman M (eds) Techniques to measure food safety and quality: microbial, chemical, and sensory. Springer International Publishing, Cham, pp 345–381

Rehmat Z, Mohammed WS, Sadiq MB et al (2019) Ochratoxin a detection in coffee by competitive inhibition assay using chitosan-based surface plasmon resonance compact system. Colloids Surf B Biointerfaces 174:569–574

Rodríguez CY, Moltó JC, Mañes J et al (2014) Development of a GC-MS/MS strategy to determine 15 mycotoxins and metabolites in human urine. Talanta 128:125–131

Rouzeau SK, Stollewerk K, Messelhäusser U et al (2020) Why be serious about emetic *Bacillus cereus*: Cereulide production and industrial challenges. Food Microbiol 85:103279

Sandvig K, Lingelem ABD, Skotland T et al (2015) 10 – Shiga toxins: properties and action on cells. In: Alouf J, Ladant D, Popoff MR (eds) The comprehensive sourcebook of bacterial protein toxins, 4th edn. Academic Press, Boston, pp 267–286

Seneca (2007) Chapter 1 – definition, typology and occurrence of alkaloids. In: Aniszewski T (ed) Alkaloids – secrets of life. Elsevier, Amsterdam, pp 1–59

Silva CJ (2018) Food forensics: using mass spectrometry to detect foodborne protein contaminants, as exemplified by Shiga toxin variants and prion strains. J Agric Food Chem 66:8435–8450

Singh AK, Lakshmi GBVS, Dhiman TK et al (2021) Bio-active free direct optical sensing of aflatoxin B1 and ochratoxin A using a manganese oxide nano-system. Front Nanotechnol 2: 621681

Soriano JM, González L, Catalá AI (2005) Mechanism of action of sphingolipids and their metabolites in the toxicity of fumonisin B1. Prog Lipid Res 44:345–356

Tevell Åberg A, Björnstad K, Hedeland M (2013) Mass spectrometric detection of protein-based toxins. Biosecur Bioterror 11(Suppl 1):S215–S226

Uhlig S, Ivanova L (2004) Determination of beauvericin and four other enniatins in grain by liquid chromatography–mass spectrometry. J Chromatogr A 1050:173–178

Wang L, Hua X, Shi J et al (2022) Ochratoxin A: occurrence and recent advances in detoxification. Toxicon 210:11–18

Yadav K, Moovendaran K, Dhenadhayalan N et al (2023) From food toxins to biomarkers: multiplexed detection of aflatoxin B1 and aflatoxin M1 in milk and human serum using PEGylated ternary transition metal sulfides. Sens Actuators Rep 5:100156

Zhang X, Li CR, Wang WC et al (2016) A novel electrochemical immunosensor for highly sensitive detection of aflatoxin B1 in corn using single-walled carbon nanotubes/chitosan. Food Chem 192:197–202

Zinedine A, Soriano JM, Moltó JC et al (2007) Review on the toxicity, occurrence, metabolism, detoxification, regulations and intake of zearalenone: an oestrogenic mycotoxin. Food Chem Toxicol 45:1–18

Detection of Adulterations in Honey Using Immunosensors

Harsh Kumar, Nidhi Sethi, Shivani Guleria, Lukas Peter, Richard Cimler, and Tabarak Malik

Abstract

Exploitation of natural sweeteners has been the prime focus of the consumers for attaining good health mainly for those suffering from diabetic problems. Honey, well known for its countless medicinal values, is appropriate for its use as a natural healthy sweetener. However, the greediness of several food industries or honey producers has led to malpractices, resulting in adulterated honey and labelling them as pure honey or as counterfeit barcodes. Taking cognizance of the adulteration, several regulatory and food safety agencies around the world have set up various parameters for defining the pure honey quality. Numerous high-ended techniques have been used for the detection of adulterants found in honey. Last few years of research have found immunosensors playing an important role in detection of harmful substances in the honey. This chapter discusses

H. Kumar · R. Cimler
Centre of Advanced Technologies, Faculty of Science, University of Hradec Kralove, Hradec Kralove, Czech Republic
e-mail: richard.cimler@uhk.cz

N. Sethi
Department of Pharmaceutical Sciences, Guru Nanak Dev University, Amritsar, India

S. Guleria
Department of Biotechnology, TIFAC-Centre of Relevance and Excellence in Agro and Industrial Biotechnology (CORE), Thapar Institute of Engineering and Technology, Patiala, India

L. Peter
Faculty of Materials Science and Technology, VSB-Technical University of Ostrava, Ostrava-Poruba, Czech Republic
e-mail: lukas.peter@vsb.cz

T. Malik (✉)
Department of Biomedical Sciences, Institute of Health, Jimma University, Jimma, Ethiopia
e-mail: tabarak.malik@ju.edu.et

K. Kuča et al. (eds.), *Ensuring Wellness: Immunosensors Monitoring Food Quality and Pathogenicity Risks*, https://doi.org/10.1007/978-3-031-97477-9_14

the quality of honey, its health benefits along with the role of diverse technologies and immunosensors for checking authenticity of honey.

Keywords

Pure honey · Mixing of sugars · Adulterants · Health risk · Detection methods

1 Introduction

Honey is defined by the *Codex alimentarius* as "the natural sweet substance produced by honey bees from the nectar of plants or from secretions of living parts of plants or excretions of plant-sucking insects on the living parts of plants, which the bees collect, transform by combining with specific substances of their own, deposit". As a result, honey can be regarded as a high-value food item that is typically consumed for its distinct flavour, high nutritional content, and health advantages that vary depending on its chemical make-up. There are notable differences in the nutritional value and biological significance of honey produced and sold around the world. Honey's antibacterial, anti-inflammatory, and antioxidant qualities are the main contributors to its health-promoting and functional qualities, and it has been used therapeutically (Zhang et al. 2021a, b). In addition to this, honey has been shown to offer a number of intriguing pharmacological qualities, including anti-HIV, anticancer, and wound-dressing capabilities (Geana and Ciucure 2020; Tsagkaris et al. 2021).

Despite being regarded as a natural food, honey, along with milk and olive oil, has also been declared to be a food that has been tampered with. As a result, honey's quality assurance and safety concerns have elevated it to the forefront of food regulatory bodies' worries about international trading (Tiwari et al. 2018). Due to its ability to be a perfect substitute for other sweeteners, particularly processed sugar, which meets the high demand in the pharmaceutical and cosmetics industries, honey is in great demand worldwide (Riswahyuli et al. 2020). Hence, this chapter discusses the nutritional profiling of honey and the various kinds of honey adulteration methodologies and related issues. Additionally, we discuss the techniques employed to detect adulterants in honey, with prime focus on the role of different immunosensors in detecting honey adulterants.

2 Nutritional Value of Honey

The nutritional composition of honey suggests that this healthy natural sweetener not only contains water, fructose, and glucose but also traces of valuable ingredients, including flavourings, minerals (Na, Ca, P, K, Mg, Se, Cu, Mn, Fe, Cr, Zn), vitamins (K, C, B9, B6, B5, B3, B2, and B1), antioxidant, and enzymes (Tables 1 and 2) (Fakhlaei et al. 2020). Honey has been one of humankind's oldest food items that preserves human health and also shields them from different diseases, such as sore

Table 1 Vitamin and mineral content of honey

Types of minerals	Amount (mg/100 g)	Types of vitamins	Amount (mg/100 g)
Sodium (Na)	1.6–17	Thiamine (B1)	0.00–0.01
Calcium (Ca)	3–31	Riboflavin (B2)	0.01–0.02
Potassium (K)	40–3500	Niacin (B3)	0.10–0.20
Magnesium (Mg)	0.7–13	Pantothenic acid (B5)	0.02–0.11
Phosphorus (P)	2–15	Pyridoxine (B6)	0.01–0.32
Selenium (Se)	0.002–0.01	Folic acid (B9)	0.002–0.01
Copper (Cu)	0.02–0.6	Ascorbic acid (C)	2.2–2.5
Iron (Fe)	0.03–4	Phyllochinon (K)	0.025
Manganese (Mn)	0.02–2		
Chromium (Cr)	0.01–0.3		
Zinc (Zn)	0.05–2		

Fakhlaei et al. (2020), under the terms and conditions of the Creative Commons Attribution [CC BY] license

Table 2 Nutritional composition of pure honey (mg/100 g)

Nutrition	Blossom honey	Honeydew honey
–	Range	Range
Water	15–20	–
Total sugars		
Monosaccharides		
Fructose	30–45	28–40
Glucose	24–40	19–32
Disaccharides		
Sucrose	0.1–4.8	0.1–4.7
Tri-saccharides		
Erlose	0.5–6.0	0.1–6.0
Melezitose	–	0.3–22
Proteins	0.2–0.4	0.4–0.7
Acids	0.2–0.8	0.8–1.5
pH value	3.2–4.5	4.5–6.5

Fakhlaei et al. (2020), under the terms and conditions of the Creative Commons Attribution [CC BY] license

throat, a cold, and cancer (Schievano et al. 2012; Ajibola et al. 2013). A study has shown that fructose present in pure honey helps reduce blood glucose in diabetic animal models (Fakhlaei et al. 2020). The mechanism of lowering blood glucose is attributed to reduction in intestinal absorption rate, extension in the emptying time of stomach, and reduction in food intake (Kellett et al. 2008; Kwon et al. 2008; Erejuwa et al. 2012; Bobiş et al. 2018). Also, fructose stimulates glucokinase enzyme in hepatocytes, which plays key role in the glucose uptake and its storage as glycogen by liver. Thus, glucose in honey, present along with fructose, enhances its absorption and facilitates its hepatic actions by enhancing its release to the liver (Meireles et al.

2013). Honey consumption as a natural sweetener or food can boost the energy levels in human body (Fakhlaei et al. 2020).

3 Methods of Honey Adulterations and Health Issues

The practice of adulterating honey is a persistent issue worldwide and has had a well-known detrimental impact on the growth of the honey business (Zhang et al. 2023). Issues with the authenticity of honey have been documented, compiled, and thus divided into two classes (Fig. 1): the direct methods of adulterating honey by adding sweeteners or mixed honeys, and misrepresenting it as premium-quality honey of botanical and geographical origins, as well as organic products; and the indirect methods of feeding honeybee colonies, harvesting unripe honey, and improperly handling and storing honey (Zhang et al. 2023).

The effects of contaminated honey on human health have been the subject of very few literature reports. Particularly, regular fake honey eating might enhance blood cholesterol and sugar levels and result in obesity. Furthermore, the functioning of internal organs such as the liver and kidney is directly impacted by the contaminated honey (Li et al. 2015; Fakhlaei et al. 2020). To test the toxicity of tainted honey, an in vitro zebrafish study was conducted. Histological examinations of the internal

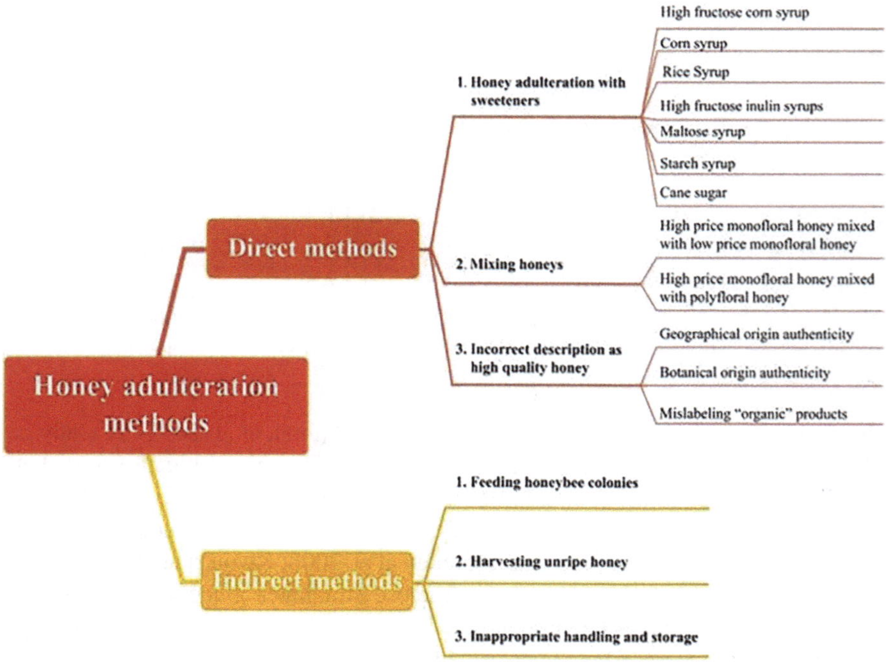

Fig. 1 Graphical representation associated with honey adulteration methods. (Zhang et al. 2023, under the CC BY-NC-ND license)

organs of the study subjects (zebrafish) treated with tainted honey revealed abnormalities in the liver, kidney, and spleen. As a result, the subjects (those exposed to honey with sugar syrup) have higher mortality rates (Fakhlaei et al. 2021). The effects of both short-term and long-term methods for assessing the consumption of contaminated honey on human health were established in a related study carried out in Malaysia. For the animal-model investigation, the researchers used male Sprague Dawley rats. The findings demonstrated that the rats' health was not significantly impacted by a brief exposure to tainted honey. However, after 2 weeks of consistent use of tainted honey, noticeable alterations were observed. Triglycerides, blood cholesterol, and glucose were all shown to be lower in the control rats (17.5 ± 1.2 mmol/L, 6.4 ± 1.4 mmol/L, and 26.5 ± 4.4 mmol/L, respectively). On the other hand, patients who regularly consumed adulterated honey for longer than 2 weeks saw substantial changes in their blood glucose levels (28.4 ± 2.5 mmol/L, 25 ± 2.6 mmol/L), tri glycerides (26 ± 1.2 mmol/L, 24.4 ± 1.8 mmol/L), and cholesterol (48.6 ± 4.8 mmol/L). According to the explanatory conclusion, test subjects died earlier than control subjects (Samat et al. 2018). As a result, the worldwide apiculture market is primarily focused on honey authenticity, which poses a barrier to food quality management at the moment (Zhang et al. 2023). Accordingly, it is crucial to constantly update and enhance relevant analytical techniques based on the physical, chemical, and biological properties of honey (Zhang et al. 2023).

4 Detection Techniques in Honey Adulterations in Brief

Table 3 lists the most often-used techniques for identifying tainted honey. Using Ultrahigh-performance liquid chromatography/quadrupole time-of-flight mass spectrometry (UHPLC/Q-TOF-MS), Du et al. (2015) developed a high resolution, quick, and high-performance analysis method that detected adulteration at even 10% of sugar syrup in less than 30 min by focussing on particular markers (di-fructose anhydrides, and polysaccharides) that were not present in the pure honey. Additionally, in order to identify aberrant saccharides, Qu et al. (2019) employed matrix-assisted laser desorption/ionization mass spectrometry (MALDI-MS) to analyse polysaccharide and oligosaccharide profiles in honey. The presence of high fructose corn syrup (HFCS) is a sign that honey has been tampered with. MALDI-MS analysis demonstrated robustness in high-throughput monitoring, offering a new method for identifying adulterants in meals high in carbohydrates. In a 2019 study, Arroyo-Manzanares and colleagues used headspace gas chromatography-ion mobility spectrometry (HS-GC-IMS) to identify pure honey with 97.4% accuracy and a 93.8% success rate when determining the amount of adulterants. When it comes to identifying various adulterants, HS-GC-IMS's analysis of volatile chemicals emitted into the headspace offers excellent efficacy and efficiency. The robustness of the analytical technique, high-performance anion exchange chromatography with pulsed amperometric detection (HPAEC-PAD) in combination with chemometric analysis for honey authentication, was further highlighted by Carabetta and

Table 3 Common techniques employed for detection of adulterants in honey

Techniques	Types of adulterants
Ultrahigh-performance liquid chromatography/ quadrupole time-of-flight mass spectrometry (UHPLC/Q-TOF-MS)	Corn syrup, high fructose corn syrup, inverted syrup, and rice syrup
Inductively coupled plasma mass spectrometry (ICP-MS)	Zn, Cu, Pb, Cd, Ni, and Cr
Cavity ring-down spectroscopy (CRDS)	High fructose corn syrup (HFCS) or cane sugar
Raman spectroscopy	Fructose, glucose, inverted sugar, hydrolysed inulin syrup, and malt
Fourier transform infrared spectroscopy (FTIR) with Attenuated Total Reflectance (ATR)	Beet sugar, and corn syrup
Matrix-assisted laser desorption/ionization mass spectrometry (MALDI-MS)	High fructose corn syrup (HFCS)
High-Performance Anion Exchange Chromatography with Pulsed Amperometric Detection (HPAEC-PAD)	Glucose, maltose, trehalose, sucrose, and palatinose
Voltammetry e-tongue	Maple, inverted sugar, corn, and rice syrups
Liquid chromatography high-resolution mass spectrometry (LC-HRMS)	Sugar syrups
High magnetic field NMR spectroscopy	Corn syrup, glucose syrup, and wheat syrup

Bose and Padmavati (2024), with licence number 6010141129424

researchers (2020), who achieved more than 90% accuracy with detailed sugar profiling. This was essential for classifying and differentiating diverse honey samples based on their botanical origin. Raman spectroscopy assisted in achieving an astounding 96.54% accuracy in differentiating between legitimate and counterfeit honey samples, as demonstrated by Oroian et al. (2018). Another reliable technique for identifying honey adulteration was FTIR-ATR (Fourier Transform Infrared Spectroscopy with Attenuated Total Reflectance), which was used by Başar and Özdemir (2018). The study analysed spectra from contaminated honey samples that contained corn syrup and beetroot sugar using partial least squares (PLS) and genetic algorithm-based inverse least squares (GILS). The FTIR method successfully detected and quantified several adulterants in honey with high accuracy and few mistakes.

In order to identify rice molasses as an adulterant in honey, Sobrino-Gregorio et al. (2019) used both traditional and reverse transcription polymerase chain reaction (RT-PCR) techniques. By simulating rice molasses at various levels of adulteration (1%, 2%, 5%, 10%, 20%, and 50%), the researchers have shown how successful PCR is at identifying and quantifying the adulterant. An alternative strategy for detecting honey adulteration combines proteomic and electrochemical techniques, which work well together to analyse and identify adulterants. Zheng et al. (2021) used chemometrics in conjunction with sequential window acquisition of all theoretical fragment ion mass spectra (SWATH-MS) to detect syrup adulteration and differentiate honey according to its botanical and geographic origins. Even at lower syrup quantities, this approach successfully detected honey adulteration. An electronic tongue system based on cyclic voltammetry has been used by Ciursa and Oroian (2021) to detect adulterated honey. This technique demonstrated its efficacy in identifying adulteration by achieving excellent validation and calibration accuracies and associating the voltametric data with physicochemical parameters such as pH and electrical conductivity. Martinello et al. (2022) used hydrophilic interaction liquid chromatography high-resolution mass spectrometry (LC-HRMS) in conjunction with non-targeted metabolomics to identify indicators of syrup-adulterated honey. Additionally, the study achieved a detection limit of about 5% and successfully identified a particular signature associated with syrup adulteration. This method demonstrated the superiority of metabolomics in honey analysis and highly accurate adulterant detection.

5 Role of Immunosensors in to Check the Honey Adulterations and Toxins

An electrochemical biosensor has been developed for detection of chloramphenicol (CAP) residues presence in honey (Gaudin et al. 2020). As per the researchers, the initial results were satisfactory as they could develop a screening test capable of detecting CAP at 0.3 ng/g or below in single honey sample. On analysis of individual honeys, this methodology was capable of discriminating blank samples from 0.3 ng/g CAP spiked samples. Therefore, the objective was fulfilled. On the other hand, as the mean honey signals of diverse botanical origins were factored, it was not possible to differentiate spiked samples and blank samples as the variation in the signal intensity due to every honey sample was too high. Yuan et al. (2008) introduced a highly sensitive surface plasmon resonance (SPR) methodology for detection of CAP which afforded a detection limit of nearly 17.5 fg/mL in spiked honey samples. The SPR-biosensor was fabricated firstly by synthesis of a separate conjugate, chloramphenicol-oligoethylene-ovalalbumin (CAP-OEG-OVA) in the laboratory. Further, mixed self-assembled monolayer (mSAM) containing 16-mercaptohexadecanoic acid (16-MHA), 11-mercaptoundecanol (11-MUOH) was formed with subsequent covalent immobilization of the CAP-OEG-OVA conjugate using OEG linker on the monolayer mSAM. The CAP detection was conducted through competitive assay format wherein mouse IgG/nanogold particles (two sizes of 10 nm and 40 nm) and anti-chloramphenicol antibody (mAb, anti-CAP) were utilized as sequential binding in rapid CAP detection. The application of 40 nm nanogold improved the signal response derived from CAP SPR-biosensor. The researchers reported that the prepared sensor was high in stability (binding regeneration cycles of ~400) and offered a ~0.74 fg/mL LOD for CAP with wide linear concentration of CAP ranging from 1 to 1000 fg/mL.

In order to identify dichlordifenyltrichlorethane (DDT) and carbaryl in honey, Cervera-Chiner et al. (2020) created piezoelectric immunosensors that operate on high fundamental frequency quartz crystal microbalance (HFF-QCM). The biorecognition was based on conjugate-coated competitive immunoassays using monoclonal antibodies as particular immunoreagents. In comparison to the standard reference technologies, the HFF-QCM immunosensors achieved a similar degree of detectability, with LODs of 0.24 μg/L for DDT and 0.05 μg/L for carbaryl. For honey samples, the realistic LOD values were 24 μg/kg for DDT and 8 μg/kg for carbaryl. Immunosensors' analytical performance enables the detection of the aforementioned pesticides in honey at EU regulatory levels with good precision (coefficients of variation calculated in the range of 9%–36%) and accuracy (recovery percentage range between 94% and 130% falling within each one's workable range of standard curve). A new integrated bio micro-electromechanical system (Bio-MEMS)-based immunosensor with eight gold microelectrodes (μWEs), one integrated platinum reference, and counter electrodes has been created by El Hassani et al. (2019). Using magnetic nanoparticles (MNPs), functionalized with poly (pyrrole-co pyrrole-2-carboxylic acid) (Py/Py-COOH/MNPs), Tetracycline (TC) was immobilized on the surface of μWEs. The developed impedimetric immunosensor had a detection limit of 1.2 pg/mL and a sensitivity of 0.208 mL/pg. When compared to similar compounds, this immunosensor platform's selectivity for tetracycline was found to be greater. The study offered a new and less expensive fabrication method for creating a high-sensing device that can identify trace amounts of the drug TC residues in honey samples without the need for a laborious pre-treatment.

The potential neurotoxic tropane alkaloids are commonly found in honey. Recently, Sun et al. (2024) have developed a user-friendly and efficient analytical method for simultaneous detection of six tropane alkaloids (L-hyoscyamine, atropine, scopolamine, anisodamine, apoatropine, and homatropine) present in honey which was based on gold nanoparticles (AuNPs) lateral flow immunoassay (AuNPs-LFIA) having broad-spectrum antibodies. The detection limits were calculated under optimal conditions and were found out to be 0.29, 0.22, 0.51, 6.34, 0.94, and 0.30 ng/mL, respectively. Dilution of the honey sample by five times led to readily detection of the contaminants using LFIA. Quantitative and semi-quantitative analyses could be completed in 17 min.

6 Conclusion

Honey as a food fulfils majority of nutritional requirements and is also a very popular natural sweetener commercially. But adulteration of honey through various means also makes it a hazardous food item for consumption of humans, especially amongst vulnerable populations. Different technologies have been employed by researchers and food industries across the world to check honey purity for its safe and healthy consumption. Although designing of immunosensors based on various working principles has been done to check honey's authenticity, yet they have not been explored fully for this food product. Till now the fabricated immunosensors mainly detect synthetic chemicals and drugs residue in honey. Hence, it is recommended to design immunosensors for detecting different sugar moieties in the near future.

References

Ajibola A, Chamunorwa JP, Erlwanger KH (2013) Dietary supplementation with natural honey promotes growth and health of male and female rats compared to cane syrup. Sci Res Essays 8(14):543–553

Arroyo-Manzanares N, García-Nicolás M, Castell A et al (2019) Untargeted headspace gas chromatography–Ion mobility spectrometry analysis for detection of adulterated honey. Talanta 205:120123

Başar B, Özdemir D (2018) Determination of honey adulteration with beet sugar and corn syrup using infrared spectroscopy and genetic-algorithm-based multivariate calibration. J Sci Food Agric 98(15):5616–5624

Bobiş O, Dezmirean DS, Moise AR (2018) Honey and diabetes: the importance of natural simple sugars in diet for preventing and treating different type of diabetes. Oxidative Med Cell Longev 2018(1):4757893

Bose D, Padmavati M (2024) Honey authentication: a review of the issues and challenges associated with honey adulteration. Food Biosci 61:105004

Carabetta S, Di Sanzo R, Campone L et al (2020) High-performance anion exchange chromatography with Pulsed Amperometric Detection (HPAEC–PAD) and chemometrics for geographical and floral authentication of honeys from Southern Italy (Calabria region). Food Secur 9(11): 1625

Cervera-Chiner L, March C, Arnau A et al (2020) Detection of DDT and carbaryl pesticides in honey by means of immunosensors based on high fundamental frequency quartz crystal microbalance (HFF-QCM). J Sci Food Agric 100(6):2468–2472

Ciursa P, Oroian M (2021) Voltammetric e-tongue for honey adulteration detection. Sensors 21(15):5059

Du B, Wu L, Xue X et al (2015) Rapid screening of multiclass syrup adulterants in honey by ultrahigh-performance liquid chromatography/quadrupole time of flight mass spectrometry. J Agric Food Chem 63(29):6614–6623

El Hassani NEA, Baraket A, Boudjaoui S et al (2019) Development and application of a novel electrochemical immunosensor for tetracycline screening in honey using a fully integrated electrochemical Bio-MEMS. Biosens Bioelectron 130:330–337

Erejuwa OO, Sulaiman SA, Ab Wahab MS (2012) Honey: a novel antioxidant. Molecules 17(4): 4400–4423

Fakhlaei R, Selamat J, Khatib A et al (2020) The toxic impact of honey adulteration: a review. Food Secur 9(11):1538

Fakhlaei R, Selamat J, Razis AFA et al (2021) In vivo toxicity evaluation of sugar adulterated heterotrigona itama honey using zebrafish model. Molecules 26(20):6222

Gaudin V, Bodin C, Hédou C et al (2020) Development and optimization of an amperometric immunosensor for the detection of banned antibiotic residues in honey. Proceedings 60:48

Geana EI, Ciucure CT (2020) Establishing authenticity of honey via comprehensive Romanian honey analysis. Food Chem 306:125595

Kellett GL, Brot-Laroche E, Mace OJ et al (2008) Sugar absorption in the intestine: the role of GLUT2. Annu Rev Nutr 28(1):35–54

Kwon S, Kim YJ, Kim MK (2008) Effect of fructose or sucrose feeding with different levels on oral glucose tolerance test in normal and type 2 diabetic rats. Nutr Res Pract 2(4):252–258

Li L, Zhao Z, Xia J et al (2015) A long-term high-fat/high-sucrose diet promotes kidney lipid deposition and causes apoptosis and glomerular hypertrophy in bama minipigs. PLoS One 10(11):e0142884

Martinello M, Stella R, Baggio A et al (2022) LC-HRMS-based non-targeted metabolomics for the assessment of honey adulteration with sugar syrups: a preliminary study. Metabolites 12(10): 985

Meireles M, Martel F, Araújo J et al (2013) Characterization and modulation of glucose uptake in a human blood–brain barrier model. J Membr Biol 246:669–677

Oroian M, Ropciuc S, Paduret S (2018) Honey adulteration detection using Raman spectroscopy. Food Anal Methods 11:959–968

Qu L, Jiang Y, Huang X et al (2019) High-throughput monitoring of multiclass syrup adulterants in honey based on the oligosaccharide and polysaccharide profiles by MALDI mass spectrometry. J Agric Food Chem 67(40):11256–11261

Riswahyuli Y, Rohman A, Setyabudi FM et al (2020) Indonesian wild honey authenticity analysis using attenuated total reflectance-fourier transform infrared (ATR-FTIR) spectroscopy combined with multivariate statistical techniques. Heliyon 6(4):e03662

Samat S, Enchang FK, Abd Razak A et al (2018) Adulterated honey consumption can induce obesity, increase blood glucose level and demonstrate toxicity effects. Sains Malays 47(2): 353–365

Schievano E, Stocchero M, Morelato E et al (2012) An NMR-based metabolomic approach to identify the botanical origin of honey. Metabolomics 8:679–690

Sobrino-Gregorio L, Vilanova S, Prohens J et al (2019) Detection of honey adulteration by conventional and real-time PCR. Food Control 95:57–62

Sun B, Wang C, Wang Z et al (2024) Strategy for accurate detection of six tropane alkaloids in honey using lateral flow immunosensors. Sensors 24(22):7265

Tiwari K, Tudu B, Bandyopadhyay R et al (2018) Voltammetric sensor for electrochemical determination of the floral origin of honey based on a zinc oxide nanoparticle modified carbon paste electrode. J Sens Syst 7(1):319–329

Tsagkaris AS, Koulis GA, Danezis GP et al (2021) Honey authenticity: analytical techniques, state of the art and challenges. RSC Adv 11(19):11273–11294

Yuan J, Oliver R, Aguilar MI et al (2008) Surface plasmon resonance assay for chloramphenicol. Anal Chem 80(21):8329–8333

Zhang XH, Qing XD, Mu ST et al (2021a) Authentication of honey of different nectar sources and antioxidant property evaluation by phenolic composition analysis with chemometrics. Food Control 124:107900

Zhang XH, Wang D, Guo WY et al (2021b) Rapid identification of adulterated honey according to the targeted analysis of phenolic compounds using chemometrics. Eur Food Res Technol 247(8):1975–1985

Zhang XH, Gu HW, Liu RJ et al (2023) A comprehensive review of the current trends and recent advancements on the authenticity of honey. Food Chem: X 19:100850

Zheng YF, Wu MC, Chien HJ et al (2021) Honey proteomic signatures for the identification of honey adulterated with syrup, producing country, and nectar source using SWATH-MS approach. Food Chem 354:129590

Detection of Adulterants in Eggs Using Immunosensors

Suresh Kumar, Chunna Yadav, Santosh Kumar Khuswaha, Veena Pande, and Anirban Dandapat

Abstract

Eggs are a key part of many diets and are an encapsulated source of animal proteins. Eggs are loved for their rich nutritional benefits and contain various bioactive macro- and micronutrients. These bioactive components have a wide range of health-enhancing properties. These properties include immunomodulatory, antioxidant, anticancer, anti-inflammatory, and antihypertensive effects. Adulteration is a type of act of intentionally debasing the natural properties of the eating substances by either adding or replacing them with artificial substances. But with the increasing issue of egg adulteration, like the use of synthetic dyes, formalin, and protein additives, it becomes a curse day by day. There are serious health concerns and food safety challenges that occurred due to the use of various artificial additives. Long-established physical methods of working for identifying these adulterants were not valuable. These methods' sensitivity, specificity, and speed are less in comparison to novel methods for discovery of adulteration. That is where immunosensors come in, utilizing the precise interactions between antibodies and antigens to effectively detect these harmful substances in eggs. This chapter provides a closer look at how immunosensors work efficiently and their applications in identifying egg adulterants. Here, novel models of

S. Kumar (✉) · V. Pande
Department of Biotechnology, Sir J. C. Bose Technical Campus, Bhimtal, Kumaun University, Nainital, Uttarakhand, India

C. Yadav · S. K. Khuswaha
Pharmaceutics and Pharmacokinetics Division, CSIR Central Drug Research Institute, Lucknow, Uttarakhand, India

A. Dandapat
University School of Automation and Robotics, Guru Gobind Singh Indraprastha University, East Delhi Campus, Surajmal Vihar, Delhi, India
e-mail: anirban.usar@ipu.ac.in

K. Kuča et al. (eds.), *Ensuring Wellness: Immunosensors Monitoring Food Quality and Pathogenicity Risks*, https://doi.org/10.1007/978-3-031-97477-9_15

immunosensors, such as electrochemical and optical ones and those enhanced with nanomaterials, demonstrate their ability to provide authentic and quick results. It is also highlighting the innovations like artificial intelligence and nanotechnology-based biosensors that help to overcome these mentioned issues. It is focusing on regulatory standards and future developments.

Keywords

Egg adulteration · Immunosensors · Biosensors · Food safety · Nanotechnology

1 Introduction

Eggs are a powerhouse of vital nutrients, playing a crucial role in supporting positively human health. Eggs are a super-versatile food packed with beneficial nutrients and a good option for tasty meals. They are an affordable source of quality protein. Plus, they provide a lot of cooking options, are enjoyable for the taste buds, and are also pocket-friendly. Including eggs in the diet can really boost the overall health (Onyenweaku et al. 2022). Eggs are packed with important nutrients, such as vitamins A, D, E, and B12, along with minerals like selenium, zinc, and iron. The yolk is especially high in choline, which is crucial for brain health and development. Plus, eggs are super versatile and easy to digest, so they are great for everyone, no matter their age (Rehault-Godbert et al. 2019). Eggs are a common allergen for kids. Both the egg white and yolk can lead to allergies, but most allergic reactions come from proteins found in the egg white, like ovomucoid, ovalbumin, ovotransferrin, and lysozyme. Because of this, it's super important to clearly label any food products that contain eggs, as mandated by EU law (1169/2011) (Paganizza 2020). Vendors often add illegal synthetic dyes to eggs because consumers judge the freshness and nutritional value of the eggs by the color of the yolk. Azo dyes such as Sudan dye and Para Red are often used in the printing industry, but they are also mixed into poultry feed to enhance the yellow or orange color of the yolk. Since darker egg yolks are preferred, sellers sometimes mix these dyes into chicken feed to make the yolk look more vibrant (He et al. 2007).

Biosensors are a game-changing technology that can really boost how we detect and monitor egg adulterants, leading to better food safety and quality. They allow for quick, precise, and on-the-spot identification of contaminants, which is a big step up from traditional methods. These adaptable tools can spot various adulterants and substances, but putting them to use can be tricky due to challenges like sensitivity and specificity, interference from the complex egg matrix, and tough regulatory standards. Sensors-based technologies play a pivotal role in detecting the egg adulterants, helping to keep the food items safe for use by detecting harmful chemicals and contaminants. Formalin, a formaldehyde solution, is a common additive found in eggs for preservation. On the other hand, sodium benzoate and boric acid are used as controlling agents for microbial growth and inducing color formation. Hydrogen peroxide is used for disinfecting the shelves. Copper and zinc have been substituted for lead oxide for pickling preserved eggs (Darji et al. 2018; Wang et al. 2021). These substances can lead to health threats like cancer, lung infections, kidney damage, breathing problems, and allergic reactions. Immunosensors, like ELISA, electrochemical sensors, and fluorescence-based sensors, are known for their quick action, sensitivity, and precise way to identify these adulterants. In contrast to traditional methods, such as chromatography and spectroscopy, immunosensors allow for real-time, on-site testing of adulterants with less sample preparation and minimum costs. Advanced technology, like smart phone-connected biosensors and AI-powered detection systems, is on the way to emergence for advanced food safety monitoring even better by enabling portable and automated testing. These advancements help to detect concealment of material impurities, ensure compliance with regulations, and safeguard public health by offering a dependable way to detect chemical preservatives in eggs (Goyal et al. 2022).

Egg adulteration detection remains a big challenge in an underresearched area, particularly on a large scale. Existing literature shows that few studies have been conducted so far, which focuses on a considerable gap in this field. The primary reason for this scarcity of research in egg adulteration is due to the complexity involved in developing consistent and standardized methodologies for detecting egg adulteration. Biodegradability, softness, and complex biological properties of the eggs make their analysis more challenging compared to other food items. The egg's composition varies naturally, which creates hindrance in detection of foreign substances or alterations. Moreover, the lack of rapid, cost-effective, and reliable detection techniques further hampers large-scale studies.

This chapter covers the few methodologies in the field of egg adulterations that show effective ways to detect egg adulteration quickly and accurately. With advancements in more authentic, reliable, and easy-to-use materials in the queue, such as artificial intelligence and portable biosensors, these techniques even become more commercial for real-time monitoring. There is still continuous research going on in different areas. By enforcing strict regulations and promoting the use of these advanced detection techniques, it can ensure better egg quality and protect consumer health in the long run.

2 Overview of Adulterants in Eggs

Egg adulteration is full of harmful substances like artificial yolk, synthetic eggs, and chemical preservatives to extend their shelf life and other morphological features. These adulterants are best to generate an economy but also serve serious health risks to society. Food poisoning and long-term toxicity is noticed in many cases. It is best for serving quality products with safety and requires strict monitoring and adherence to food safety regulations (Eddin et al. 2019).

2.1 Synthetic Dyes in Yolk

The color, texture, and look of food play a big role in what consumers choose to buy. Foods that are visually appealing tend to sell better and boost profits. Because of this, both natural and artificial dyes are often used on various food products. Unfortunately, many colorful foods can fall victim to this practice. Fruits, veggies, eggs and egg-based foods, spices, sweeteners, and candy are some of the main targets. Even processed meats and fish get a splash of coloring agents added to them (Mendez-Vilas 2017). Traders sometimes mix artificial azo dyes into eggs because many folks think the color of the yolk indicates how fresh and nutritious the egg is. Sudan dyes, which are synthetic azo dyes, are typically used in industries like printing rather than in food. There is also another dye called Para Red that is intended for printing purposes. Since people tend to flavor eggs with vibrant yellow-orange yolks, some sellers add these dyes to chicken feed to boost the yolk's color. However, using these dyes in food is both illegal and unsafe (He et al. 2007).

2.2 Preservatives

Formalin is applied to tackle external parasites that can invade fish eggs. These pesky parasites can damage the eggs and lower their chances of survival. By using formalin in right doses, fish farmers can safeguard the eggs from these infections. But, if used too much, formalin can be detrimental to both the eggs and the surrounding environment. So, it is crucial to stick to proper guidelines when using it in fish hatcheries (Rahman et al. 2012). Keeping egg and larval samples intact is super important, for studying ichthyoplankton means the eggs and larvae of fish, especially when looking into how mothers affect egg quality. Old egg collections that have been stored in formalin can give us great insights, but there are worries about how it impacts egg size and health risks. This chapter also indicates how formalin preservation, which is used for 30–36 months can, affected the egg sizes of Japanese sardines, chub mackerels, and Japanese anchovies. The studies also showed a slight drop in egg size during the first few months, but then the sizes leveled off and grew much the same as fresh eggs. There was some variation in anchovy eggs, but no significant differences have been seen between the 5% and 10% formalin solutions. Still, it is important to consider that formalin is toxic and can create health issues for people, like breathing problems and possible cancer risks. Proper safety measures are implemented when working with formalin in research. Overall, this study shows that measurements from formalin-preserved eggs can be trusted for long-term studies, but caution is required in both scientific integrity and safety in mind.

2.3 Protein Adulterations in Eggs and Egg-Based Products

Melamine is a heterocyclic organic nitrogen-containing rich compound that is often used to mimic proteins in eggs (Caldara et al. 2022). It contains a high nitrogen content of about 66.7% (Wang et al. 2016). This is used as a preferred adulterant to synthetically increase protein levels. Similarly, egg powder and other egg-based food products are also categorized as melamine contamination products. This unethical practice is done to enhance market value, but it poses serious health risks. Kidney damage and other toxic effects are noticed in some cases. Strict monitoring and food safety regulations are essential to control such contamination and protect consumer health (Rodriguez Mondal et al. 2010).

2.4 Other Chemical Adulterants

Adulterated eggs are treated with various substances to improve their morphological features, such as shape, size, weight, and shelf life. Artificial dyes and chemicals are used to make eggs appear fresh and gain the attention of the customers. Coating with different ingredients such as wax, mineral oil, or paraffin wax provides the eggs with a glossy finish and extends their shelf life (Wandrey et al. 2010). Sodium alginate, gelatin, paraffin wax, calcium chloride, and artificial egg yolk coloring are among the substances that are also used in this market (Joshi et al. 2022; Pasquali et al. 2012). Formalin is used as a preservative, but it is highly toxic. Some eggs may contain lead and other heavy metals due to contamination of poultry feed (Stadelman 2017).

3 Impact of Adulterants on Health and the Egg Industry

Egg adulteration means serious health risks to both consumers and the egg industry as a whole. Contaminants in eggs can lead to different toxic reactions that lead to allergic issues. Adulteration in eggs leads to antibiotic resistance due to harmful substances (Adegbeye et al. 2024). Plus, this kind of adulteration can also minimize the nutritional value of eggs, which might cause negative overall health. The consequences for the egg industry are just as severe.

Adulteration in egg industries can lead to major financial crises, stricter regulations, and damage at the level of trust and recognition. Companies could face hefty fines, risk termination of business, and deal with trade barriers if their products do not meet safety standards. To handle these problems, producers might standardize the quality control and testing methods. Techniques like ELISA are required for detecting adulterants and ensuring food safety (Guan et al. 2022). By regularly testing eggs, companies can prevent harmful products from penetrating the market, maintain industry standards, and recoup consumer trust. Effective testing is also crucial for smooth international trade by meeting safety regulations (Vogel 2009). To overcome adulteration, the egg industry must actively participate in the improvement of quality assurance, using emerging testing technologies, and working with food safety organizations. Educating consumers about the importance of buying eggs from trusted sources and regular monitoring can also help reduce the risk of future contamination. By prioritizing food safety, the egg industry can protect public health, enhance consumer confidence, and ensure its long-term success (Lin et al. 2021).

4 Role of Immunosensors

Khanal et al. (2024) developed screen-printed gold electrodes (Au-SPEs) as a label-free point-of-care detection of colistin (Polymyxin B) in chicken eggs. The sensor attached a specific colistin antibody to a conventional Au-SPE that had been polyethyleneimine (PEI) modified by a carbodiimide-mediated amide coupling procedure. Results showed that limit of detection (LOD) of 1.539 μg/kg of egg with high specificity toward antigen against other antibiotics was observed. An amperometric immunosensor was developed to spot ovotransferrin (Gal d 3), which is a key allergen present in egg whites, including various food products. It is noticed that small exposure to egg allergens through the food chain, skin contact, or inhalation can trigger severe allergic reactions. This sensor is designed on the basis of a sandwich method for detecting this allergen. It has screen-printed carbon electrodes (SPCE) that are immobilized with antibodies that specifically attach to Gal d 3. The detection part of this sensor has another set of anti-Gal d 3 antibodies that are equipped with horseradish peroxidase (HRP). The colorimetric substrate TMB (3, 3′, 5, 5′-tetramethylbenzidine) is used for the oxidation with hydrogen peroxide (H_2O_2) by using horseradish peroxide (HRP), generating an electrochemical signal that can be noticed and recorded. Both raw and cooked egg samples were analyzed with this technique. It is able to explain the allergen amount in the food items. It was also used to check 25 different commercial food products for compliance with EU labeling rules, and their detection mechanism is similar to those from an ELISA test. Additionally, the sensor's prepared surface stayed stable for 20 days, which is useful to reduce the need for reagents with lower expenses (Freitas et al. 2022).

5 Advances in Immunosensor Technology for Egg Adulteration

Egg adulteration is no doubt a serious concern in the food industry due to their abundant use, making it important to design quick, sensitive, and dependable egg adulteration detection methods. Immunosensor technology has stepped up as a strong remedy here for spotting contaminants, harmful substances, and any fraudulent changes in the internal as well as external morphology of the eggs. Recent advancements in this area are based on the implementation of nanomaterials for the design of these immunosensors to enhance their sensitivity and specificity for finding out the adulterants in a short period of time (He et al. 2021; Thiruvengadam et al. 2024).

5.1 Role of Nanomaterials in Enhancing Sensitivity

Nanomaterials are game changers in immunosensor technology equipped with suitable physicochemical traits like a large surface area, better biocompatibility, and splendid electrical conductivity (Nasrollahpour et al. 2023). These features really help to boost sensitivity when it comes to detecting adulterants in eggs. Gold nanoparticles (AuNPs) are great for amplifying signals due to their great ability of electron transfer. Here, antibodies are used to make antigen–antibody interactions for better identification results. This is very useful in colorimetric assays, allowing for quick and clear detection of contaminants (Anh et al. 2022). Quantum dots (QDs) show strong fluorescence properties, which makes them valuable with highly sensitive optical detection. Multiple adulterants are also identified in a single test, and they are stable enough for long-term use in biosensors (Oliveira et al. 2025). Carbon nanotubes (CNTs) are known for boosting electron transfer, which enhances the strength of electrochemical signals. They have a large surface area for attaching biomolecules, making detection more efficient. They are the best option for label-free immunosensors, providing quick and pocket-friendly analysis (Saguin et al. 2024). Graphene and graphene oxide have great electrical conductivity, making them ultra-sensitive. They help with the efficient attachment of antibodies and other bioreceptors, plus they offer stability and reusability in biosensing setups (Raghavan et al. 2021). Magnetic nanoparticles (MNPs) help concentrate analytes, which improves detection limits. They are used in magnetic immunoassays to effectively separate target molecules, cutting down on background noise and boosting sensor specificity (Socas-Rodriguez et al. 2020).

5.2 Nanoparticles as a Unit of Sensors

Sensors have two compartments: one is a recognition element, and another one is a transducer element (Xing et al. 2017). The recognition elements are merged with DNA, peptides, aptamers, or antibodies. This element is known for high specificity due to their selective attachment to the target analytes (Koyun et al. 2012). On the other hand, the transducer element takes the benefits of biological interaction and turns it into a detectable signal. It is based on utilizing specific chemical or physical properties. Nanoparticles (NPs) are formed in the range of 1–100 nm, like silver, gold, magnetic particles, and quantum dots, and have become the first choice as effective signal transducers. Their large surface area, magnetic features, excellent electrical conductivity, and unique physical and chemical traits make them perfect for boosting sensor performance. Nanosensors are generally divided into electrochemical and optical types based on how they transmit signals, giving them advanced detection abilities for a variety of applications (Tarannum et al. 2024).

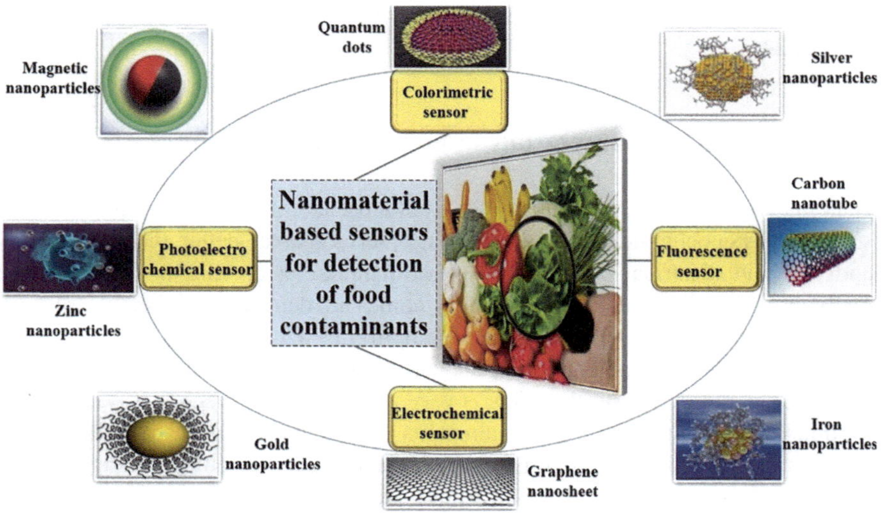

Fig. 1 Nanomaterial-based sensors for detection of food contaminants. (Tarannum et al. 2024, under the CC BY license)

Figure 1 illustrates the advanced use of nanomaterial-based sensors for detecting food contaminants, highlighting their high sensitivity, specificity, and rapid detection capabilities.

5.3 Smartphone-Enabled Immunosensors

Smartphone-based immunosensors use mobile technology to make portable, user-friendly, and cost-effective products that are useful for detecting egg adulterants. These systems use the phone's camera or specialized attachments to detect immunosensor signals. They provide immediate results through selected applications. The integration of machine learning algorithms further enhances accuracy by improving pattern recognition and minimizing errors in detection (Lu et al. 2019).

6 International Standard for Egg Safety, Such as FSSAI, FDA, Codex Alimentarius

International egg safety standards are set by different organizations, bodies *established* jointly, such as the Food Safety and Standards Authority of India (FSSAI) and the US Food and Drug Administration (FDA) in association with the Codex Alimentarius Commission (CAC), which lay out explicit guidelines to ensure eggs are ready to table from the farm with safety and high quality. These standards lay down science-based standards for articles of everything from egg production and handling to storage, aiming to keep food safe and reduce the risk of foodborne illnesses. The main areas of research include maintaining farm hygiene, providing optimum temperature conditions during transport and storage, and following specific labeling rules (Fayaz et al. 2022; Singh 2023).

Eggs freshness during storage and transportation can be maintained for an extended period by applying temperature ranges of 4 °C and 7 °C, or 39 °F to 45 ° F. Humidity levels should ideally be between 70% and 80% relative humidity (Akter et al. 2014). Eggs can last about 4–5 weeks in deep cooling from the date of packing. Good hygiene practices can significantly overcome the bacterial contamination. According to available data, implementing international standards for egg production has helped to reduce the foodborne illnesses linked to eggs over the years (Chousalkar et al. 2021). Additionally, labeling contains the date of production to the date of best-before use. Traceability information also ensures the consumer's safety (Chatzopoulou et al. 2020). To keep up with international standards, the regular conduct of inspections or audits are also applicable. These parameters provide compliance with an emphasis on the pivotal role in the development or rising economy. Therefore, it also allows for how to tackle new market opportunities and increases the shelf life of the products. In egg production, specific pathogens like Salmonella enteritidis are a big concern, but implementing these standards has led to a noticeable drop in their prevalence. It is also a priority to set authentic and easy-to-understand quantitative goals for quality indicators, such as acceptable levels of microbial contamination and the maximum time allowed between collecting eggs and refrigerating them (Guyonnet 2023; Solis et al. 2023). When it comes to layer eggs, the presence of tetracycline residues raises some red flags regarding human health risks from consuming eggs with these residues. While finding chlortetracycline and tetracycline levels above the maximum residue limit (MRL) is worrying, the potential health risks for consumers who eat these eggs cannot be ignored (Bishnoi et al. 2025).

7 Recommended Limits for Adulterants in Eggs

Regulatory bodies such as FSSAI, FDA, and EFSA designed some norms related to food safety such as egg safety from harmful substances (Baliyan et al. 2024). Azo dyes, including Sudan dyes and Para Red, are completely avoided because of their toxic and cancer-causing properties. Formalin, which is sometimes used as a preservative, is also not allowed. There are specific limits for heavy metals like lead ($\leq$0.1 mg/kg), cadmium ($\leq$0.05 mg/kg), and arsenic ($\leq$0.1 mg/kg) to avoid contamination. Antibiotics like tetracycline in the amount of $\leq$0.2 mg/kg and sulfonamides having an amount $\leq$0.1 mg/kg need to be used within safe limits, while nitrofurans, due to resistance, are banned altogether (Sharma et al. 2023). Pesticide residues in eggs cannot exceed 0.01–0.05 mg/kg, depending on the type (Ahmad et al. 2010). These parameters are in place to minimize health risks and ensure that the eggs on the plate are safe to eat (Balali-Mood et al. 2021).

8 AI-Driven (Artificial Intelligence) Immunosensors for Rapid Data Analysis

Artificial Intelligence (AI) is the game changer in the world of immunosensor technology by providing data analysis that is faster, more accurate, and fully automated. With AI-powered platforms, interpreting complex biosensor signals becomes easier, boosting detection sensitivity and helping with better decision-making (Wasilewski et al. 2024). Deep learning and neural networks, by using machine learning (ML) algorithms, transfer the detection system from large amounts of biosensor data to spot patterns and minimize the false results (Rafique et al. 2024). Cloud-based AI (Artificial Intelligence) analytics are valuable in remote access to immunosensor data. They speed up the clinical decisions and the tracking of pandemics. Artificial Intelligence (AI) analytics is useful in forecasting the disease progression through immunosensor readings, which is useful for personalized medicine (Nashruddin et al. 2024).

9 Conclusion

Egg adulteration is responsible for mild to severe health impacts that threaten food safety, consumer trust, and financial damage. Rapid and accurate detection of fraudulent addition of adulterants is crucial to ensuring the integrity of the egg supply. In recent years, researchers have explored the use of antibody-based assays as a means of detecting proteins and other added products in eggs. These methods anchor the specificity of antibodies to detect the specific foreign proteins. It is known for providing a sensitive and accurate approach to egg adulteration detection. Enzyme-linked immunosorbent assays (ELISA) is another approach that is implemented for the detection of the presence of proteins in egg samples in a better way. These assays are based on antibodies specific to proteins considered as antigens. It allows for rapid and reliable identification of adulteration. Furthermore, the development of paper-based PCR assays also has provided a simple and cost-effective method for the detection of bacterial contaminants, which can also serve as indicators of egg adulteration. Therefore, some advanced methods, like nanomaterials-based biosensors, smart and artificial techniques, and implemented equipment, are also suitable for the detection of adulteration with full accuracy. Instead of that, further research is still ongoing to optimize and validate them for widespread use in the poultry industry. Continued development of rapid, sensitive, and reliable egg adulteration detection methods will be introduced to maintain the quality and safety of the global egg supply.

Egg adulteration detection is still a relatively unexplored area in large aspects. The available data show few studies conducted so far. This limited research is mainly due to the challenges faced by the researchers with developing reliable and standardized methods for detecting adulterants in eggs. Additionally, eggs are easily spoilt and complex biological materials, which make their detailed studies more difficult compared to other food products. The lack of awareness is also one of the reasons for possible adulteration practices in eggs. That leads to a contribution to the limited research. As a result, more comprehensive studies are required in upcoming time to establish effective detection techniques. The innovative detection methods development ensures the safety and quality of eggs available to consumers.

References

Adegbeye MJ, Adetuyi BO, Igirigi AI et al (2024) Comprehensive insights into antibiotic residues in livestock products: distribution, factors, challenges, opportunities, and implications for food safety and public health. Food Control 163:110545

Ahmad R, Nida MS, Estaitieh H (2010) Occurrence of organochlorine pesticide residues in eggs, chicken and meat in Jordan. Chemosphere 78(6):667–671

Akter Y, Kasim A, Omar H et al (2014) Effect of storage time and temperature on the quality characteristics of chicken eggs. J Food Agric Environ 12(3-4):87–92

Anh NH, Doan MQ, Dinh NX (2022) Gold nanoparticle-based optical nanosensors for food and health safety monitoring: recent advances and future perspectives. RSC Adv 12(18): 10950–10988

Pasquali F, Manfreda G, Olivi P, Rocculi P, Sirri F, Meluzzi A (2012) Modified-atmosphere packaging of hen table eggs: Effects on pathogen and spoilage bacteria. Poult Sci 91 (12):3253–3259

Balali-Mood M, Naseri K, Tahergorabi Z (2021) Toxic mechanisms of five heavy metals: mercury, lead, chromium, cadmium, and arsenic. Front Pharmacol 12:643972

Baliyan N, Kumari R, Kumari A (2024) Food safety and detection of bacterial foodborne pathogens in India. In: Handbook of public health nutrition: international, national, and regional perspectives. Springer Nature Switzerland, Cham, pp 1–24

Bishnoi K, Moudgil P, Soni D (2025) Occurrence and risk assessment of tetracycline residues in layer eggs in Haryana, India. J Food Prot 88:100449

Caldara M, Lowdon JW, Royakkers J (2022) A molecularly imprinted polymer-based thermal sensor for the selective detection of melamine in milk samples. Food Secur 11(18):2906

Chatzopoulou S, Eriksson NL, Eriksson D (2020) Improving risk assessment in the European food safety authority: lessons from the European Medicines Agency. Front Plant Sci 11

Chousalkar KK, Khan S, McWhorter AR (2021) Microbial quality, safety and storage of eggs. Current Opinion in Food Science 38:91–95

Darji MA, Lalge RM, Marathe SP (2018) Excipient stability in oral solid dosage forms: a review. AAPS PharmSciTech 19:12–26

Eddin AS, Ibrahim SA, Tahergorabi R (2019) Egg quality and safety with an overview of edible coating application for egg preservation. Food Chem 296:29–39

Fayaz S, Naqash F, Morab S et al (2022) Food quality and safety regulation issues. In: Shelf life and food safety, pp 375–390. https://doi.org/10.1201/9781003091677-20

Freitas M, Del Rio M, Nouws HP (2022) Tracking a major egg allergen to assess commercial food label compliance: towards a simple and fast immunosensing device. Biosensors 12(12):1109

Goyal K, Kumar P, Verma K (2022) Food adulteration detection using artificial intelligence: a systematic review. Arch Comput Methods Eng 29(1):397–426

Guan T, Xu Z, Wang J et al (2022) Multiplex optical bioassays for food safety analysis: toward on-site detection. Compr Rev Food Sci Food Saf 21(2):1627–1656

Guyonnet V (2023) Regulations for shell eggs in different countries. In: Handbook of egg science and technology. CRC Press, pp 323–340. https://doi.org/10.1201/9781003254430-23

He L, Su Y, Shen X (2007) Determination of Sudan dye residues in eggs by liquid chromatography and gas chromatography-mass spectrometry. Anal Chim Acta 594(1):139–146

He Y, Bai X, Xiao Q et al (2021) Detection of adulteration in food based on nondestructive analysis techniques: a review. Crit Rev Food Sci Nutr 61(14):2351–2371

Joshi R, Baek I, Joshi R (2022) Detection of fabricated eggs using Fourier transform infrared (FT-IR) spectroscopy coupled with multivariate classification techniques. Infrared Phys Technol 123:104163
Khanal S, Taneja A, Kumar H et al (2024) Detection of colistin (Polymyxin B) a restricted drug in chicken eggs using a label-free immunosensor based on Au screen-printed electrodes. Chem Pap 78:1009–1020
Koyun A, Ahlatcolu E, Koca Y et al (2012) Biosensors and their principles. In: A roadmap of biomedical engineers and milestones, vol 5, pp 42–117. https://doi.org/10.5772/48824
Lin X, Chang SC, Chou TH et al (2021) Consumers' intention to adopt blockchain food traceability technology towards organic food products. Int J Environ Res Public Health 18(3):912
Lu Y, Shi Z, Liu Q (2019) Smartphone-based biosensors for portable food evaluation. Curr Opin Food Sci 28:74–81
Mendez-Vilas A (2017) Science within food: up-to-date advances on research and educational ideas. Formatex Research Center, Badajoz
Nashruddin SNABM, Salleh FHM, Yunus RM (2024) Artificial intelligence-powered electrochemical sensor: recent advances, challenges, and prospects. Heliyon 10(18):e37964
Nasrollahpour H, Sanchez BJ, Sillanpaa M (2023) Metal nanoclusters in point-of-care sensing and biosensing applications. ACS Appl Nano Mater 6(14):12609–12672
Oliveira S, Sharifuzzaman M, Moro G (2025) Advances in optical devices for the detection of contaminants in food and water. TrAC Trends Anal Chem 184:118139
Onyenweaku EO, Akah LU, Kesa H et al (2022) Protein quality evaluation of some commonly consumed bird egg varieties using amino acid scores. Biochem Res Int 2022(1):6536826
Paganizza V (2020) A European overview on Regulation (EU) No 1169/2011 after the entry into force. Riv Dir Aliment 14(1):11–30
Rafique SH, Abdallah A, Musa NS (2024) Machine learning and deep learning techniques for internet of things network anomaly detection-current research trends. Sensors 24(6):1968
Raghavan VS, O'Driscoll B, Bloor JM, Li B, Katare P, Sethi J, Gorthi SS, Jenkins D (2021) Emerging graphene-based sensors for the detection of food adulterants and toxicants–a review. Food Chem 355:129547
Rahman MM, Ahmed S, Hosen MM (2012) Detection of formalin and quality characteristics of selected fish from wet markets at Sylhet city in Bangladesh. Bangladesh Res Pub J 7(2):161–169
Rehault-Godbert S, Guyot N, Nys Y et al (2019) The golden egg: nutritional value, bioactivities, and emerging benefits for human health. Nutrients 11(3):684
Rodriguez Mondal AM, Desmarchelier A, Konings (2010) Liquid chromatography-tandem mass spectrometry (LC−MS/MS) method extension to quantify simultaneously melamine and cyanuric acid in egg powder and soy protein in addition to milk products. J Agric Food Chem 58(22):11574–11579
Saguin NSG, Maulik G, Cao X (2024) CNTs-based biosensors for enzyme detection. Sens Actuators A: Phys 377:115753
Sharma A, Pant K, Brar D (2023) A review on Api-products: current scenario of potential contaminants and their food safety concerns. Food Control 145:109499
Singh M (2023) Nanosensor platforms for detection of milk adulterants. Sens Actuators Rep 5: 100159
Socas-Rodriguez B, Herrera-Herrera AV, Asensio-Ramos M (2020) Recent applications of magnetic nanoparticles in food analysis. Processes 8(9):1140
Solis D, Cordero N, Quezada-Reyes M (2023) Prevalence of salmonella in eggs from conventional and cage-free egg production systems and the role of consumers in reducing household contamination. Foods 12(23):4300
Stadelman WJ (2017) Quality identification of shell eggs. In: Egg science and technology. CRC Press, Boca Raton, pp 39–66
Tarannum N, Gautam A, Chauhan T (2024) Nanomaterial based sensors for detection of food contaminants: a prospect. Sens Technol 2(1):2373196

Thiruvengadam M, Kim JT, Kim WR (2024) Safeguarding public health: advanced detection of food adulteration using nanoparticle-based sensors. Crit Rev Anal Chem:1–21. https://doi.org/10.1080/10408347.2024.2399202

Vogel D (2009) Trading up: consumer and environmental regulation in a global economy. Harvard University Press, Cambridge

Wandrey C, Bartkowiak A, Harding SE (2010) Materials for encapsulation. In: Encapsulation technologies for active food ingredients and food processing. Springer, New York, pp 31–100

Wang Y, Xuan H, Lin G (2016) A melamine-assisted chemical blowing synthesis of N-doped activated carbon sheets for supercapacitor application. J Power Sources 319:262–270

Wang Y, Xiong C, Luo W (2021) Effects of packaging methods on the quality of heavy metals–free preserved duck eggs during storage. Poult Sci 100(5):101051

Wasilewski T, Kamysz W, Gębicki J (2024) Artificial Intelligence-assisted detection of biomarkers by sensors and biosensors for early diagnosis and monitoring. Biosensors 14(7):356

Xing C, Jing X, Zhang X et al (2017) Ultrasensitive indirect competitive ELISA and strip sensor for detection of furazolidone metabolite in animal tissues. Food Agric Immunol 28(6):1269–1282

Application of Immunosensors in Detecting Fungi and Their Toxins in Different Food Matrices

Harshvadan Patel, Ratnakar Chitte, Jitendra Kumar Pal, and Jinal Naik

Abstract

The detection of common fungal contaminants such as *Aspergillus*, *Penicillium*, and *Fusarium* and their mycotoxins like aflatoxins, ochratoxins, and fumonisins in food matrices is a critical concern in food safety due to the health risks. These mycotoxins cause carcinogenic, nephrotoxic, and immunosuppressive effects. Several traditional analytical methods are used for the detection of these fungi and their toxins. Traditional methods are accurate, complex, time-consuming and require high resources. Because of the above, their use for rapid and large-scale testing of fungi and associated toxins are limited. To reduce these limitations, a powerful alternative is the use of immunosensors because they are based on the specificity of antigen–antibody interactions for sensitive, rapid, and cost-effective mycotoxin detection. Some of the latest discoveries in nanotechnology and miniaturization further enhanced sensor performance and made on-site rapid testing feasible. Developing proper regulatory guidelines for food safety, immunosensors exist as an effective and rapid method to detect and qualify fungal contamination and their toxins. The use of immunosensors for food safety has potential to enhance intensive care, reduce risks of contamination during storage, and act as a safeguard for public health.

Keywords

Fungi · Mycotoxin · Immunosensor · Fungal toxin detection · Aflatoxin · Ochratoxin · Fumonisin

H. Patel (✉) · R. Chitte
Department of Microbiology, Vanita Vishram Women's University, Athwagate, Surat, Gujarat, India

J. K. Pal · J. Naik
Department of Biotechnology, Vanita Vishram Women's University, Athwagate, Surat, Gujarat, India

K. Kuča et al. (eds.), *Ensuring Wellness: Immunosensors Monitoring Food Quality and Pathogenicity Risks*, https://doi.org/10.1007/978-3-031-97477-9_16

1 Introduction

Contamination in food matrices by fungal species represents a considerable global issue affecting food safety, public health, and economic stability. Fungi are ubiquitous in the environment and eagerly contaminate a wide range of food products, especially cereals, nuts, fruits, and spices. The environmental conditions, like warm temperatures and high humidity, as well as improper storage, may lead to rapid fungal growth and production of mycotoxin, a secondary metabolite harmful to humans and animals (Pitt and Hocking 2009; Klich 2007). The most explored mycotoxins are aflatoxins, ochratoxins, and fumonisins, and their occurrence and significant toxicological effects in food supplies are reported by various researchers. The fungi such as *Aspergillus flavus* and *Aspergillus parasiticus* primarily produced aflatoxins, which are potent carcinogens and reported to be particularly responsible for liver cancer (IARC 2012). Prolonged exposure of aflatoxins caused hepatocellular carcinoma, immunosuppression, and was also responsible for stunted growth in children, particularly in regions in which there is no proper food monitoring system or the system may be underdeveloped (Gong et al. 2004; Wild and Gong 2010).

Ochratoxins, produced by various *Aspergillus* and *Penicillium* species, are nephrotoxic and cause kidney damage and increased cancer risk in humans (Pfohl-Leszkowicz and Manderville 2007). The most common form of ochratoxin is ochratoxin A, which is mainly responsible for kidney disease and also has immunosuppressive and teratogenic effects, increasing concerns about its contamination in foods, mainly coffee, cereals, and dried fruits (Marquardt and Frohlich 1992; EFSA 2020). Fumonisins, produced by *Fusarium* species, predominantly found in maize and maize products, cause neural tube defects and esophageal cancer. This toxin is mainly responsible for disruption of sphingolipid metabolism, which can lead to cell membrane instability and multiple organ damage (Rheeder et al. 1992; van der Westhuizen et al. 2011). One of the less discussed mycotoxins is zearalenone, produced by *Fusarium* species, mainly *Fusarium graminearum*. These mycotoxins are potent estrogenic compounds and mainly affect the reproductive systems of both humans and animals and cause infertility, hormonal imbalances, and are also responsible for other related health issues (Pestka et al. 2008). This is mainly found in cereals such as wheat, barley, and corn (Haug et al. 2014).

Exposure to mycotoxin leads to serious health issues, not only cancer but also damaged organs, create neurological effects and immunosuppression as well as developmental issues. Chronic mycotoxin exposure leads to neurotoxicity, particularly fumonisins, which is an emerging area of concern because these toxins are responsible for impairing brain function and altering neurotransmitter pathways in animals (Marin et al. 2013). The risks of mycotoxins are not restricted to human health; they also impact livestock and food industries worldwide. Animals consuming infected feed may suffer from reduced growth rates, reproductive issues, and immune deficiencies, which may indirectly affect food security and agricultural economies (Bennett and Klich 2003; Richard 2007).

Restriction or supervising mycotoxin is an area of interest among current researchers aiming to reduce contamination in food matrices by improved agricultural practices, storage conditions, and regulatory interferences. Various innovative approaches like biological control using antagonistic microorganisms, method of chemical detoxification and using various adsorbent materials in which mycotoxins are bound in the gastrointestinal tract of animals are being explored to mitigate mycotoxin risks (Mannaa and Kim 2017; Bullerman and Bianchini 2007). Because of the expansion of global food trade, ensuring robust regulatory frameworks and adopting integrated mycotoxin management strategies across food production chains are essential to minimize health risks and protect food supplies (Codex Alimentarius 2021). Global regulatory bodies such as the European Union, the Food and Drug Administration (FDA), and the Codex Alimentarius Commission, have recommended maximum allowable levels of various mycotoxins in food products (FAO 2004). But control of contamination remains a challenge due to the resilience of fungal spores and the favorable environmental conditions that affect mycotoxin production.

2 Importance of Fungal Detection in Food Safety

The reduction of economic losses in food industries and increasing the detection of fungi and their associated toxins present in food are highly essential for public health. A number of species of *Aspergillus*, *Penicillium*, and *Fusarium* produce harmful mycotoxins, which may be responsible for serious health risks to consumers like humans and animals (Zain 2011; Benkerroum 2020). These mycotoxins cause carcinogenic, immunosuppressive, and hepatotoxic effects. Because of the above, it is highly essential for early detection of these toxins to ensure food safety (Alshannaq and Yu 2017).

2.1 Traditional Detection Methods

Chromatography and mass spectrometry, like traditional analytical methods, are widely used for the detection of mycotoxin. These traditional methods are extensively used for detection because of their high accuracy and sensitivity. For effective identification and quantification of mycotoxins, techniques such as high-performance liquid chromatography (HPLC) and gas chromatography mass spectrometry (GC-MS) have been developed (Turner et al. 2009; Rahmani et al. 2009). These advanced methods are not feasible for large-scale or real-time testing because they require complex sample preparation, expensive laboratory equipment, and skilled technicians (Pereira et al. 2015). Moreover, their time-consuming nature renders them unfeasible for rapid screening, particularly in scenarios requiring immediate results, such as food processing plants or import/export inspections.

2.2 Advances in Rapid Detection: Immunosensors

Innovative mycotoxin detection techniques, such as immunosensors, were developed because of the limitations of traditional detection methods. Immunosensors have gained attention as promising alternatives to conventional techniques due to their rapidity, sensitivity, and cost-effectiveness (Paniel and Radoi 2014). Immunosensors work on the specificity of antigen–antibody interactions for detection of mycotoxins. Because of less extensive requirement of sample preparation and complex equipment as well as real-time detection capability, biosensors are beneficial in high-throughput environments as they allow for efficient screening, enhancing food safety, and reducing waste (Goryacheva et al. 2007; Scognamiglio 2013).

A number of immunosensors have been developed for mycotoxin detection. Biosensors based on electrochemical and optical principles are also used for the detection of various mycotoxins with high sensitivity even at low concentrations (Pohanka 2018; Kavita and Prasad 2020). The detection of mycotoxin on-site is becoming more feasible, allowing for rapid action and prevention of contaminated products from reaching consumers, which requires increasing the availability of portable biosensors (Rodríguez-Mozaz et al. 2006).

2.2.1 Structure and Working Principle

Mainly three components (bioreceptor, transducer, and signal processor) are used for immunosensor development. These components play an important role in the detection process. The bioreceptor is generally an antibody or antigen that binds selectively and specifically to the target analyte or biomolecule, like a protein, virus, or bacterium. This specific binding capability allows the immunosensor to achieve high specificity in complex samples. For example, antibodies are frequently used as bioreceptors due to their strong binding affinity for antigens, making them highly effective for selective detection in clinical and environmental settings (Gugleva et al. 2017).

The transducer is responsible for converting the interaction of bioreceptor–analyte into a measurable signal. It converts in electrochemical, optical, or piezoelectric signals based on the type of immunosensor used. Electrochemical immunosensors use changes in current, potential, or impedance resulting from the antibody–antigen interaction, which may allow quantitative measurements with high sensitivity (Yola et al. 2014). Optical transducers use change in light properties such as fluorescence or absorbance into signals in the presence of target analyte. The signal processor interprets the data generated by the transducer and produces output data that can be read and analyzed by the user. Technology advancements in digital signal processing increase the accuracy and reliability of immunosensors. Because of that, immunosensors are being used at point-of-care testing and other rapid diagnostic applications. Signal processors allow immunosensors to integrate with data storage and display systems, enabling real-time monitoring and easy data retrieval (Luong et al. 2008). Each component of the immunosensor must work in harmony to ensure sensitivity, specificity, and reliability. Interaction between the bioreceptor and the target analyte triggers a measurable response. The transducer then converts the response into a quantifiable signal that reflects the analyte concentration. The high sensitivity of immunosensors makes them suitable for detecting very low concentrations of biomolecules in various samples.

3 Types of Immunosensors

Immunosensors are developed on the basis of their signal transduction mechanisms. Mainly optical, electrochemical, and piezoelectric immunosensors are used for the detection process. Each immunosensor offers unique advantages for different applications.

3.1 Optical Immunosensors

Changes in fluorescence, absorbance, and refractive index like optical properties are detected by what is known as an optical immunosensor. As a result of interactions between antigens and antibodies these changes may occur. Changes in optical signals can be monitored and quantified to determine the presence and concentration of target analytes. Different optical techniques are used for the detection, and among them, surface plasmon resonance (SPR) is particularly prominent due to its label-free detection capability. Because of that, direct monitoring of antigen–antibody binding events in real time is possible (Homola 2008). Hence, SPR is more suitable in applications where labeling might interfere with the biomolecular interaction. Recent advancements in fluorescence-based immunosensors have led to increased sensitivity using low concentrations of analytes that play a crucial role in fields such as clinical diagnostics and environmental monitoring (Kim et al. 2020). Because of compatibility with reduced platforms, optical immunosensors are suitable for incorporation into microfluidic systems with increasing applicability for point-of-care diagnostics (Wang and Niessner 2019).

3.2 Electrochemical Immunosensors

The detection of changes in electrical properties like current, voltage, and resistance due to antigen–antibody reactions on the immunosensor surface is known as electrochemical immunosensors. These sensors are mainly used because of their high sensitivity, selectivity, and low cost. The simplicity of the electrochemical transduction process makes it suitable for easy integration into portable devices. Hence, these immunosensors are ideal for applications that require rapid and on-site testing (Pemberton et al. 2014). Electrochemical sensors can also detect a wide range of biomolecules through modification of sensor electrodes with specific antibodies, making them suitable for the development of multiplexed platforms (Labib et al. 2016). Incorporation of nanomaterials like gold nanoparticles and carbon nanotubes in electrochemical immunosensors enhances signal sensitivity, thereby improving detection limits and enabling the identification of ultralow concentrations of target analytes.

The detection of mass changes on the immunosensor surface due to binding of antigens with immobilized antibodies on piezoelectric crystals is known as piezoelectric immunosensors. These sensors offer real-time and label-free detection, making them suitable for continuous monitoring applications like environmental pollutant detection and clinical diagnostics (Brahmbhatt et al. 2016). Quartz Crystal Microbalance (QCM) is a commonly used piezoelectric immunosensor that detects minute mass changes with high accuracy, making them suitable for quantitative analysis of biomolecular interactions (Marx 2003). As compared to optical and electrochemical sensors, these piezoelectric sensors do not require complex labeling procedures, play a crucial role in simplifying the detection process, and reduce the overall analysis costs (Su et al. 2021). Current QCM technology development increases sensitivity and stability as well as expanding its applications in areas like pathogen detection and drug discovery (Ricci et al. 2019). Each type of immunosensor offers different advantages depending on the application requirements like real-time monitoring, portability, and high sensitivity. The selection of immunosensor type always depends on factors like target analyte concentration range, sample matrix, and desired response time (Lavanya et al. 2018). These immunosensor technologies continue to be in the development phase, focusing on advancements in nanotechnology and material science which are expected to further enhance their performance and broaden their application scope.

3.3 Detection of Fungi in Food Matrices

3.3.1 Common Fungi in Food Products

One of the important food contaminants are fungi due to their capability to grow under diverse environmental conditions and produce a variety of harmful secondary metabolites known as mycotoxins. Among the most commonly found fungi in food products are *Aspergillus*, *Penicillium*, and *Fusarium*, each associated with specific mycotoxins that present serious health risks to consumers. These mycotoxins can have carcinogenic, immunotoxic, and hepatotoxic effects (Frisvad et al. 2006; Marin et al. 2013). Table 1 summarizes some common fungi, their associated mycotoxins, and the typical food matrices in which they are found with their optimal growth condition, detection method, and regulatory limits.

3.3.1.1 *Aspergillus*: Aflatoxins in Nuts, Seeds, and Cereals

Aspergillus flavus and *Aspergillus parasiticus* are well known for producing potentially carcinogenic, mutagenic, and immunosuppressive mycotoxin known as aflatoxins (Klich 2007; Eaton and Gallagher 1994). Aflatoxin B1 is classified as most dangerous due to its high carcinogenicity and prevalence in contaminated food sources (IARC 2002). These fungi frequently infect crops such as peanuts, maize, and other nuts under conditions of high warmth and humidity. Improper storage or exposure to fluctuating environmental conditions can raise the moisture content

Table 1 Common fungi, their associated mycotoxins, and the typical food matrices in which they are found with their optimal growth condition, detection method, and regulatory limits

Fungi	Mycotoxins	Food matrices	Health effects	Control methods	Optimal growth conditions	Detection methods	Regulatory limits (ppm)	References
Aspergillus	Aflatoxins	Nuts, seeds, cereals	Liver damage, cancer, immune suppression	Proper drying, biocontrol agents, good storage	25–30 °C, humid conditions	ELISA, HPLC, LC-MS	EU: 2–10, US: 20	Klich (2002) and EFSA (2020)
Penicillium	Ochratoxins	Fruits, nuts, cheese	Kidney damage, neurotoxic effects	Temperature control, irradiation, antifungal agents	15–20 °C, moderate humidity	TLC, LC-MS, GC-MS	EU: 5, US: No specific limit	Pitt and Hocking (2009) and EFSA (2020)
Fusarium	Fumonisins	Cereals, grains	Esophageal cancer, birth defects	Crop rotation, resistant varieties, fungicides	20–25 °C, moderate to high humidity	HPLC, PCR, LC-MS	EU: 1–4, US: 2–4	Desjardins (2006) and US FDA (2018)
Claviceps	Ergot alkaloids	Rye, barley, wheat	Vasoconstriction, hallucinations	Ergot-free seeds, crop monitoring, mechanical sorting	10–25 °C, moderate humidity	HPLC, GC-MS, fluorescence	EU: 0.5, US: No specific limit	Richard (2007)
Alternaria	Alternariol	Tomatoes, fruits, grains	DNA damage, cytotoxicity	Fungicides, proper harvesting, reducing field moisture	20–25 °C, high humidity	HPLC, ELISA, LC-MS	EU: No specific limit, US: No specific limit	Ostry (2008)

within stored grains, creating a conducive environment for *Aspergillus* growth and aflatoxin production (Magan and Aldred 2007; Hell et al. 2008). Contamination because of aflatoxin is particularly concerning in tropical and subtropical regions where warm climates facilitate fungal proliferation. Unfortunately, aflatoxins are heat-stable and difficult to eliminate through standard cooking or processing methods, posing persistent risks to food safety (Williams et al. 2004). Regulatory agencies like the FDA and the EU have established strict limits for aflatoxins in foods to mitigate the health risks, yet contamination persists in both developed and developing countries due to limitations in storage infrastructure and contamination monitoring (Wu and Khlangwiset 2010; Wild and Gong 2010).

3.3.1.2 *Penicillium*: Ochratoxins in Fruits, Nuts, and Cheese

Penicillium verrucosum and *Penicillium nordicum* are prominent in the production of ochratoxin A (OTA). OTA is a nephrotoxin linked to renal disorders and has been associated with carcinogenic and teratogenic effects in animals (Pitt and Hocking 2009; Varga et al. 2015). OTA exposure in humans is primarily through the consumption of contaminated foods like dried fruits, coffee, wine, beer, and aged cheese in which fungal contamination often occurs under improper storage conditions with high humidity (Bayman and Baker 2006; Ostry et al. 2013). Ochratoxins are comparatively stable and persevere through food processing which makes detection and prevention challenging. Exposure to OTA is a significant concern in Europe. Various studies have found detectable levels in blood samples, indicating chronic low-dose exposure through diet. Efforts to reduce ochratoxin contamination include implementing good manufacturing practices (GMPs) and hazard analysis and critical control points (HACCP) systems to control moisture levels and temperature during storage (Bennett and Klich 2003; FAO 2004).

3.3.1.3 *Fusarium*: Fumonisins in Cereals and Grains

Fusarium verticillioides and *Fusarium proliferatum* predominantly produce fumonisin B1 which is responsible for esophageal cancer and neural tube defects (Munkvold 2017; Rheeder et al. 1992). These fungi frequently infect cereals, especially maize, where the presence of cool, moist conditions during harvest and storage promotes fungal growth (Desjardins 2006; Shephard 2008). Fumonisins interfere with lipid metabolism by inhibiting ceramide synthase, which disrupts cell signaling pathways and ultimately poses a risk to human and animal health (Voss et al. 2007). Fumonisin contamination is especially widespread in countries like South Africa and parts of Latin America where maize is a main food ingredient. Regulatory agencies like the FDA set limits on fumonisins, yet these continue to pose a risk due to environmental and agricultural practices that facilitate fungal growth (Bullerman and Bianchini 2007).

4 Immunosensors for Fungal Detection

Use of immunosensors for detection of specific fungi in diverse food matrices has increased and play an essential role for food safety and quality control. By combining biological recognition with physical transducers, immunosensors offer precise, rapid, and sensitive detection of pathogenic fungi that can lead to food spoilage and health hazards, if undetected. The advancements in immunosensor technology have enabled the detection of various fungi such as *Aspergillus, Penicillium*, and *Fusarium* species which can produce harmful mycotoxins affecting human health and the food industry.

4.1 *Aspergillus* Detection

Aspergillus species, the most common contaminants in nuts and cereals, are well-known for producing aflatoxins which are highly toxic compounds with significant health risks (Iha et al. 2021). Recent studies have suggested optical immunosensors for detecting *Aspergillus* species, especially due to their ability to use fluorescence and surface plasmon resonance (SPR) for improved sensitivity. These sensors work by immobilizing antibodies specific to *Aspergillus* antigens on the sensor surface. When exposed to a sample containing *Aspergillus*, binding between antigens and antibodies triggers an optical signal. Rosi et al. (2011) demonstrated an effective application of SPR immunosensors in detecting *Aspergillus* with high selectivity in cereal samples. The fluorescence-based sensors, on the other hand, offer quantitative assessment with a lower detection limit and essential for identifying even trace amounts of fungi that could pose health risks (Eivazi et al. 2018).

4.2 *Penicillium* Detection

Penicillium species are mainly responsible for spoilage of fruits and vegetables which may lead to significant economic losses. Because of fast response times and minimal sample preparation, electrochemical immunosensors are mainly used for the detection of *Penicillium*. These sensors measure changes in electrical signals upon antibody–antigen interactions providing real-time monitoring with high sensitivity. Marin et al. (2013) demonstrated that electrochemical immunosensors could detect low levels of *Penicillium* in fruit samples making them valuable in preventing post-harvest spoilage. Moreover, improvements in nanomaterial enhanced electrodes have led to even greater signal amplification, which allows for detection at lower limits (Hernández-Hernández et al. 2020).

4.3 *Fusarium* Detection

The spoilage in grains is often found because of growth of *Fusarium* species, which are notorious for producing mycotoxins such as fumonisins, which are harmful to both humans and livestock (Munkvold and Arias 2019). Piezoelectric immunosensors are effectively used for real-time *Fusarium* detection in grains because it can measure mass changes upon antibody–antigen binding. The piezoelectric effect interprets mechanical distortion into an electrical signal and allows these sensors for the detection even at minor concentrations of *Fusarium* with outstanding sensitivity. Sun et al. (2014) reported the effectiveness of piezoelectric sensors for the early detection of fungal contaminants in grains under storage conditions, helping in the prevention of large-scale contamination. Recent advances in piezoelectric sensor materials and miniaturization further increase sensor sensitivity and operational efficiency (Li et al. 2021). Table 2 shows a comparison of common fungal contaminants and main techniques used for identification in various food matrices.

5 Immunosensors for Mycotoxin Detection

The use of immunosensors for mycotoxin detection has significantly been enhanced over the past two decades. Because of that, sensitive and selective platforms for rapid detection of harmful mycotoxins in food and agricultural products have been created. Immunosensors are particularly advantageous because of their high specificity, sensitivity, and the ability to offer real-time analysis. Mycotoxins like aflatoxins, ochratoxins, and fumonisins are among the most studied because of their prevalence in food contamination and toxicological effects on both human health and livestock.

5.1 Aflatoxin Detection

For the detection of aflatoxins, the most carcinogenic and prevalent mycotoxins found in food items (nuts, cereals, and spices), electrochemical immunosensors have been developed. Electrochemical immunosensors are designed to be highly specific and offer rapid results which is crucial for ensuring food safety in commercial and field settings. Ngundi et al. (2005) used electrochemical sensors for the detection of aflatoxins at the low nanogram per liter range. Moreover, these systems have been adapted for on-site detection, making them suitable for field-based testing and minimizing the need for complex laboratory procedures (Zhao et al. 2018).

Table 2 Comparative table summarizing the different types of immunosensors used for fungal detection in food matrices

Fungi	Immunosensors	Technique	Food matrices	Advantages	Regulatory limits	References
Aspergillus	Optical immunosensors	Fluorescence, Surface Plasmon Resonance (SPR)	Nuts, cereals	High sensitivity, quantitative, rapid detection	EU: 2–10 μg/kg	Rosi et al. (2011) and Eivazi et al. (2018)
Penicillium	Electrochemical immunosensors	Electrochemical impedance spectroscopy	Fruits	Fast response time, low detection limits	EU: 50 μg/kg	Marin et al. (2013) and Hernández-Hernández et al. (2020)
Fusarium	Piezoelectric immunosensors	Piezoelectric effect (mass change detection)	Grains	Real-time monitoring, high sensitivity	EU: 200–2000 μg/kg	Sun et al. (2014) and Li et al. (2021)

5.2 Ochratoxin Detection

Ochratoxins, particularly ochratoxin A (OTA), are common contaminants in coffee, wine, and dried fruit. The use of optical immunosensors, including surface plasmon resonance (SPR) and fluorescence-based platforms, has shown promising results for detecting OTA with high sensitivity and specificity. SPR-based sensors can detect changes in the refractive index near the sensor surface upon binding with the target mycotoxin, offering real-time, label-free detection (Pires et al. 2014). However, in fluorescent-based methods, fluorescent probes are used for the emission of light as binding with ochratoxins makes the process rapid and efficient with lesser sample preparation (Cai et al. 2012). These platforms are particularly useful for monitoring food safety in products like wine and coffee where traditional chemical analysis methods may be time-consuming and require expensive equipment.

5.3 Fumonisin Detection

Fumonisins, a group of mycotoxins primarily found in maize, are associated with various health problems, including neural tube defects in humans and liver and kidney damage in animals. To address the need for rapid and reliable detection, piezoelectric immunosensors have been developed for the detection of fumonisins in food. These sensors utilize the changes in the resonance frequency of a quartz crystal microbalance (QCM) sensor upon the binding of fumonisin molecules to the immunoreceptor. Such systems offer high sensitivity, low detection limits, and rapid responses, making them suitable for both laboratory and field applications. Ricci et al. (2013) used QCM-based sensors for fumonisin detection and the potential of QCM sensors reported in agricultural and food industries where quick screening is needed. Moreover, piezoelectric immunosensors can be integrated with portable devices for on-site testing, further expanding their practical applications (Gao et al. 2014). Various immunosensors used for detection of mycotoxins in different food matrices are shown in Table 3.

6 Regulatory Considerations

An authoritative framework/guideline of standardization and validation process require for the extensive application of immunosensors in food safety monitoring and regulation. Regulatory bodies like the US Food and Drug Administration (FDA) and European Food Safety Authority (EFSA) are essential to develop protocols that ensure the reliability and accuracy of immunosensor-based testing methods (Gaudin et al. 2020). Application of immunosensors requires laborious validation processes, including comparative studies against traditional laboratory methods to demonstrate their effectiveness and consistency in detecting a range of foodborne pathogens and contaminants (Goh et al. 2021). Standardization process is crucial for the confirmation that immunosensor technologies are universally applicable across different

Table 3 Comparative table summarizing the different types of immunosensors used for mycotoxins detection in food matrices

Mycotoxin	Immunosensors	Detection mechanism	Detection platform	Food matrices	Advantages	References
Aflatoxins	Electrochemical immunosensor	Utilizes antibody–antigen binding to generate an electrochemical signal. Common methods include amperometric or voltammetric detection	Electrochemical sensors amplified signals through enzyme-linked systems to increase sensitivity. The enzyme labels, such as horseradish peroxidase (HRP), amplify the signal for more sensitive detection at low concentrations	Nuts, cereals, spices (e.g., groundnut, maize, rice)	Rapid field-based detection and screening for aflatoxin contamination	Ngundi et al. (2005), Zhao et al. (2018)
Ochratoxins	Optical immunosensor	Based on changes in optical properties (e.g., SPR or fluorescence) upon binding of mycotoxin to immobilized antibodies. SPR detects changes in the refractive index when the mycotoxin binds to the sensor surface, while fluorescence measures light emitted upon excitation of bound toxins	Surface Plasmon Resonance (SPR), fluorescence-based detection platforms. Label-free or low-cost detection methods that can be integrated with portable devices	Coffee, wine, dried fruit, spices (e.g., pepper, raisins)	Used for monitoring food safety and ensuring compliance with food safety regulations	Pires et al. (2014) and Cai et al. (2012)
Fumonisins	Piezoelectric immunosensor	Works by detecting changes in frequency or mass of a quartz crystal	Quartz Crystal Microbalance (QCM)	Maize, corn products (e.g.,	High sensitivity and rapid results for agricultural industries.	Ricci et al. (2013), Gao et al. (2014),

(continued)

Table 3 (continued)

Mycotoxin	Immunosensors	Detection mechanism	Detection platform	Food matrices	Advantages	References
		oscillator due to antigen–antibody binding. QCM-based sensors detect the mass shift from the binding of fumonisins to antibodies, which changes the frequency of the crystal oscillator	and Surface Acoustic Wave (SAW) sensors	tortillas, popcorn)	Useful for quick screening in food processing plants and field testing	and Zhao et al. (2018)
Zearalenone	Electrochemical immunosensor	Antibody–antigen interaction is detected by changes in current or potential at the sensor surface, often with signal amplification techniques such as mediated electrochemistry. Amplified signals using mediators such as ferrocene to increase sensitivity in the detection of low concentrations of zearalenone	Electrochemical sensors (amperometric or potentiometric)	Cereals, maize, wheat (e.g., corn-based products)	Can be adapted for on-site rapid testing and portable detection. Particularly useful in preventing livestock contamination and ensuring crop safety	Wang and Niessner (2019) and Zhao et al. (2018)
Trichothecenes (e.g., DON)	Optical and electrochemical immunosensor	Optical immunosensors utilize fluorescence or SPR, while electrochemical sensors	Fluorescence and SPR-based sensors; Amperometric electrochemical sensors.	Wheat, barley, oats (e.g., wheat flour, bread)	Widely used in agriculture and food production, especially in grain products	Wang and Niessner (2019)

		detect DON by binding to specific antibodies. Fluorescence sensors use fluorescently labeled antibodies to detect the toxin in food and beverages, providing high sensitivity even at trace levels	Multiple assay formats including lateral flow assays for field detection			

regions and regulatory environments. Establishment of guidelines for the quality control of immunosensor devices addressing issues like sensor calibration, sensor lifespan, and environmental influences need to be developed by regulatory bodies (López et al. 2021). The development of such standardized protocols not only facilitates the combination of immunosensors into existing food safety systems but also improves public confidence in these novel technologies. In addition, future efforts' focus on ensuring interoperability among different immunosensor platforms which enable data sharing and collaboration across regulatory agencies, food industries, and public health organizations are required.

7 Conclusion

Immunosensors are recognized as revolutionary tools for the detection of fungi and their associated mycotoxins in food matrices with significant advantages over traditional methods. Immunosensors are capable for rapid, sensitive, and highly specific detection of fungi and their mycotoxins and because of that, immunosensors are invaluable in ensuring food safety. Immunosensors are particularly useful for monitoring food products (grains, fruits, and dairy) in which contamination of fungi and their mycotoxins are often found and create a serious health risk. The real-time detection capability of immunosensors allows for faster identification of contamination, enabling quick action to prevent foodborne illnesses. Use of immunosensors provides many advantages, but some challenges, like sensor stability, accuracy in complex food matrices, and cross-reactivity are still to be tackled. Continuous research and technological advancement are essential for addressing these challenges and improving the sensor performance will play a crucial role in expanding immunosensors applicability across a broader range of food products.

References

Alshannaq A, Yu JH (2017) Occurrence, toxicity, and analysis of major mycotoxins in food. Int J Environ Res Public Health 14(6):632

Bayman P, Baker JL (2006) Ochratoxins: a global perspective. Crit Rev Food Sci Nutr 46(7): 493–519

Benkerroum N (2020) Mycotoxins in dairy products: a review. Food Chem Toxicol 62:63–75

Bennett JW, Klich M (2003) Mycotoxins. Clin Microbiol Rev 16(3):497–516

Brahmbhatt H, Bhardwaj N, Mehta J, Kim KH, Deep A (2016) Piezoelectric immunosensors for real-time and label-free detection: principles, applications, and challenges. Biosens Bioelectron 89:645–656

Bullerman LB, Bianchini A (2007) Stability of mycotoxins during food processing. Int J Food Microbiol 119(1–2):140–146

Cai W, Wang Y, Zhu X, Zhang J, Chen Q, Wei Q, Xu Y (2012) Fluorescence-based biosensors for mycotoxin detection. Sensors 12(5):5686–5702

Codex Alimentarius (2021) General standard for contaminants and toxins in food and feed. CODEX STAN 193-1995 (FAO and WHO)

Desjardins AE (2006) *Fusarium mycotoxins:* chemistry, genetics, and biology. American Phytopathological Society Press, St. Paul. ISBN: 978-0890543354

Eaton DL, Gallagher EP (1994) Mechanisms of aflatoxin carcinogenesis. Annu Rev Pharmacol Toxicol 34(1):135–172

EFSA Panel on Contaminants in the Food Chain (CONTAM) (2020) Risk assessment of ochratoxin A in food. EFSA J 18(5):6113

Eivazi S, Gholizadeh S, Yari K (2018) Fluorescence-based optical biosensors for foodborne pathogen detection. Anal Biochem 550:58–66

FAO (2004) Worldwide regulations for mycotoxins in food and feed in 2003. Food and Agriculture Organization

Frisvad JC, Thrane U, Samson RA, Pitt JI (2006) Fungal contaminants in food. Mycopathologia 162:215–223

Gao Y, Zhang Y, Yang Z, Liu X (2014) Piezoelectric biosensors for mycotoxin detection: developments and future trends. Microchim Acta 181(11–12):1359–1367

Gaudin P, Meyer J, Nguyen T (2020) Standardization and regulatory concerns in the use of immunosensors for food safety monitoring. Food Control 109:106–954

Goh PS, Tan PT, Lau KS (2021) Advancements in immunosensor validation for foodborne pathogen detection. Trends Anal Chem 140:116–283

Gong YY, Hounsa A, Egal S, Turner PC, Sutcliffe AE, Hall AJ, Cardwell KF, Zhou X, Wild CP (2004) Post-weaning exposure to aflatoxin results in impaired child growth: a longitudinal study in Benin. Trends West Afr Environ Health Perspect 112(13):1334–1338

Goryacheva IY, De Saeger S, Eremin SA (2007) Immunochemical methods for rapid mycotoxin detection: evolution from single to multiple analyte screening. A review. Anal Bioanal Chem 389(2):1763–1765

Gugleva V, Ivanova K, Dimitrov I, Petrov T (2017) The application of antibodies in biosensing. Anal Bioanal Chem 409(21):5021–5035

Haug K, Smeds AI, Andersen B, Eriksen GS (2014) Occurrence of zearalenone in cereal grains: a review on the influence of environmental factors. Int J Food Microbiol 190:10–19

Hell K, Cardwell KF, Setamou M, Poehling HM (2008) The influence of storage practices on aflatoxin contamination in maize in four agroecological zones of Benin, West Africa. J Stored Prod Res 44(4):552–556

Hernández-Hernández A, Valdés MG, Calzón JAG (2020) Nanomaterial-enhanced electrochemical immunosensors for food safety: application to mycotoxin detection. Food Chem 312:126053

Homola J (2008) Surface plasmon resonance sensors for detection of chemical and biological species. Chem Rev 108(2):462–493

IARC (2002) Some traditional herbal medicines, some mycotoxins, naphthalene and styrene. International Agency for Research on Cancer

IARC (2012) Aflatoxins. In: IARC monographs on the evaluation of carcinogenic risks to humans, vol 100F. IARC, pp 225–249

Iha MH, Barbosa CB, Heck AR, Trucksess MW (2021) Aflatoxins and ochratoxin A in commercial breakfast and infant cereals from Brazil. Food Control 120:107491

Kavita SK, Prasad S (2020) Electrochemical and optical biosensors for aflatoxin detection. Anal Bioanal Chem 408(1):29–40

Kim H, Lee J, Park S, Choi Y, Lee H (2020) Fluorescence-based immunosensors: advances and applications. Sens Actuators B Chem 307:127–137

Klich MA (2002) Identification of common *Aspergillus species*. CBS-KNAW Fungal Biodiversity Centre, Utrecht. ISBN: 978-9070351443

Klich MA (2007) *Aspergillus flavus*: the major producer of aflatoxin. Mol Plant Pathol 8(6): 713–722

Labib M, Zhang Y, Tan Y, Liu X, Wang L (2016) Nanomaterial-based electrochemical sensors for biomedical applications. Biosens Bioelectron 77:845–856

Lavanya N, Sundararajan V, Rajendran S, Rao GN (2018) Immunosensors and their applications: a review. Trends Anal Chem 98:28–41

Li J, Chen Q, Zhang J (2021) Advances in piezoelectric immunosensors for detecting foodborne pathogens. Trends Food Sci Technol 109:269–281

López L, Martín C, Vázquez C (2021) Integrating machine learning with immunosensors for food safety applications. Food Control 123:107739

Luong JHT, Xie SH, Dung DT (2008) Data processing techniques in biosensors. Biosens Bioelectron 23(10):1817–1830

Magan N, Aldred D (2007) Conditions of mycotoxin production by *Fusarium* and *Aspergillus spp.* Microbiol Cereal Prod 6:41–55

Mannaa M, Kim KD (2017) Influence of temperature and water activity on deleterious fungi and mycotoxin production during grain storage. Mycobiology 45(4):240–254

Marin S, Ramos AJ, Cano-Sancho G, Sanchis V (2013) Mycotoxins: occurrence, toxicology, and exposure assessment. Food Chem Toxicol 60:218–237

Marquardt RR, Frohlich AA (1992) A review of recent advances in understanding ochratoxicosis. J Anim Sci 70(12):3968–3988

Marx KA (2003) Quartz crystal microbalance: a useful tool for studying thin polymer films and complex biomolecular systems at the solution-surface interface. Biomacromolecules 4(5): 1099–1120

Munkvold GP (2017) *Fusarium* species and their associated mycotoxins. J Agric Food Chem 45(8): 302–310

Munkvold GP, Arias S (2019) *Fusarium* and mycotoxins in cereal grains. Cereal Foods World 64(6):1–7

Ngundi M, Wanjau RW, Muriuki J (2005) Development of electrochemical immunosensors for aflatoxin detection in food. Anal Chim Acta 529(1):45–50

Ostry V (2008) Alternaria mycotoxins: an overview of chemical characterization, producers, toxicity, analysis, and occurrence in foodstuffs. World Mycotoxin J 1(3):251–274

Ostry V, Malíková D, Toman J, Červenka J (2013) A review of ochratoxin A contamination and its role in human disease. World Mycotoxin J 6(3):237–250

Paniel N, Radoi A (2014) Immunosensors for rapid mycotoxin detection in food matrices. Trends Food Sci Technol 36(1):52–63

Pemberton RM, Turner APF, D'Souza SF (2014) Electrochemical immunosensors for food and environmental monitoring: progress and challenges. Anal Chem 86(3):798–812

Pereira V, Fernandes J, Lima N (2015) Mycotoxins in cereals and related foodstuffs: a review on occurrence and recent methods of analysis. Trends Food Sci Technol 61:75–90

Pestka JJ, Lee PP, Akins RB (2008) Zearalenone: a review of its effects on human and animal health. Environ Toxicol Pharmacol 25(2):231–238

Pfohl-Leszkowicz A, Manderville RA (2007) Ochratoxin A: an overview on toxicity and carcinogenicity in animals and humans. Mol Nutr Food Res 51(1):61–99

Pires JR, Araujo RF, Ferreira DL, Costa MR (2014) Optical immunosensors for ochratoxin A detection: surface plasmon resonance and fluorescence-based platforms. Biosens Bioelectron 55:335–340

Pitt JI, Hocking AD (2009) Fungi and food spoilage, 3rd edn. Springer Science & Business Media. ISBN 0387922075, 9780387922072

Pohanka M (2018) Overview of piezoelectric biosensors, immunosensors, and their applications. J Immunol Methods 459:1–9

Rahmani A, Jinap S, Soleimany F (2009) Qualitative and quantitative analysis of mycotoxins. Compr Rev Food Sci Food Saf 8(2):202–220

Rheeder JP, Marasas WFO, Vismer HF, Sydenham EW (1992) Fumonisin contamination in corn and its association with esophageal cancer in South Africa. J Agric Food Chem 40(5):954–958

Ricci F, Minunni M, Tombelli S (2013) Piezoelectric immunosensors for fumonisin detection in maize. Sens Actuators B Chem 176:835–842

Ricci F, D'Aloia A, DeCesare F, Mazzocchetti L (2019) Advances in piezoelectric biosensing technologies for pathogen detection. Sensors 19(4):877

Richard JL (2007) Some major mycotoxins and their mycotoxicosis-an overview. Int J Food Microbiol 119(1–2):3–10

Rodríguez-Mozaz S, Marco MP, López de Alda MJ, Barceló D (2006) Biosensors for environmental monitoring of endocrine disruptors: a review article. Anal Bioanal Chem 386(4):1025–1041

Rosi P, Ricci F, Amine A (2011) Immunosensors for aflatoxin detection in food: a review. Trends Anal Chem 30(6):782–791

Scognamiglio V (2013) Nanotechnology in the development of biosensors for food safety: challenges and perspectives. Trends Anal Chem 62:60–73

Shephard GS (2008) Impact of mycotoxins on human health in developing countries. Food Addit Contam 25(2):146–151

Su H, Zhang L, Wang Y, Zhang Y, Li W, Li J, Liu Z (2021) Recent advances in piezoelectric biosensors: fundamentals and applications. Biosens Bioelectron 172:112770

Sun J, He H, Liu X (2014) Advances in piezoelectric immunosensors for *Fusarium* detection in grains. Biosens Bioelectron 51(12):3534–3540

Turner NW, Subrahmanyam S, Piletsky SA (2009) Analytical methods for determination of mycotoxins: a review. Anal Chim Acta 632(2):168–180

US FDA (2018) Guidance for industry: fumonisin levels in foods and feeds. U.S. Food and Drug Administration, Center for Food Safety and Applied Nutrition

van der Westhuizen L, Shephard GS, Rheeder JP, Burger HM, Gelderblom WC, Wild CP, Gong YY (2011) Fumonisin biomarkers in maize eaters and implications for human disease. World Mycotoxins J 4(3):257–265

Varga J, Tóth B, Kocsubé S, Kocsis B, Rácz M, Kozakiewicz Z, Ossa F (2015) Ochratoxin production and taxonomy of *Aspergillus* section Circumdati. Stud Mycol 69:121–139

Voss KA, Bowers RM, Gelder SW, Nunn PD, McCully MT, Larsen JR (2007) Fumonisin contamination in maize and its impact on human health: a review. Mycopathologia 164(3): 95–102

Wang J, Niessner R (2019) Miniaturized optical sensors for point-of-care diagnostics: a review. Sensors 19(3):583

Wild CP, Gong YY (2010) Mycotoxins and human disease: a largely ignored global health issue. Carcinogenesis 31(1):71–82

Williams JH, Phillips TD, Jolly PE, Stiles JK, Jolly CM, Aggarwal D (2004) Human aflatoxicosis in developing countries: a review of toxicology, exposure, potential health consequences, and interventions. Am J Clin Nutr 80(5):1106–1122

Wu F, Khlangwiset P (2010) Health economic impacts and cost-effectiveness of aflatoxin reduction strategies in Africa: case studies in biocontrol and postharvest interventions. Food Addit Contam A 27(4):496–509

Yola ML, Atar N, Eren T, Yilmaz AF (2014) Electrochemical biosensors for antigen detection. Sens Actuators B Chem 192:121–134

Zain ME (2011) Impact of mycotoxins on humans and animals. J Saudi Chem Soc 15(2):129–144

Zhao L, Zhang Y, Wang X, Li M, Zhang H (2018) Recent advances in electrochemical immunosensors for food safety detection. Trends Anal Chem 105:1–12

Innovative Techniques and Future Trends in Immunosensor Technology

Suresh Kumar, Megha Pant, Veena Pande, and Anirban Dandapat

Abstract

Immunosensors represent a dynamic and rapidly developing field of biosensor technology that uses specific antigen—antibody interactions to detect biomolecules. In this area, recent innovations have produced immunosensors with advanced immunosensing capabilities, very high sensitivity, and specificity. Key advances include various nanomaterials, such as gold nanoparticles, carbon nanotubes (CNTs), and quantum dots (QDs), which promote signal transduction and facilitate miniaturisation. The advent of microfluidic systems and lab-on-chip devices have enabled rapid on-site analysis of specimens. Furthermore, the emergence of advanced transmission techniques, including electrochemical, optical, and piezoelectric techniques, has opened up a new avenue in the applications of immunosensors. The immunosensors have been utilised as diagnostic instruments for a number of disorders brought on by bacteria, viruses, prions, and unchecked cell proliferation. As immunosensor technology continues to develop, it holds great promise for personalised medicine, environmental monitoring, and public health monitoring. Artificial intelligence and machine learning techniques can be used for real-time data analysis, making the process faster and easier. The potential benefits of these immunosensor devices are quite promising in the realm of their ability to support early identification of various diseases and thus can play a critical role in preventing outbreaks of diseases. In this way, advanced immunosensor technology would certainly enhance the quality of our

S. Kumar · M. Pant · V. Pande
Department of Biotechnology, Sir J. C. Bose Technical Campus, Bhimtal, Kumaun University, Nainital, Uttarakhand, India

A. Dandapat (✉)
University School of Automation and Robotics, Guru Gobind Singh Indraprastha University, East Delhi Campus, Surajmal Vihar, Delhi, India
e-mail: anirban.usar@ipu.ac.in

K. Kuča et al. (eds.), *Ensuring Wellness: Immunosensors Monitoring Food Quality and Pathogenicity Risks*, https://doi.org/10.1007/978-3-031-97477-9_17

lives and overcome the global health challenges. In this chapter, we explore how advanced immunosensor technology has the potential to significantly boost the quality of our lives and help overcome global health challenges.

Keywords

Immunosensors · Antigen–antibody interactions · Nanomaterials · Advanced transmission techniques · Global health

1 Introduction

Trace detection and molecular recognition are the major areas of extensive research over the last 20 years. These fields become more sensitive for the safety and security of individuals globally, especially in developed nations (Bhatia et al. 2024). There are lots of advancements required for further investigation to improve the effectiveness of detection and diagnostic systems (Khalifa and Albadawy 2024). The technology also helps to reduce the vulnerabilities to various threats. Numerous chemicals, including biochemical and biophysical methods and devices are in progress for the detection and diagnosis of various problems. They have been used for the detection of infectious microorganisms, hazardous substances, and environmental pollutants (Franco-Duarte et al. 2019). The dedicated research efforts of academic professionals, scientists, and engineers in both industry and government sectors are in progress for the development of these areas. Furthermore, these technologies are invaluable as they enhance detection and recognition capabilities in each prospect (Durai et al. 2024). At present, certain biochemical and biophysical analytical techniques are linked with immunosensors to make significant results. The advanced bioimmunosensors are the area of interest to biomedical scientists, biotechnologists, and the life sciences community (Chen et al. 2023).

An immunosensor is a biosensor detection system designed for the sensing of an analyte by utilising an antibody or antigen as its sensing component. The antibodies and receptors employed in the component of immunosensors. These antibodies may be derived from different sources, such as animals, and also produced through genetic engineering. Additionally, cell-based sensing methods are another one of these purposes. Antibody pairs are designed in such a manner that they perform well in research settings. It may not yield the same high-quality results in field analysis using an immunosensor (Aboul-Ella et al. 2024). Sometimes, the critical performance factors, such as stability, response time, and limit of quantification, are not adequately addressed. Most animals and plants possess immune systems that are inherently capable of recognising and responding to biological pathogens. The human immune system, including the skin, is particularly adept at detecting both chemical and biological agents (Kant et al. 2021).

Antigen–antibody interactions are the basic event of immunosensing. This is a super-specific technique used in biosensor technology. The strong affinity and selectivity of antibodies for their specific antigens allow for accurate detection of target molecules. The high affinity and selectivity of antibodies for their corresponding antigens enable the precise detection of target molecules. This is crucial for reliable diagnostic applications. It allows the identification of analytes ranging from pathogens and toxins to disease-associated biomarkers (Lara and Perez-Potti 2018). The strength of antigen–antibody binding means that biosensors can maintain high sensitivity and specificity. They also deal with tiny amounts of analytes. This is crucial for early disease detection and monitoring. It is helpful in spotting the low levels of biomarkers. It is useful in clinical outcomes (Wani et al. 2023). By incorporating these interactions into biosensor systems; they enhance the diagnostic performance, ensuring accuracy, consistency, and minimal cross-reactivity. Immunosensors have become key players in modern diagnostics due to their quick response times and ability to perform tests at the point of care (Hemdan et al. 2024). They are used in a variety of areas, including detecting infectious diseases like COVID-19 rapid tests, identifying cancer biomarkers, such as HER-2 for breast cancer (Gamble et al. 2021), and monitoring chronic diseases like insulin or HbA1c for diabetes (Horber et al. 2020). It also ensures food safety, allergen detection, and checking for environmental pollutants and toxins. These uses show their importance and how versatile and important immunosensors are in improving healthcare, food safety, and environmental protection (Arya et al. 2023).

Micro- and nanofluidics are known as game changers in analytical systems. This technology lays the foundations for lab-on-a-chip (LOC) devices that are based on the "sample-in-answer-out" principle (Behera et al. 2023). The biological samples are now directly analysed by using these cutting-edge microfluidic tools. They are capable of processing the whole or lightly biological samples (Wu et al. 2022). They are applicable because of their qualitative and quantitative analysis of disease markers and other given bio-analytes in the sample. These are compact devices that analyse the samples in various clinical steps, such as sample preparation, manipulation, separation, reaction, and detection. This technology provides a smart and efficient way to diagnose things in modern aspects. The micro- and nanofluidics work in a collaborative way to enhance the experience of lab-on-a-chip diagnosis. It is pocket-friendly technology compared to traditional clinical setups (Shi et al. 2021). This chapter mainly focuses on the advancement in bioanalytical applications that work on the principle of integrated microfluidic devices, merged with advanced optical and electrochemical techniques. Amperometry, cyclic voltage, potentiometry, and impedance spectroscopy are some of the advanced techniques. Artificial intelligence-powered immunosensors are making waves in early cancer detection by utilising biomarkers. Some models with deep learning networks have achieved an impressive 95% drop in false positives for prostate and breast cancer markers by examining patterns in immuno-assay data (Bhattacharya et al. 2022). Plus, artificial intelligence has been equipped with immunosensors for quick detection of COVID-19 antigens. Machine learning systems also enhance the identification of different respiratory infections with great accuracy.

In point-of-care environments, artificial intelligence-enhanced immunosensors are being used for diseases like HIV and malaria, allowing for speedy diagnoses with minimal resources (Parihar et al. 2023). These devices have cut down diagnosis time significantly, going from hours to just minutes. Moreover, artificial intelligence-driven wearable devices equipped with immunosensors continuously monitor health conditions, sending data to cloud platforms for real-time analysis. They utilise artificial intelligence to catch early signs of inflammation or metabolic issues (Khan et al. 2023). In this chapter, we take a closer look at the latest advancements in immunosensor technology. We will explore how cutting-edge innovations such as microfluidics, artificial intelligence, and nanotechnology are transforming bioanalytical applications. These advancements are leading to highly accurate, efficient, and user-friendly diagnostic tools that enhance healthcare, food safety, and environmental protection.

2 Advances in Immunosensor Materials

2.1 Nanomaterials in Immunosensors

The advanced immunosensor technology has been all about using cutting-edge materials to enhance the working of these sensors. These new materials play a significant role in the enhancement of sensitivity, specificity, and overall performance. All these features make these sensors proficient diagnostic tools. The reliability and performance with efficiency are all greatly blissed. The cutting-edge materials profoundly increase the working of these immunosensors. These new materials play a significant role in the enhancement of sensitivity, specificity, and overall performance. These are all features that make this diagnostic tool more reliable and efficient than ever (Ramesh et al. 2022).

2.1.1 Gold Nanoparticles

Gold nanoparticles are known for their optical and electronic properties, like surface plasmon resonance. Due to this, they are capable of scattering and absorbing light (Choosang et al. 2021; Lee et al. 2018). Due to its micro size and large surface area, it has the potential to make a bond with a tonne of biomolecules (carbohydrates, proteins, lipids, and nucleic acids) (Mikhailova 2021). AuNPs are used as important ingredients in different immunoassays. They enhance the colorimetric detection methods, in which a visible colour change appears due to the presence of target analytes (Chang et al. 2019). It is also a valuable tool in lateral flow assays, providing immediate and authenticated results. AuNPs are eco-friendly, easy to use, easily modified, and can be manufactured with antibodies for better performance. Due to their optical properties, it enhances the sensitivity. It is valuable in both direct and indirect detection methods.

Severe acute respiratory syndrome Coronavirus 2 creates chaos worldwide by causing serious respiratory illnesses. SARS-CoV-2 first appeared in 2019 and caused a full range of symptoms. Mild cough and fever and severe pneumonia are threat issues that arise due to this. Scientific society developed various diagnostic tools against them. These tools are real-time polymerase chain reactions, including point-of-care tests to detect the virus and antibodies (Lai et al. 2020). Point-of-care tests are preferable tests because they are easily applicable, portable; everyone can afford, and can be done immediately on site without any advance training. SARS-CoV-2 variants are on their emerging peaks, so the healthcare system required quick action to manage the epidemic effectively (Fistera et al. 2023). Among these tests, gold nanoparticle (GNP)-based lateral flow assays (LFAs) are one of the advanced diagnostic tools that have emerged as a fighter against them. These tools identified the COVID-19 proteins and antibodies associated with this (Ardekani and Thulstrup 2022).

2.1.2 Carbon Nanotubes (CNTs)

Carbon nanotubes (CNTs) are getting a lot of interest in the biosensor world because of their amazing electrical, mechanical, and chemical traits (Chawla et al. 2023). These tube-shaped carbon structures are made from one or more rolled-up graphene sheets. It will be helpful in achieving a high surface area-to-volume ratio, which is advantageous for enhancing sensor sensitivity (Rauti et al. 2019). Due to these marvellous features, they are perfectly used for creating sensitive, quick, and selective biosensors. These carbon nanotubes (CNTs)-based biosensors can be used in areas like medical diagnostics, environmental monitoring, and the food industry for safety. There are various kinds of carbon nanotubes (CNTs)-based glucose sensors that can be designed to make electrochemical sensors. They designed to check the issues as either minimally invasive or completely non-invasive. The carbon nanotubes-based biosensor can identify more than just glucose, tapping into other key markers for diabetes (Yuwen et al. 2023). The conversation dives into the different methods carbon nanotubes (CNTs) use for glucose detection. It covers both enzymatic sensors that utilise enzymes and non-enzymatic sensors that do not require enzymes. Although there are still some hurdles to tackle, this field is set to expand and innovate, potentially making a huge difference for millions of people with diabetes globally (Sheoran et al. 2022).

In the last few years, there has been a lot of advancement in creating carbon nanotubes (CNTs)-based materials for these electrochemical glucose sensors. These advancements are single-walled carbon nanotubes (SWCNTs) and multi-walled carbon nanotubes (MWCNTs), as shown in Fig. 1a. A top-notch glucose biosensor was developed by creating a bio-nanohybrid film with great electro activity on a glassy carbon electrode. These advanced carbon nanotubes (CNTs) were designed by a straightforward one-step electrochemical polymerisation process. This biosensor is formed by a nanocomposite made up of polypyrrole (PPy), Nafion (Nf), and functionalised multi-walled carbon nanotubes (fMWCNTs). Once the nanocomposite film was ready, chitosan-glucose oxidase (CH-GOx) was attached to its surface. This blend led to a highly effective biosensor for detecting glucose. The advanced combination resulted in an effective biosensor for glucose detection, as shown in Fig. 1b.

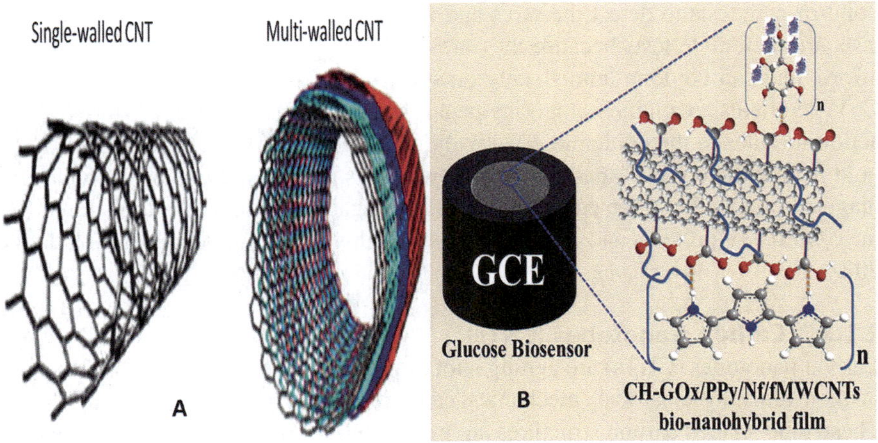

Fig. 1 (**a**) Molecular structure of single-walled carbon nanotubes (SWCNTs) and multi-walled carbon nanotubes (MWCNTs). (Zhang et al. (2010), License number 5887130376568). (**b**) A top-notch glucose biosensor was successfully created using a one-step in situ electro-polymerisation of a PPy-Nf-fMWCNTs nanocomposite right on a glassy carbon electrode (GCE), and then CH-GOx was immobilised on it. (Shrestha et al. (2016), License number 5887131318092)

2.1.3 Quantum Dots (QDs)

Quantum dots (QDs) are tiny semiconductor crystals. They have amazing optical properties which are more suitable for design of immunosensors. Usually, the size is measured between 2 and 10 nm. The quantum dots (QDs) show light emission which can be adjusted based on their size (Singh et al. 2020). Plus, they are super stable even when they are under light exposure. This quality makes them more useful in biosensing, especially for detecting multiple targets over a long period (Mercy et al. 2023). In immunosensors, quantum dots (QDs) are also merged with biomolecules like antibodies, antigens, or aptamers to specifically latch onto certain analytes. These biosensors detect the pathogens or disease indicators. When these quantum dots (QDs) come into contact with their target, they emit fluorescence. The fluorescence is utilised to detect the target with great sensitivity. When quantum dots (QDs) are compared with regular fluorophores, they are brighter and less prone to photo bleaching. This property enhances the detection sensitivity and analyses the analyte at low concentrations as well (Sun et al. 2021; Şahin et al. 2021). Fluorescence Resonance Energy transfer is another popular approach utilised in quantum dots (QDs) immunosensors development. In this approach, quantum dots (QDs) act as a donor that provides energy. This energy is given to the acceptor molecules during antigen–antibody binding. This provides a fluorescent signal at the end of the reaction. This fluorescent signal is used for real-time tracking of immune responses. It is also helpful to identify the pathogen with high specificity (Verma et al. 2023).

Multiplexity is another feature of these quantum dots (QDs). Due to their wide excitation and narrow emission spectra, these biosensors can detect multiple targets at once in a single test. This feature is super useful in diagnostics, where we need to

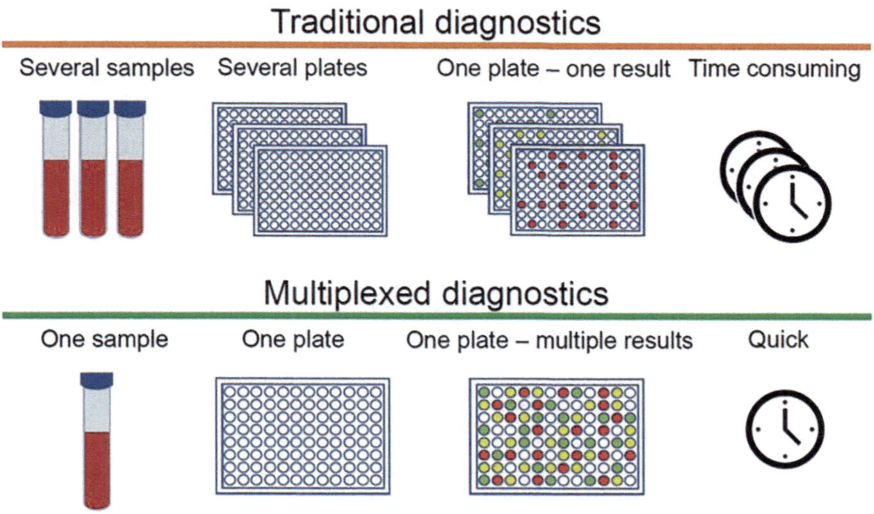

Fig. 2 Traditional diagnostics usually aim to find one biomarker or analyte at a time. The multiplexed diagnostics can detect several biomarkers all at once in a single test. (Jarockyte et al. (2020), under the CC BY license)

identify several biomarkers or pathogens at the same time. For instance, quantum dots (QDs)-based multiplexed immunoassays have been created to detect various infectious diseases that are caused by the human immunovirus, Zika virus, and hepatitis, offering quick, precise, and scalable diagnostic solutions (Mousavi et al. 2022; Jarockyte et al. 2020). Multiplexed analysis is a beneficial approach as compared to traditional approach shown in Fig. 2, because it identifies several biomarkers qualitatively or quantitatively in one sample. This method comes with a bunch of perks, like getting more information from each sample, cutting down costs for each data point, reducing mistakes by needing fewer samples, and boosting overall efficiency (Jarockyte et al. 2020).

2.2 Signal Transduction Enhancements

Signal transduction in immunosensors is used to check the signals developed during biological recognition events, such as antigen–antibody interactions. These immunosensors are sensitive and specific. By merging the different nanomaterials, such as quantum dots (QDs), carbon nanotubes (CNTs), and graphene, with immunosensors, it will help to boost the signal transmission. This advancement can detect things better and respond faster (Aizawa 1994; Ramesh et al. 2022).

2.2.1 Mechanisms of Signal Amplification

Signal amplification is an important part of advanced immunosensors and will help to find out the low-abundance biomarkers. The advanced version of the immunosensor developed by using nanomaterials enhances the immune recognition process (Chen et al. 2023). Gold nanoparticles and carbon nanotubes (CNTs) are showing great efficacy to generate signals in electrochemical and optical immunosensors. They improve the electron transfer and optical responses. This results in the formation of a strong output signal. In electrochemical immunosensors, gold nanoparticles coated with antibodies increase the number of sites for electron transfer. This association enhances the sensor's sensitivity (Fu et al. 2024; Tessaro et al. 2021). Quantum dots (QDs) linked with enzymes can enhance the immune signals through catalytic processes. An enzyme horseradish peroxidase is linked with an immunosensor to trigger the reactions. These immunosensors produce measurable signals. They are also amplifying the response in the presence of an antigen. These enzymes are utilised for getting multiple product molecules from a single reaction by boosting the overall signal (Popov et al. 2021; Liu et al. 2022).

Polyaniline and polypyrrole-conducting polymers-based immunosensors can also apply for the amplification of electrochemical signals. These polymer-based nanomaterials help to facilitate electron transfer between the sensing elements and electrodes. It results in improving the quality of signal strength and allows for the detection of low levels of antigens or pathogens (Drobysh et al. 2023; Thadathil et al. 2022). Quantum dots (QDs) are also merging with FRET-based systems to enhance signal amplification. Their high quantum yield and adjustable optical properties give us efficient energy transfer and fluorescence detection.Quantum dots (QDs) coated with antibodies or antigens can improve the immune interactions. This is converted into amplified optical signals. It significantly enhances the sensitivity of the immunosensors. Carbon quantum dots (CQDs) are utilised for detection and imaging, whether inside living organisms or in lab settings. They are also involved in drug delivery systems aimed at targeted treatments, including photodynamic and photothermal therapies, immunotherapy, and gene therapy (Kaurav et al. 2023; Yang et al. 2023), as shown in Fig. 3.

2.3 Impact of Nanomaterials on Sensitivity and Specificity

Nanomaterials are the game changers when compiled with immunosensors. They may help in boosting the ability to detect tiny amounts of substances that were impossible to identify earlier. Due to the large surface area property of the nanomaterials, unique electronic features, and easy customisation, they are perfect for taking biosensing technology to the next level. Carbon nanotubes (CNTs), gold nanoparticles, and graphene are known for their amazing conductivity and electron transfer capabilities. These features greatly improve the sensitivity of electrochemical immunosensors. They expand the surface area for antigen–antibody interactions (Joseph et al. 2023). Easy-to-spot low-level biomarkers in clinical samples are possible with immunosensors. Plus, nanomaterials help to reduce the background noise in optical immunosensors. It leads to more accurate and reliable detection. Functionalising nanomaterials with biomolecules that specifically attach to target

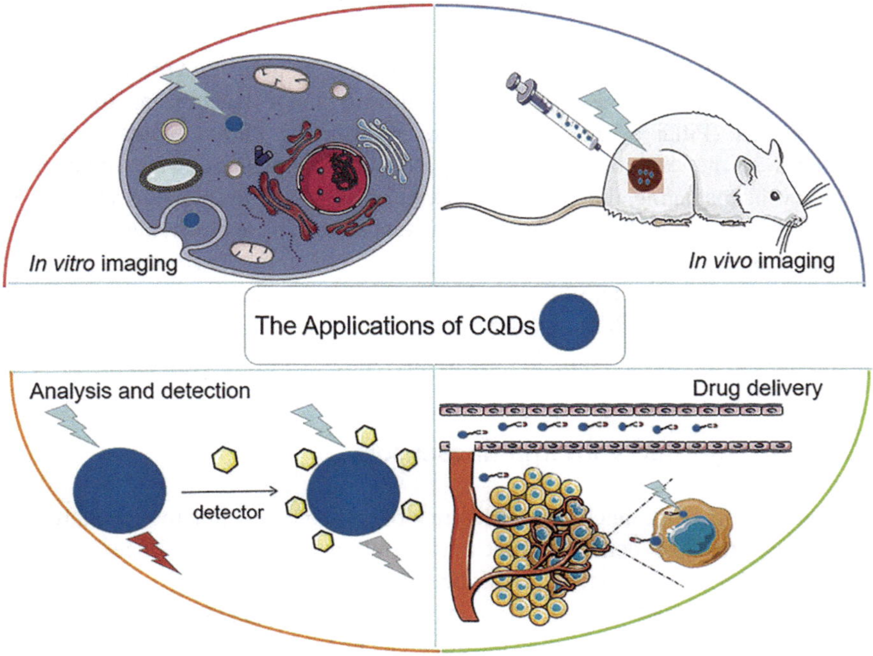

Fig. 3 Carbon quantum dots (CQDs) are utilised for detection and imaging, whether inside living organisms or in lab settings. (Yang et al. (2023), under the CC BY license)

antigens or antibodies enhances their results in a significant upgrade (Lohcharoenkal et al. 2021).

The careful design of nanomaterial surfaces allows for selective binding of analytes. It reduces the non-specific interactions and also overcomes the false results. This is useful in medical diagnostics, such as detecting disease markers. It is vital for patient care. Nanomaterials have paved the way for immunosensors that can monitor immune responses in real time (Dhahi et al. 2024). Quantum dots (QDs) and nanowires are useful in continuous monitoring the antigen–antibody interactions. It provides valuable insights into the dynamics of immune processes. This characteristic is essential in many applications, such as analysis of vaccine effectiveness, and also useful in tracking the progression of infectious diseases (Mousavi et al. 2022). Signal transduction faced many hurdles with nanomaterials. The biocompatibility of certain nanomaterials, especially those with heavy metals, causes problems in the detection (Thakur and Kumar 2022).

3 Microfluidic Systems and Lab-on-Chip Devices

Microfluidics analysis can be used to find out the biomarkers and proliferating their efficiency to revolutionise diagnostics. It also tailors healthcare to personal requirements (Pillai et al. 2023). These microfluidic chip systems handle the obstacles in the way of sensitivity, speed, and sample size while maintaining accuracy in managing fluids (Pattanayak et al. 2021). The miniaturisation of the chips will provide small pores that are sufficient for complete biochemical processes, starting from loading reagents to last step-down streaming, performed by a single chip. The pros of this microfluidic technology require the lowest amount of reagent; using surface forces such as capillary action to move liquids, along with the stability that laminar flow offers, really boosts sensitivity.

3.1 Integration with Immunosensors

Immunosensors work much better and are more reliable now that they use advanced design features like better biological recognition parts, cutting-edge transducer systems, and new ways to treat the surface. Plus, making these devices tiny in nature brings uncountable advantages (Mollarasouli et al. 2019). These enhance the sensitivity, are easy to carry out, and generate quicker results in a short period of time. These properties of immunosensors make them easy to use in real-life situations. As the industrialisation demand increases, merging miniaturised immunosensors with advanced technology like microfluidics and wearable gadgets will keep expanding the possibilities in diagnostics and biosensing (Apoorva et al. 2024). It is useful in developments in point-of-care testing (Cho et al. 2018), personal medication, neurodegenerative diseases (Dhahi et al. 2024), and environmental monitoring.

3.1.1 Design and Functionality

Immunosensors are a beautiful type of biosensor that is designed on the basis of interaction between an antigen and its related antibody (Mollarasouli et al. 2019). To make them specific for interactions, the researchers must ensure their sensitivity, specificity, and the durability of the product. These immunobiosensors are used in healthcare and medical testing, environment monitoring, and the food safety industry (Ghasemi-Varnamkhasti et al. 2018).

3.1.2 Advantages of Miniaturisation

Minimising immunosensors has a bunch of privileges that are effective in many ways (Wang et al. 2008). The miniature immunosensors are easier to carry. The compact immunosensors are super portable, perfect for on-the-spot testing. For these tiny immunosensors, a smaller amount of sample is needed, which means tiny sensors use smaller amounts of reagents and samples. These are designed to reduce costs and wastage. They provide better sensitivity due to their higher surface-to-volume ratio. It makes them easier to detect low-concentration substances. It provides quicker results due to its smaller size. It means faster reactions and signal processing, and we will get results in a short period of time (Huang et al. 2021; Zambry et al. 2022).

3.2 Rapid On-Site Analysis

Rapid on-site analysis is all about quickly testing biological or chemical samples at the collected site (Choi et al. 2017). This analysis shows no need to send the sample to a lab. This method is very useful in point-of-care situations because it provides instant results. It helps healthcare professionals to take decisions and start treatments in the perfect manner (Rasmi et al. 2021).

3.2.1 Techniques for Specimen Handling

Specimens handling with proper care is very important for keeping samples intact. It reduces the chances of contamination and provides accurate results in immunosensor testing (Ahmed et al. 2014). Sterile methods are used when the samples are collected from different areas to make them less contaminated, especially in sensitive immunosensor tests. The samples, such as proteins or enzymes, are stored at the right temperature to avoid degradation. Cryopreservation also uses it to preserve biological samples in liquid nitrogen if they cannot be processed right away. Proper packaging and cold chain logistics are important to maintain the viability of the samples during transportation (Moon 2023). Stability and specificity are maintained by preventing changes in biomarker levels. A microfluidic system is another technique for in situ sample preparation. Microfluidic systems are common in point-of-care devices. It provides efficient handling of small sample volumes and overcomes the need for manual work (Niculescu et al. 2021). Samples with high analyte concentrations will need to be diluted on-site before processing. Buffering helps in keeping the pH levels stable for accurate immunosensor readings. All sampling tools and containers must be sterile to avoid cross-contamination (De Paoli 2005; Bharti et al. 2022).

3.2.2 Applications in Point-of-Care Testing

Immunosensors play a pivotal role in point-of-care testing because they allow for quick diagnostics and real-time health tracking. These sensors perform unique interactions between antigens and antibodies to spot different biomarkers. They provide instant results that help with clinical decisions (Janik-Karpinska et al. 2022). Modern point-of-care immunosensors are built to be easy to use and compact. Modern design is so familiar that it can be utilised in a range of places, whether it is a remote clinic or at home (Heidt et al. 2020). Plus, their connection with smartphones and digital apps has really boosted their usefulness. The advanced technology enables real-time data sharing and telemedicine consultations (Zhang et al. 2023). The diagram (Fig. 4) demonstrates the multifaceted nature of smartphone-based biosensors, emphasising their capability to process data using a variety of methods such as optical and electrochemical detection, with AI playing a role in interpretation.

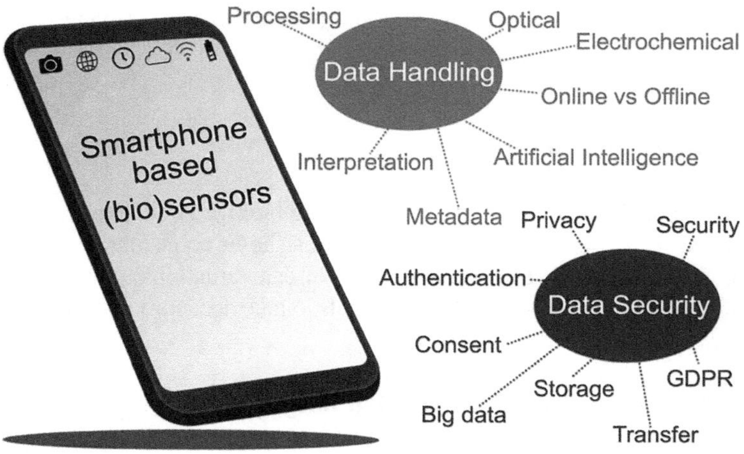

Fig. 4 Diagram illustrating key aspects of smartphone-based biosensors, highlighting data handling processes (including optical, electrochemical, and AI integration) and data security considerations (such as privacy, authentication, and big data management). (Ross et al. (2023), under the CC BY license)

Research showing that an electrochemical immunosensor has been designed in such a way that uses anti-Gly m TI IgG/Gly m TI. This immunosensor can detect the amount of soybean in complicated food items. It can detect concentrations ranging from 1 ag mL^{-1} to 0.1 mg mL^{-1}, with a limit of detection under 1.0 ag mL^{-1}. In this context, "ag" refers to *attogram*, which is a unit of mass in the metric system. An attogram is equal to 10^{-18} g. This sensor is applicable to perform different food types and processing methods. This immunosensor is allowing it to detect down to 0.1 mg kg^{-1}. This capability is crucial for managing food allergens and safeguarding 99% of people with allergies in various food items (Dias et al. 2024).

4 Advanced Transduction Techniques

4.1 Electrochemical Immunosensors

Electrochemical immunosensors are categorised as analytical tools. They are affinity-based bio-sensing equipments, which are used to convert the recognition of an antigen by an antibody into quantifiable electrical signals (Felix and Angnes 2018). These are a class of immunosensors profoundly used in clinical diagnostics as point-of-care devices (Justino et al. 2016). The principle is based on measuring changes in electrical properties, such as current, potential, and impedance. This technology has great potential due to its high sensitivity, portability, and cost-effectiveness (Felix and Angnes 2018).

Electrochemical immunosensors utilise different approaches, such as amperometric, potentiometric, and impedimetric techniques (Aydin et al. 2019). These biosensors are well suited for the identification of the disease biomarkers. Infectious diseases, such as HIV, hepatitis, and more recently, COVID-19, are identified with these immunosensors (Drobysh et al. 2022). Successful applications feature enzyme-linked electrochemical immunoassays for detecting HIV in its early stages (Farzin et al. 2020) and label-free impedimetric sensors for spotting respiratory viruses (Shoute et al. 2023). These studies showcase how effectively they deliver authenticated real-time results in clinical environments.

4.2 Optical Immunosensors

Optical immunosensors rely on light to spot interactions between antigens and antibodies. They are well-known for being super-sensitive and can deliver both quantitative and qualitative results.

4.2.1 Types of Optical Techniques (e.g., Fluorescence, SPR)

Fluorescence-based immunosensors are equipped with fluorophore-labelled antibodies. These advanced immunosensors find their target site and bind with them, upon binding with them, they start to emit light. This allows for highly sensitive detection (Huang et al. 2022). Surface plasmon resonance is another common optical technique that detects changes in refractive index at the sensor surface. They facilitate the label-free detection of viral and bacterial antigens (Capelli et al. 2023). Photoluminescence has the properties of fluorescence and phosphorescence. This shows how a molecule emits light after it gets excited by absorbing photons from the visible or UV light spectrum. In a simple application, this property was used in an immunosensor. One study showed that the Titanium dioxide (TiO_2) nanoparticles were deposited on glass substrates and treated with antibodies that specifically target an opportunistic bacteria *Salmonella typhimurium*. When the optical immunosensors are attached with bacteria, there was a noticeable drop in the intensity of photoluminescence. This sensor was able to detect *Salmonella typhimurium* in concentrations ranging from 10^3 to 10^5 cells/mL (Viter et al. 2017).

Surface plasmon resonance (SPR) biosensors are label-free optical biosensors. They measure how the refractive index changes when molecules bind to biomolecules immobilised on the sensor (Piliarik et al. 2009). They are super-popular biosensors that measure the refractive index changes during binding of biomolecules immobilised on the surface of biosensors. It is used for detecting various foodborne pathogens. They detect the pathogens individually or in groups without needing labels. They are known for their impressive sensitivity and ability to monitor binding interactions in real time (Kourti et al. 2023). The first recorded use of a surface plasmon resonance (SPR) sensor for spotting bacteria goes back to 1998. In this study, a sandwich immunoassay was used to detect *Escherichia coli* O157 in a buffer solution. This setup was improved by adding protein A or protein G and then attaching either a mouse monoclonal or rabbit polyclonal antibody. The detection limit ranged from 5 to 7×10^7 colony-forming units per milliliter (CFU/mL), depending on the type of antibody used. Plus, there was also an SPR biosensor with a protein G layer that was used to detect *Vibrio cholerae* O1 in a buffer (Jyoung et al. 2006).

4.3 Piezoelectric Immunosensors

Piezoelectric immunosensors are analytical devices used to recognise various large molecules and microorganisms, such as bacteria. These biosensors are defined by using antibodies as a key component for the identification of specific antigen as target. Specificity of the biosensors depends on the function and ability of the antibodies to detect the antigens (Castillo-Henríquez et al. 2020). In simple terms, piezoelectric immunosensor working is based on the antibodies they are specifically used with. The remaining components of the sensors are not playing any significant role in the reaction with unrelated substances present in the detection medium (Mehrotra 2016). Typically, immunosensors are designed under the consideration of their antigen–antibody binding affinities. However, the immunosensors are also designed by using fixed specific antigens to identify the specific antibody in the sample (Gan et al. 2023). In this case, antibodies are used as target molecules. These types of piezoelectric immunosensors containing immobilised antigens are used to detect various infectious diseases (Pohanka 2018).

5 Applications of Immunosensors

Immunosensors are the most demanding and indispensable tools for the detection and monitoring of a wide range of microorganisms that are responsible for various diseases (Janik-Karpinska et al. 2022). These tools are revolutionising the fields of medical diagnostics for various infections caused by bacteria, viruses, and prions. Environment monitoring and food safety are also diagnosed by rapid, reliable results that are used to enhance public health responses and mitigate risks associated with infection outbreaks (Rios et al. 2024; Castillo-Henríquez et al. 2020). As the need for exact and quick diagnostic methods keeps going up, new progress in immunosensor tools uses new ideas supplemented with smart technology and machine systems, increasing the speed of finding pathogens in a more right and fast way. This helps with addressing the growing challenges of increasing problems from sicknesses that spread from bugs that don't respond to treatment. Antibacterial resistance is one of the major issues today (Janik-Karpinska et al. 2022; Sin et al. 2014). The increase in demand has shown the importance of constantly watching and finding new ways to solve medical issues. It shows how the way immunosensors work could change the diagnosis patterns of diseases by giving us quick and useful information about a patient's health (Kim et al. 2019). In Fig. 5, the smart wearable sensors show how they are preferably utilised in different health issues. These issues start from skin problems, heart conditions, and unusual movements to metabolic disorders. Figure 5 also points out the challenges that are faced during applying these technologies into

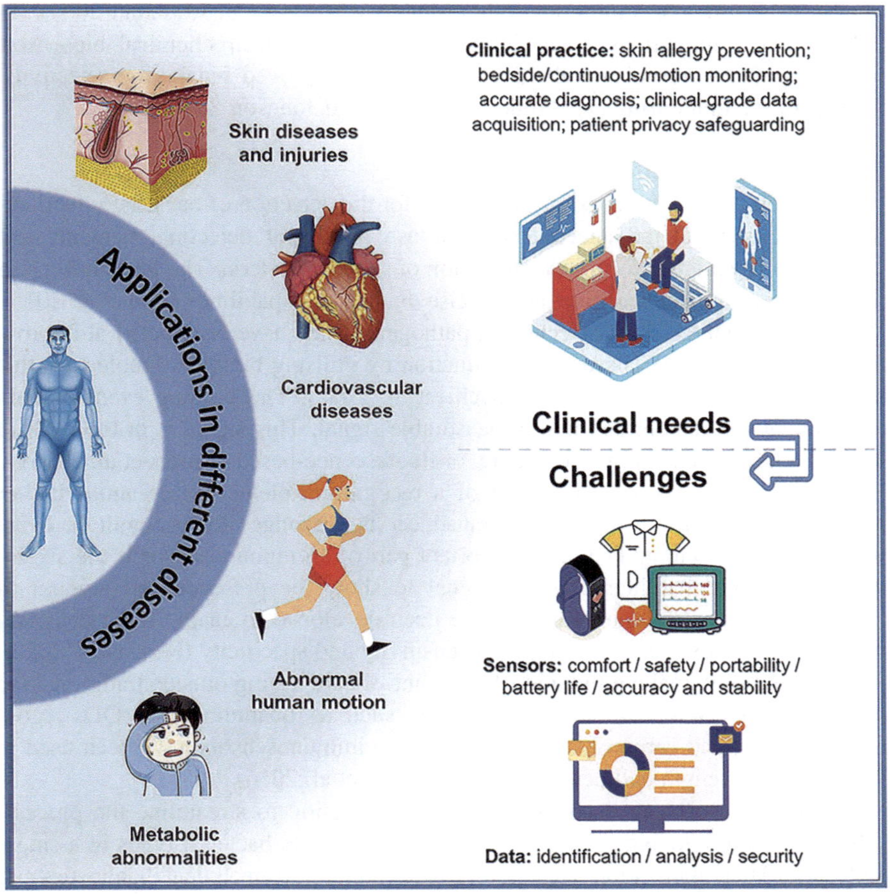

Fig. 5 This figure shows an overview of smart wearable sensors used for different health issues. These sensors track and measure important body signals linked to these conditions, sending live data to connected gadgets. The illustration also points out the hurdles in using these technologies in healthcare, like technology challenges and the need for teamwork between medical and engineering experts. (Qingxiao et al., under the CC BY license)

practice, stressing the need to tackle technical hurdles and encourage teamwork between healthcare and engineering experts.

5.1 Detection of Pathogens

Detecting pathogens quickly and accurately is super important for diagnosing and treating infectious diseases early on. Bacterial and viral infections, along with prion diseases, really challenge the healthcare system because they can spread fast and often become resistant to treatments (Abbas et al. 2023). Established techniques such as polymerase chain reaction have been widely used for pathogen identification, but they can be time-consuming and labour-intensive (Yang and Rothman 2004). To address these limitations, microfluidic platforms and electrochemical biosensors have emerged as promising alternatives, offering improved sensitivity, selectivity, and real-time monitoring capabilities (Cesewski and Johnson 2020).

5.1.1 Bacterial Infections

Immunosensors are designed in such a way for the detection of bacterial infections. Immunosensors are advanced tools that use biological detecting elements like antibodies or antigens for the identification of specific bacteria. The immunosensors are known for quick, sensitive, and precise diagnostic capabilities (Zhu et al. 2015). They are crucial for early detection of pathogens. They have been better at improving treatment results. These sensors function by utilising biological molecules that attach to bacterial components (Nnachi et al. 2022). The binding event of both components is transformed into a measurable signal. This signal is in the form of optical, electrochemical, piezoelectric, or fluorescence-based (Ansari et al. 2016).

Immunosensors designed consist of a recognition element, like antibodies or engineered proteins. It also has a transducer that changes the recognition event into a readable output. The most important part of an immunosensor is the signal-processing unit that interprets the signal to show the presence and amount of detecting bacteria. Immunosensors have been developed to enable rapid detection within minutes to hours, offering high sensitivity and specificity (Rosa et al. 2022). Good portability, making them ideal for point-of-care testing outside traditional lab environments. Nanomaterial advancements such as quantum dots (QDs), gold nanoparticles, and carbon nanotubes (CNTs) in immunosensors have been used to boost signal strength and detection limits (Wang et al. 2016).

Modern sensors can even offer label-free detection to streamline the process. They have multiplexing capabilities to identify multiple bacterial types in a single detection. Their application range is very wide, ranging from clinical diagnostics and food safety to environmental monitoring (Haleem et al. 2021). The detection system allows the quick identification of bacterial infections, contaminated food items, contaminated environments with different pathogens, and spotting bacterial diseases to public health (Ali et al. 2020). However, there are still lots of things that need to improve, like the stability of immunosensors over time and work to inhibit cross-reactivity and false results. It is therefore increasingly important to make the technology both affordable and scalable. Despite these challenges, ongoing research is required. The nanotechnology involvement to develop these immunosensors shows great potential for enhancing bacterial infection diagnostics (Ramesh et al. 2022).

5.1.2 Viral Infections

Immunosensors are being used more widely in clinical environments for quick diagnosis of viral infections. They play a key role in detecting illnesses like influenza, dengue, HIV, and, most recently, COVID-19. Point-of-care devices have evolved into more advanced tools that are portable and easy to use. They allow for testing right at the site without requiring a full lab setup. However, there are still

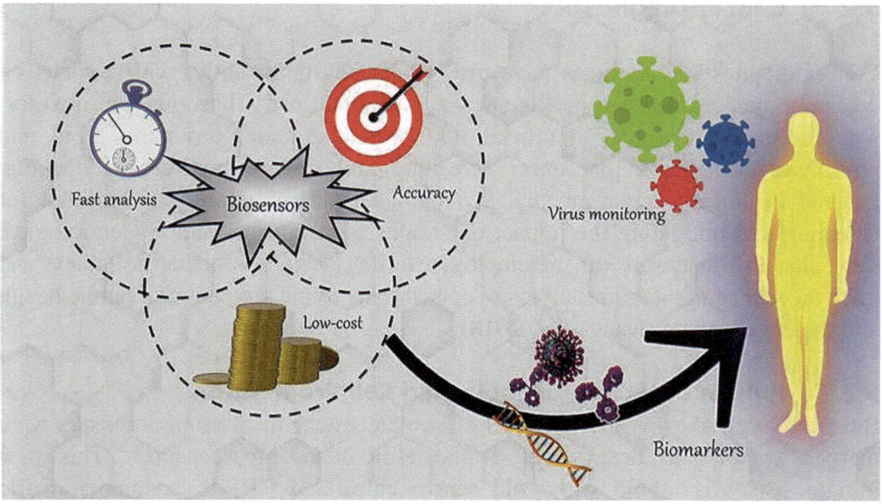

Fig. 6 Biosensors play an integral role in modern diagnostics due to their fast analysis, high accuracy, and cost-effectiveness. (Brazaca et al. (2021), License number 5903610490630)

issues that must be overcome. To make sensors stable over time, avoiding cross-reactivity to get false-positive results, and easy affordability to everyone is needed to improve (Kabir et al. 2021; Brazaca et al. 2021). Figure 6 highlights key attributes of biosensors, including fast analysis, high accuracy, and low-cost operation, emphasising their role in virus monitoring and biomarker detection for human health.

5.1.3 Prion Diseases

Prion diseases are a unique set of rare and deadly neurodegenerative disorders that can be passed from one individual to another. They mainly show up with symptoms like worsening dementia, muscle jerks, and coordination issues. This category includes illnesses like scrapie in sheep, mad cow disease in cattle, chronic wasting disease in deer and similar animals, along with several prion diseases that impact humans (Geschwind 2015). Transmissible spongiform encephalopathies (TSEs) are deadly brain disorders caused by misfolded prion proteins (PrPSc), which lead to quick nerve cell damage. Although they are pretty rare in people, prion diseases like Creutzfeldt-Jakob disease and Kuru can cause rapid onset of dementia and coordination issues, with most patients passing away within a year. New approaches have been developed, like nanoparticles and immunotherapy. Important approaches like antibody vaccines, dendritic cell vaccines, and transferring $CD4^+$ T lymphocytes are some of the useful therapies that are introduced to overcome the symptoms of these diseases and enhance the immune system (Burchell and Panegyres 2016).

5.2 Cancer Diagnostics

Cancer diagnostic techniques have also advanced significantly, with a focus on detecting uncontrolled cell proliferation and enabling early detection and monitoring. Biosensors can play a crucial role in this domain, providing rapid and non-invasive methods for cancer screening and disease progression tracking. Beyond healthcare, environmental and public health monitoring is essential for detecting and mitigating the impact of various contaminants and epidemiological surveillance. Biosensors can be employed to detect environmental pollutants and track the spread of infectious diseases, contributing to the protection of public health and the environment (Kang et al. 2018).

5.2.1 Role in Detecting Uncontrolled Cell Proliferation

Research work showed that the uses of two electrochemical immunosensors were merged together to detect HER-1 and HER-2 for breast cancer. This new immunosensor is inbuilt with gold sensor chips and HRP-based amperometric detection. They employed an indirect sandwich ELISA method with monoclonal antibodies. They show the achieving limits of detection of 1.06 and 0.95 ng mL^{-1} in buffer. Gold nanoparticles (AuNPs) equipped immunosensors showed improved sensitivity. Therefore, it shows LODs as low as 30 pg mL^{-1}. In serum samples, the HER-2 detection had a LOD of 50 pg mL^{-1}. It showed excellent specificity and minimal cross-reactivity. These biosensors are great for early detection and monitoring, which can lead to better patient outcomes (Wignarajah et al. 2023).

6 The Role of Artificial Intelligence and Machine Learning

6.1 Integration with Immunosensors

Artificial intelligence and machine learning are the superheroes in the world of immunosensors by boosting data analysis experiences. These techniques enhance the precision diagnostic tests and make them available for quick decision-making (Flynn and Chang 2024). Data processing and interpretation with artificial intelligence and machine learning generate huge amounts of data by refining it from immunosensors. It quickly accesses the target molecules, pathogens, or biomarkers without causing any delay (Nashruddin et al. 2024).

6.1.1 Real-Time Data Analysis

Pattern recognition in the feature of machine learning provides deep learning. This excels in identifying intricate data patterns (Manakitsa et al. 2024). These learnings through machine may be helpful in identifying the antigen–antibody interactions. It also noticed the biological response accompanying through different condition (Zhang et al. 2024). The noisy data can be cured by using these techniques. Neural networks is a type of artificial intelligence method that is useful in sifting the noisy data to authenticate form in immunosensors. They can filter out the irrelevant information and make the detection more reliable. Artificial Intelligence can clean up background noise in fluorescence or electrochemical signals (Zhao et al. 2023). The result is to boost sensitivity. Adaptive learning through Artificial Intelligence-driven systems can adjust to changing datasets (Aldoseri et al. 2023).

6.1.2 Automation in Diagnostics

Artificial intelligence-powered immunosensors can take over the decision-making process by automation. It eases the data interpretation by reducing the manual work. It is used to figure out positive or negative results. Biomarker-level-based measurement is helpful in suggesting next steps for diagnosis (Manickam et al. 2022). The robotics and automated systems utilise the data and make a streamlined workflow. Artificial-based immunosensors are useful in handling the high screening of the data. The large samples can be treated parallelly without causing any disturbance in their results (Wasilewski et al. 2024).

Remote diagnosis and telemedicine are possible by using artificial intelligence-designed immunosensors. It is also helpful to evaluate immunosensor data on mobile devices and in the cloud. The results are accessible quickly through this technology. These fast-obtaining results are helpful to both patients and healthcare providers (Kuziemsky et al. 2019). Artificial Intelligence-Enhanced User Interfaces are super to overcome the complicated diagnostic data. These data are easily converted into understandable formats. This technology offers visual aids and recommendations for additional tests and treatments. These advanced features are useful to make immunosensor data more accessible for non-specialists and healthcare workers in areas with limited resources (Wang et al. 2023).

6.2 Enhancing Speed and Accuracy

Artificial intelligence and machine learning are enhancing the speed and precision of immunosensors. Both the technologies make the diagnostics more effective and valuable.

6.2.1 AI-Based Predictive Models

Artificial intelligence-driven predictive models can be used to estimate the chances of disease or infection. These advanced immunosensors can analyse the history and track disease development. It also proposes possible outcomes based on previous data. Machine learning algorithms and techniques, such as random forests, support vector machines, and neural networks, can accurately classify samples (Dixon et al. 2024). They get information from previous datasets to effectively distinguish between positive and negative samples and identify different development stages of diseases. Early detection performed by artificial learning-powered models can improve the sensitivity of immunosensors (Bhaiyya et al. 2024). They allow for the detection of diseases in their early stages and sometimes before symptoms appear. This is especially crucial for illnesses like cancer, where identifying it early can significantly enhance treatment success (Hunter et al. 2022). Customisable thresholds in artificial learning systems can flexibly modify detection thresholds based on testing context, patient history, and specific diagnostic requirements; personalised and accurate diagnostics.

6.2.2 Reducing False Positives/Negatives

Machine learning models can easily separate out the false positives and negatives by constantly learning from labelled data. Artificial learning algorithms are great at spotting potential errors. It makes real-time adjustments that boost overall reliability. Artificial learning can also generate synthetic data to enhance training datasets. It provides solutions for unusual situations. This plays a pivotal role in minimising false positives and negatives, especially in those areas where certain conditions are less common (Javaid et al. 2022).

Additionally, artificial learning also provides clear insight into diagnostic data that could lead to wrong results. The algorithms can adapt to changes in the data. These are in the form of sensor drift and sample differences. It provides consistent outcomes. Artificial learning shows combined readings from different immunosensors (Shajari et al. 2023). It also has been used to cross-check the results to verify diagnostic accuracy. For instance, optical and electrochemical immunosensor data can be merged together to lead to a stronger analysis than just using one type of sensor (Wang et al. 2023). Lastly, artificial learning can help standardise the diagnostic criteria used by immunosensors. It reduces the differences in interpretation among various operators and labs. This provides a uniform diagnostic result across different platforms.

7 Future Trends in Immunosensor Technology

7.1 Personalised Medicine and Next-Generation Immunosensor Devices

The use of immunosensors in personalised medicine is a game changer for designing healthcare products to provide for individual needs. 3D-printed immunosensors are becoming an emerging field in the electrochemical sensor world. This innovative approach is used efficiently to make highly accurate and unique sensors using biocompatible materials (Sharafeldin et al. 2020).

The flexibility of 3D printing features makes these devices come in all sorts of shapes and sizes. The unique sizes and shapes of these will help boost their detection capabilities. There is a lot of promise for 3D-printed immunosensors in personalised medicine. As they can be designed in such a way as to fulfil the specific diagnostic requirement. SARS-CoV-2 viruses are also monitored through this advanced technology. Bioprinting is the cutting edge of 3D printing. This is used to create organs and tissues from specialised materials. These materials are in the form of intestinal microvilli that help to enhance electrochemical sensors for detecting gliadin (Ali et al. 2022). Plus, wearable immunosensors can be integrated into clothing, medical devices, or personal items. This will help to allow for real-time health monitoring by detecting biomarkers on the skin (Polli et al. 2024).

7.2 Immunosensors in Pharmacogenomics

Incorporating immunosensors into pharmacogenomics allows for a deeper understanding of how genes influence a person's immune system to react to medications (Qahwaji et al. 2024). These uniquely designed immunosensors help healthcare providers identify genetic and protein markers. This is helpful to predict the treatments and dosages according to the needs of the patient. It may help to optimise the treatment plans and minimise the chances of negative side effects (Oates and Lopez 2018).

Immunosensors utilise the biomarkers for each specific issue of the patient. The designed treatment plans are framed based on the instant changes in the individual body. This means therapies can be perfectly matched to a patient's issues. The end result is more effective and fruitful. They mainly focused on the treatment plan (Chen et al. 2023). The cancer biomarkers can also be useful to help tweak treatment strategies to boost effectiveness while reducing side effects (Gadade et al. 2024).

8 Challenges and Limitations

8.1 Stability and Reproducibility

Immunosensors are able to encounter issues with signal stability. They reproduce the results by using nanomaterials. These issues are surface instability, variability in binding efficiency, and the impact of environmental factors, such as temperature and pH changes. Additionally, the working of quantum dots and other nanomaterials does not stay the same throughout their lives. It can be impacting the reliability of sensor responses (Khazaei et al. 2023).

8.1.1 Miniaturisation and Scalability

Miniaturisation enhances portability and makes it easy to handle for point-of-care applications. It also creates difficulties in terms of signal strength, precision, and sample handling. During the scaling down of immunosensors, maintaining sensitivity and specificity without producing noise becomes a significant challenge (Roudsari et al. 2024). Furthermore, scalable manufacturing methods for producing high-quality, reproducible miniaturised sensors are still a research area under development. Scaling up in an authentic way may help in impacting commercial viability (Jelva Hussan et al. 2024).

8.2 Regulatory and Ethical Considerations

8.2.1 Approval Pathways

Integrating advanced nanomaterials is one of the hurdle areas to take permission for. Regulatory bodies have been following the comprehensive testing and validation. This has been used to ensure the safety, efficacy, and reliability (Allan et al. 2021). For instance, the use of quantum dots (QDs) in healthcare areas may raise additional examination or observation. Because, it is directly related to their impacts on the environment and their toxicity. Due to this, it causes complications in the path to approval (Reshma and Mohanan 2019).

8.2.2 Privacy and Data Security

Digital technology linked with the development of immunosensors has led to the collection of data and analysis. The linked equipped technology must be checked to prove the security of patient data. The safety of personal data of a patient is a priority. Privacy concerns is the thinking area because in point-of-care settings, data might be transmitted over networks (Ross et al. 2023). It has been ensured that compliance with data protection regulations creates complexity (e.g. GDPR, HIPAA). Security measures must be robust to prevent unauthorised access and ensure patient data confidentiality (Act 1996; Perry 2021).

9 Conclusion

Immunosensor technologies are merged with significant advancements. The integration of innovative materials such as gold nanoparticles and quantum dots (QDs) with these immunosensors can enhance the sensitivity, specificity, and response time of these sensors. Electrochemical immunosensors development with improved signal transduction has allowed for the detection of low-abundance biomarkers. They are facilitating the early diagnosis and effective disease monitoring. Future prospects for immunosensor technology include the incorporation of artificial intelligence for data monitoring, data analysis, miniaturised point-of-care platforms, and easily wearable or implantable sensors that improve continuous health monitoring. Additionally, microfluidic systems are expected to streamline sample handling. It is contributing to more efficient diagnostic tools. These advancements hold substantial promise for global health. The improvement has been done in low-resource areas where portable and cost-effective immunosensors can provide rapid and accurate results. By supporting personalised medicine, such technologies enable tailored treatments based on individual biomarker profiles. It improves patient outcomes, reducing healthcare costs.

Artificial intelligence-powered immunosensors are making waves in early cancer detection by utilising biomarkers. Some models with deep learning networks have achieved an impressive 95% drop in false positives for prostate and breast cancer markers by examining patterns in immunoassay data. Plus, artificial intelligence has been combined with immunosensors for quick detection of COVID-19 antigens.

Machine learning systems ameliorated the ability to distinguish between similar respiratory infections, boosting accuracy. In point-of-care environments, artificial intelligence-enhanced immunosensors are being used for diseases like HIV and malaria, allowing for speedy diagnoses with minimal resources. These devices have cut down diagnosis time significantly, going from hours to just minutes. Moreover, artificial intelligence-driven wearable devices equipped with immunosensors continuously monitor health conditions, sending data to cloud platforms for real-time analysis. They utilise artificial intelligence to catch early signs of inflammation or metabolic issues. The broad accessibility and user-friendly nature of modern immunosensors have the potential to bridge healthcare disparities and enhance public health management on a global scale. The latest breakthroughs in immunosensor technology, driven by AI and cutting-edge materials, are set to change the game in healthcare. They promise to make diagnostics more accessible, efficient, and tailored to individual needs. It could lead to better health outcomes worldwide and reshape the future of medical care.

References

Abbas N, Song S, Chang MS et al (2023) Point-of-Care diagnostic devices for detection of Escherichia coli O157: H7 using microfluidic systems: a focused review. Biosensors 13(7):741

Aboul-Ella H, Gohar A, Ali AA et al (2024) Monoclonal antibodies: from magic bullet to precision weapon. Mol Biomed 5(1):1–44

Act A (1996) Health insurance portability and accountability act of 1996. Public Law 104:191

Ahmed A, Rushworth JV, Hirst NA et al (2014) Biosensors for whole-cell bacterial detection. Clin Microbiol Rev 27(3):631–646

Aizawa M (1994) Immunosensors for clinical analysis. Adv Clin Chem 31:247–275

Aldoseri A, Al-Khalifa KN, Hamouda AM (2023) Re-thinking data strategy and integration for artificial intelligence: concepts, opportunities, and challenges. Appl Sci 13(12):7082

Ali AA, Altemimi AB, Alhelfi N et al (2020) Application of biosensors for detection of pathogenic food bacteria: a review. Biosensors 10(6):58

Ali MA, Hu C, Zhang F et al (2022) N protein-based ultrasensitive SARS-CoV-2 antibody detection in seconds via 3D nanoprinted, microarchitected array electrodes. J Med Virol 94(5):2067–2078

Allan J, Belz S, Hoeveler A et al (2021) Regulatory landscape of nanotechnology and nanoplastics from a global perspective. Regul Toxicol Pharmacol 122:104885

Ansari MH, Hassan S, Qurashi A et al (2016) Microfluidic-integrated DNA nanobiosensors. Biosens Bioelectron 85:247–260

Apoorva S, Nguyen NT, Rajan SK et al (2024) Recent developments and future perspectives of microfluidics and smart technologies in wearable devices. Lab Chip 24:1833

Ardekani LS, Thulstrup PW (2022) Gold nanoparticle-mediated lateral flow assays for detection of host antibodies and COVID-19 proteins. Nanomaterials 12(9):1456

Arya SS, Dias SB, Jelinek HF (2023) The convergence of traditional and digital biomarkers through AI-assisted biosensing: a new era in translational diagnostics. Biosens Bioelectron 235:115387

Aydin EB, Aydin M, Sezginturk MK (2019) Advances in electrochemical immunosensors. Adv Clin Chem 92:1–57

Behera PP, Kumar N, Kumari M (2023) Integrated microfluidic devices for point-of-care detection of bio-analytes and disease. Sens Diagn 2:1437

Bhaiyya M, Panigrahi D, Rewatkar P (2024) Role of machine learning assisted biosensors in point-of-care-testing for clinical decisions. ACS sens 9(9):4495–4519

Bharti B, Li H, Ren Z et al (2022) Recent advances in sterilization and disinfection technology: a review. Chemosphere 308:136404

Bhatia D, Paul S, Acharjee T et al (2024) Biosensors and their widespread impact on human health. Sens Int 5:100257

Bhattacharya I, Khandwala YS, Vesal S et al (2022) A review of artificial intelligence in prostate cancer detection on imaging. Ther Adv Urol 14:17562872221128791

Brazaca LC, Dos Santos PL, de Oliveira PR et al (2021) Biosensing strategies for the electrochemical detection of viruses and viral diseases–a review. Anal Chim Acta 1159:338384

Burchell JT, Panegyres PK (2016) Prion diseases: immunotargets and therapy. ImmunoTargets Ther:57–68

Capelli D, Scognamiglio V, Montanari R (2023) Surface plasmon resonance technology: recent advances, applications and experimental cases. TrAC Trends Anal Chem 163:117079

Castillo-Henríquez L, Brenes-Acuña M, Castro-Rojas A et al (2020) Biosensors for the detection of bacterial and viral clinical pathogens. Sensors 20(23):6926

Cesewski E, Johnson BN (2020) Electrochemical biosensors for pathogen detection. Biosens Bioelectron 159:112214

Chang CC, Chen CP, Wu TH et al (2019) Gold nanoparticle-based colorimetric strategies for chemical and biological sensing applications. Nanomaterials 9(6):861

Chawla S, Bansal I, Rai P (2023) Modified carbon nanomaterials for diagnosis, drug delivery and stem cell therapy. In: Functionalized carbon nanomaterials for theranostic applications. Elsevier, pp 331–350

Chen H, Zhang J, Huang R et al (2023) The applications of electrochemical immunosensors in the detection of disease biomarkers: a review. Molecules 28(8):3605

Cho IH, Lee J, Kim J et al (2018) Current technologies of electrochemical immunosensors: perspective on signal amplification. Sensors 18(1):207

Choi J, Seong TW, Jeun M et al (2017) Field-effect biosensors for on-site detection: recent advances and promising targets. Adv Healthc Mater 6(20):1700796

Choosang J, Khumngern S, Thavarungkul P et al (2021) An ultrasensitive label-free electrochemical immunosensor based on 3D porous chitosan–graphene–ionic liquid–ferrocene nanocomposite cryogel decorated with gold nanoparticles for prostate-specific antigen. Talanta 224:121787

De Paoli P (2005) Biobanking in microbiology: from sample collection to epidemiology, diagnosis and research. FEMS Microbiol Rev 29(5):897–910

Dhahi TS, Dafhalla AKY, Al-Mufti AW et al (2024) Application of nanobiosensor engineering in the diagnosis of neurodegenerative disorders. Results Eng:102790

Dias C, Costa J, Mafra I et al (2024) Electrochemical immunosensor for point-of-care detection of soybean Gly m TI allergen in foods. Talanta 268:125284

Dixon D, Sattar H, Moros N et al (2024) Unveiling the influence of AI predictive analytics on patient outcomes: a comprehensive narrative review. Cureus 16(5):e59954

Drobysh M, Ramanaviciene A, Viter R et al (2022) Biosensors for the determination of SARS-CoV-2 virus and diagnosis of COVID-19 infection. Int J Mol Sci 23(2):666

Drobysh M, Ramanavicius A, Baradoke A (2023) Polyaniline-based electrochemical immunosensor for the determination of antibodies against SARS-CoV-2 spike protein. Sci Total Environ 862:160700

Durai S, Manoharan G, Ashtikar SP (2024) Harnessing artificial intelligence: pioneering sustainable solutions for a greener future. In: Social and ethical implications of AI in finance for sustainability. IGI Global, pp 89–117

Farzin L, Shamsipur M, Samandari L et al (2020) HIV biosensors for early diagnosis of infection: the intertwine of nanotechnology with sensing strategies. Talanta 206:120201

Felix FS, Angnes L (2018) Electrochemical immunosensors–a powerful tool for analytical applications. Biosens Bioelectron 102:470–478

Fistera D, Kikull K, Risse J et al (2023) Point-of-care PCR testing of SARS-CoV-2 in the emergency department: influence on workflow and efficiency. PLoS One 18(8):e0288906

Flynn CD, Chang D (2024) Artificial intelligence in point-of-care biosensing: challenges and opportunities. Diagnostics 14(11):1100

Franco-Duarte R, Cernakova L, Kadam S (2019) Advances in chemical and biological methods to identify microorganisms—from past to present. Microorganisms 7(5):130

Fu Y, Liu T, Wang H et al (2024) Applications of nanomaterial technology in biosensing. J Sci Adv Mater Devices:100694

Gadade DD, Jha H, Kumar C et al (2024) Unlocking the power of precision medicine: exploring the role of biomarkers in cancer management. Future J Pharm Sci 10(1):5

Gamble P, Jaroensri R, Wang H et al (2021) Determining breast cancer biomarker status and associated morphological features using deep learning. Commun Med 1(1):14

Gan SY, Tye GJ, Chew AL et al (2023) Linker-mediated oriented antibody immobilisation strategies for a more efficient immunosensor and diagnostic applications: a review. Biosens Bioelectron X 14:100379

Geschwind MD (2015) Prion diseases. Continuum (Minneap Minn) 21(6):1612–1638

Ghasemi-Varnamkhasti M, Apetrei C, Lozano J et al (2018) Potential use of electronic noses, electronic tongues and biosensors as multisensor systems for spoilage examination in foods. Trends Food Sci Technol 80:71–92

Haleem A, Javaid M, Singh RP (2021) Biosensors applications in medical field: a brief review. Sens Int 2:100100

Heidt B, Siqueira WF, Eersels K et al (2020) Point of care diagnostics in resource-limited settings: a review of the present and future of PoC in its most needed environment. Biosensors 10(10):133

Hemdan M, Ali MA, Doghish AS et al (2024) Innovations in biosensor technologies for healthcare diagnostics and therapeutic drug monitoring: applications, recent progress, and future research challenges. Sensors (Basel) 24(16):5143

Horber S, Achenbach P, Schleicher E (2020) Harmonization of immunoassays for biomarkers in diabetes mellitus. Biotechnol Adv 39:107359

Huang X, Zhu Y, Kianfar E (2021) Nano biosensors: properties, applications and electrochemical techniques. J Mater Res Technol 12:1649–1672

Huang J, Wei F, Cui Y et al (2022) Fluorescence immunosensor based on functional nanomaterials and its application in tumor biomarker detection. RSC Adv 12(48):31369–31379

Hunter B, Hindocha S, Lee RW (2022) The role of artificial intelligence in early cancer diagnosis. Cancers 14(6):1524

Janik-Karpinska E, Ceremuga M, Niemcewicz M et al (2022) Immunosensors—the future of pathogen real-time detection. Sensors 22(24):9757

Jarockyte G, Karabanovas V, Rotomskis R et al (2020) Multiplexed nanobiosensors: current trends in early diagnostics. Sensors 20(23):6890

Javaid M, Haleem A, Singh RP et al (2022) Significance of machine learning in healthcare: features, pillars and applications. Int J Intell Netw 3:58–73

Jelva Hussan KS, Subramaniam MP, Mohammed Kenz KT et al (2024) Fabrication and challenges of 3D printed sensors for biomedical applications-comprehensive review. Results Eng 21: 101867

Joseph TM, Kar Mahapatra D, Esmaeili A et al (2023) Nanoparticles: taking a unique position in medicine. Nanomaterials 13(3):574

Justino CI, Duarte AC, Rocha-Santos TA (2016) Immunosensors in clinical laboratory diagnostics. Adv Clin Chem 73:65–108

Jyoung JY, Hong S, Lee W et al (2006) Immunosensor for the detection of Vibrio cholerae O1 using surface plasmon resonance. Biosens Bioelectron 21(12):2315–2319

Kabir MA, Zilouchian H, Younas MA et al (2021) Dengue detection: advances in diagnostic tools from conventional technology to point of care. Biosensors 11(7):206

Kang EB, Mazrad ZAI, Robby AI et al (2018) Alkaline phosphatase-responsive fluorescent polymer probe coated surface for colorimetric bacteria detection. Eur Polym J 105:217–225

Kant L, Immanuel G, Narain JP (2021) Disease pandemics and the threat of microbial emergence. In: Climate change and the health sector. Routledge India, pp 47–57

Kaurav H, Verma D, Bansal A et al (2023) Progress in drug delivery and diagnostic applications of carbon dots: a systematic review. Front Chem 11:1227843

Khalifa M, Albadawy M (2024) AI in diagnostic imaging: Revolutionising accuracy and efficiency. Comput Methods Programs Biomed Update:100146

Khan AI, Khan M, Khan R (2023) Artificial intelligence in point-of-care testing. Ann Lab Med 43(5):401–407

Khazaei M, Hosseini MS, Haghighi AM et al (2023) Nanosensors and their applications in early diagnosis of cancer. Sens Bio-Sens Res 41:100569

Kim J, Campbell AS, de Avila BEF et al (2019) Wearable biosensors for healthcare monitoring. Nat Biotechnol 37(4):389–406

Kourti D, Angelopoulou M, Petrou P (2023) Optical immunosensors for bacteria detection in food matrices. Chemosensors 11(8):430

Kuziemsky C, Maeder AJ, John O et al (2019) Role of artificial intelligence within the telehealth domain. Yearb Med Inform 28(01):035–040

Lai CC, Shih TP, Ko WC et al (2020) Severe acute respiratory syndrome coronavirus 2 (SARS-CoV-2) and coronavirus disease-2019 (COVID-19): the epidemic and the challenges. Int J Antimicrob Agents 55(3):105924

Lara S, Perez-Potti A (2018) Applications of nanomaterials for immunosensing. Biosensors 8(4):104

Lee JH, Cho HY, Choi HK (2018) Application of gold nanoparticle to plasmonic biosensors. Int J Mol Sci 19(7):2021

Liu J, Ruan G, Ma W et al (2022) Horseradish peroxidase-triggered direct in situ fluorescent immunoassay platform for sensing cardiac troponin I and SARS-CoV-2 nucleocapsid protein in serum. Biosens Bioelectron 198:113823

Lohcharoenkal W, Abbas Z, Rojanasakul Y (2021) Advances in nanotechnology-based biosensing of immunoregulatory cytokines. Biosensors 11(10):364

Manakitsa N, Maraslidis GS, Moysis L et al (2024) A review of machine learning and deep learning for object detection, semantic segmentation, and human action recognition in machine and robotic vision. Technologies 12(2):15

Manickam P, Mariappan SA, Murugesan SM et al (2022) Artificial intelligence (AI) and internet of medical things (IoMT) assisted biomedical systems for intelligent healthcare. Biosensors 12(8):562

Mehrotra P (2016) Biosensors and their applications–a review. J Oral Biol Craniofac Res 6(2):153–159

Mercy DJ, Girigoswami K, Girigoswami A (2023) A mini review on biosensor advancements-emphasis on quantum dots. Results Chem 7:101271

Mikhailova EO (2021) Gold nanoparticles: biosynthesis and potential of biomedical application. J Funct Biomater 12(4):70

Mollarasouli F, Kurbanoglu S, Ozkan SA (2019) The role of electrochemical immunosensors in clinical analysis. Biosensors 9(3):86

Moon S (2023) Extending the shelf-life of immunoassay-based microfluidic chips through freeze-drying sublimation techniques. Sensors 23(20):8524

Mousavi SM, Hashemi SA, Yari Kalashgrani M et al (2022) The pivotal role of quantum dots-based biomarkers integrated with ultra-sensitive probes for multiplex detection of human viral infections. Pharmaceuticals 15(7):880

Nashruddin SNABM, Salleh FHM, Yunus RM et al (2024) Artificial intelligence– powered electrochemical sensor: recent advances, challenges, and prospects, vol 10. Heliyon, p e37964

Niculescu AG, Chircov C, Birca AC et al (2021) Fabrication and applications of microfluidic devices: a review. Int J Mol Sci 22(4):2011

Nnachi RC, Sui N, Ke B et al (2022) Biosensors for rapid detection of bacterial pathogens in water, food and environment. Environ Int 166:107357

Oates JT, Lopez D (2018) Pharmacogenetics: an important part of drug development with a focus on its application. Int J Biomed Investig 1(2):111

Parihar A, Yadav S, Sadique MA et al (2023) Internet-of-medical-things integrated point-of-care biosensing devices for infectious diseases: toward better preparedness for futuristic pandemics. Bioeng Transl Med 3:e10481

Pattanayak P, Singh SK, Gulati M et al (2021) Microfluidic chips: recent advances, critical strategies in design, applications and future perspectives. Microfluid Nanofluidics 25:1–28

Perry PY (2021) Meeting data compliance with a wave of new privacy regulations: GDPR, CCPA, PIPEDA, POPI, LGPD, HIPAA, PCI-DSS, and More. Cloud Netapp

Piliarik M, Vaisocherova H, Homola J et al (2009) Surface plasmon resonance biosensing. Biosens Biodetect:65–88

Pillai S, Kwan JC, Yaziji F et al (2023) Mapping the potential of microfluidics in early diagnosis and personalized treatment of head and neck cancers. Cancers 15(15):3894

Pohanka M (2018) Overview of piezoelectric biosensors, immunosensors and DNA sensors and their applications. Materials 11(3):448

Polli F, Simonetti F, Surace L et al (2024) Nanoparticles in electrochemical immunosensors–a concept and perspective. ChemElectroChem 11(2):e202300408

Popov A, Brasiunas B, Kausaite-Minkstimiene A et al (2021) Metal nanoparticle and quantum dot tags for signal amplification in electrochemical immunosensors for biomarker detection. Chemosensors 9(4):85

Qahwaji R, Ashankyty I, Sannan NS et al (2024) Pharmacogenomics: a genetic approach to drug development and therapy. Pharmaceuticals 17(7):940

Ramesh M, Janani R, Deepa C et al (2022) Nanotechnology-enabled biosensors: a review of fundamentals, design principles, materials, and applications. Biosensors 13(1):40

Rasmi Y, Li X, Khan J et al (2021) Emerging point-of-care biosensors for rapid diagnosis of COVID-19: current progress, challenges, and future prospects. Anal Bioanal Chem 413(16): 4137–4159

Rauti R, Musto M, Bosi S et al (2019) Properties and behavior of carbon nanomaterials when interfacing neuronal cells: how far have we come? Carbon 143:430–446

Reshma VG, Mohanan PV (2019) Quantum dots: applications and safety consequences. J Lumin 205:287–298

Rios TB, Maximiano MR, Feitosa GC et al (2024) Nanosensors for animal infectious disease detection. Sens Bio-Sens Res 43:100622

Rosa BG, Akingbade OE, Guo X et al (2022) Multiplexed immunosensors for point-of-care diagnostic applications. Biosens Bioelectron 203:114050

Ross GMS, Zhao Y, Bosman AJ et al (2023) Best practices and current implementation of emerging smartphone-based (bio) sensors–part 1: data handling and ethics. TrAC Trends Anal Chem 158: 116863

Roudsari ZO, Khoramrooz SS, Rouhi S et al (2024) Electrochemical and optical biosensors for the detection of *E. Coli*. Clin Chim Acta 565:119984

Şahin S, Unlu C, Trabzon L (2021) Affinity biosensors developed with quantum dots in microfluidic systems. Emerg Mater 4:187–209

Shajari S, Kuruvinashetti K, Komeili A et al (2023) The emergence of AI-based wearable sensors for digital health technology: a review. Sensors 23(23):9498

Sharafeldin M, Kadimisetty K, Bhalerao KS et al (2020) 3D-printed Immunosensor arrays for cancer diagnostics. Sensors 20(16):4514

Sheoran K, Kaur H, Siwal SS et al (2022) Recent advances of carbon-based nanomaterials (CBNMs) for wastewater treatment: synthesis and application. Chemosphere 299:134364

Shi Y, Ye P, Yang K et al (2021) Application of microfluidics in immunoassay: recent advancements. J Healthcare Eng 2021(1):2959843

Shoute LC, Abdelrasoul GN, Ma Y et al (2023) Label-free impedimetric immunosensor for point-of-care detection of COVID-19 antibodies. Microsyst Nanoeng 9(1):3

Shrestha BK, Ahmad R, Mousa HM et al (2016) High-performance glucose biosensor based on chitosan-glucose oxidase immobilized polypyrrole/Nafion/functionalized multi-walled carbon nanotubes bio-nanohybrid film. J Colloid Interface Sci 482:39–47

Sin ML, Mach KE, Wong PK et al (2014) Advances and challenges in biosensor-based diagnosis of infectious diseases. Expert Rev Mol Diagn 14(2):225–244

Singh S, Dhawan A, Karhana S et al (2020) Quantum dots: an emerging tool for point-of-care testing. Micromachines 11(12):1058

Sun F, Zhang J, Yang Q et al (2021) Quantum dot biosensor combined with antibody and aptamer for tracing food-borne pathogens. Food Qual Safety 5:019

Tessaro L, Aquino A, de Carvalho APA et al (2021) A systematic review on gold nanoparticles based-optical biosensors for Influenza virus detection. Sens Actuators Rep 3:100060

Thadathil A, Pradeep H, Joshy D et al (2022) Polyindole and polypyrrole as a sustainable platform for environmental remediation and sensor applications. Mater Adv 3(7):2990–3022

Thakur A, Kumar A (2022) Recent advances on rapid detection and remediation of environmental pollutants utilizing nanomaterials-based (bio) sensors. Sci Total Environ 834:155219

Verma AK, Noumani A, Yadav AK et al (2023) FRET based biosensor: principle applications recent advances and challenges. Diagnostics 13(8):1375

Viter R, Tereshchenko A, Smyntyna V et al (2017) Toward development of optical biosensors based on photoluminescence of TiO2 nanoparticles for the detection of Salmonella. Sens Actuators B Chem 252:95–102

Wang H, Shen G, Yu R (2008) Aspects of recent development of immunosensors. In: Electrochemical sensors, biosensors and their biomedical applications. Academic Press, p 237

Wang H, Sugiarto S, Li T et al (2016) Advances in nanomaterials and their applications in point of care (POC) devices for the diagnosis of infectious diseases. Biotechnol Adv 34(8):1275–1288

Wang C, He T, Zhou H et al (2023) Artificial intelligence enhanced sensors-enabling technologies to next-generation healthcare and biomedical platform. Bioelectron Med 9(1):17

Wani AK, Akhtar N, Katoch V et al (2023) Advancing biological investigations using portable sensors for detection of sensitive samples. Heliyon 9:e22679

Wasilewski T, Kamysz W, Gębicki J (2024) AI-assisted detection of biomarkers by sensors and biosensors for early diagnosis and monitoring. Biosensors 14(7):356

Wignarajah S, Chianella I, Tothill IE (2023) Development of electrochemical immunosensors for HER-1 and HER-2 analysis in serum for breast cancer patients. Biosensors 13(3):355

Wu K, He X, Wang J et al (2022) Recent progress of microfluidic chips in immunoassay. Front Bioeng Biotechnol 10:1112327

Yang S, Rothman RE (2004) PCR-based diagnostics for infectious diseases: uses, limitations, and future applications in acute-care settings. Lancet Infect Dis 4(6):337–348

Yang HL, Bai LF, Geng ZR et al (2023) Carbon quantum dots: preparation, optical properties, and biomedical applications. Mater Today Adv 18:100376

Yuwen T, Shu D, Zou H et al (2023) Carbon nanotubes: a powerful bridge for conductivity and flexibility in electrochemical glucose sensors. J Nanobiotechnol 21(1):320

Zambry NS, Obande GA, Khalid MF et al (2022) Utilizing electrochemical-based sensing approaches for the detection of SARS-CoV-2 in clinical samples: a review. Biosensors 12(7): 473

Zhang Y, Bai Y, Yan B (2010) Functionalized carbon nanotubes for potential medicinal applications. Drug Discov Today 15(11–12):428–435

Zhang Y, Hu Y, Jiang N et al (2023) Wearable artificial intelligence biosensor networks. Biosens Bioelectron 219:114825

Zhang G, Kuang X, Zhang Y et al (2024) Machine-learning-based structural analysis of interactions between antibodies and antigens. BioSystems 243:105264

Zhao Y, Wang X, Sun T et al (2023) Artificial intelligence-driven electrochemical immunosensing biochips in multi-component detection. Biomicrofluidics 17(4):041301

Zhu C, Yang G, Li H et al (2015) Electrochemical sensors and biosensors based on nanomaterials and nanostructures. Anal Chem 87(1):230–249

Index

K. Kuča et al. (eds.), *Ensuring Wellness: Immunosensors Monitoring Food Quality and Pathogenicity Risks*, https://doi.org/10.1007/978-3-031-97477-9

GPSR Compliance

The European Union's (EU) General Product Safety Regulation (GPSR) is a set of rules that requires consumer products to be safe and our obligations to ensure this.

If you have any concerns about our products, you can contact us on ProductSafety@springernature.com

In case Publisher is established outside the EU, the EU authorized representative is:

Springer Nature Customer Service Center GmbH
Europaplatz 3
69115 Heidelberg, Germany

Batch number: 10755332

Printed by Printforce, the Netherlands